T0182137

NEUROMETHODS

Series Editor:
Wolfgang Walz
University of Saskatchewan
Saskatoon, Canada

For further volumes:
http://www.springer.com/series/7657

Receptor-Receptor Interactions in the Central Nervous System

Edited by

Kjell Fuxe

Department of Neuroscience, Karolinska Institutet, Stockholm, Sweden

Dasiel O. Borroto-Escuela

Department of Neuroscience, Karolinska Institutet, Stockholm, Sweden

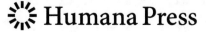 Humana Press

Editors
Kjell Fuxe
Department of Neuroscience
Karolinska Institutet
Stockholm, Sweden

Dasiel O. Borroto-Escuela
Department of Neuroscience
Karolinska Institutet
Stockholm, Sweden

ISSN 0893-2336   ISSN 1940-6045   (electronic)
Neuromethods
ISBN 978-1-4939-9331-4   ISBN 978-1-4939-8576-0   (eBook)
https://doi.org/10.1007/978-1-4939-8576-0

Series Preface

Experimental life sciences have two basic foundations: concepts and tools. The *Neuromethods* series focuses on the tools and techniques unique to the investigation of the nervous system and excitable cells. It will not, however, shortchange the concept side of things as care has been taken to integrate these tools within the context of the concepts and questions under investigation. In this way, the series is unique in that it not only collects protocols but also includes theoretical background information and critiques which led to the methods and their development. Thus, it gives the reader a better understanding of the origin of the techniques and their potential future development. The *Neuromethods* publishing program strikes a balance between recent and exciting developments like those concerning new animal models of disease, imaging, in vivo methods, and more established techniques, including immunocytochemistry and electrophysiological technologies. New trainees in neurosciences still need a sound footing in these older methods in order to apply a critical approach to their results.

Under the guidance of its founders, Alan Boulton and Glen Baker, the *Neuromethods* series has been a success since its first volume published through Humana Press in 1985. The series continues to flourish through many changes over the years. It is now published under the umbrella of Springer Protocols. While methods involving brain research have changed a lot since the series started, the publishing environment and technology have changed even more radically. *Neuromethods* has the distinct layout and style of the Springer Protocols program, designed specifically for readability and ease of reference in a laboratory setting.

The careful application of methods is potentially the most important step in the process of scientific inquiry. In the past, new methodologies led the way in developing new disciplines in the biological and medical sciences. For example, physiology emerged out of anatomy in the nineteenth century by harnessing new methods based on the newly discovered phenomenon of electricity. Nowadays, the relationships between disciplines and methods are more complex. Methods are now widely shared between disciplines and research areas. New developments in electronic publishing make it possible for scientists who encounter new methods to quickly find sources of information electronically. The design of individual volumes and chapters in this series takes this new access technology into account. Springer Protocols makes it possible to download single protocols separately. In addition, Springer makes its print-on-demand technology available globally. A print copy can therefore be acquired quickly and for a competitive price anywhere in the world.

Saskatoon, Canada   *Wolfgang Walz*

Preface

G protein-coupled receptor (GPCR) can form homo- and heteroreceptor complexes with allosteric receptor–receptor interactions representing a novel molecular integrative mechanism in the central nervous system (CNS) [1, 2]. This takes place through direct physical interactions between co-expressed GPCRs either by physical association between the same receptor (homoreceptor complex) or different GPCR types (heteroreceptor complex) and with or without the participation of adapter proteins [3]. There are many impressive examples demonstrating GPCR–GPCR heteromerization, which through allosteric receptor–receptor interactions alters the GPCR protomer recognition, signaling, and trafficking. These changes in the receptor protomers produce marked increases in diversity and bias of GPCR function.

One emerging concept in neuropsychopharmacology is that a dysfunction of the allosteric receptor–receptor interactions contributes to disease progression in mental and neurological disorders [4–10]. So far their stoichiometry and topology are unknown within the heteromer formed as well as the number of adapter proteins participating, including their architecture. The heteromers in the CNS should therefore be described as heteroreceptor complexes [2].

The overall architecture of the global GPCR heterodimer network shows that inter alia D2R are hub components forming more than ten heterodimer pairs [11]. It should be underlined that GPCR can also form complexes with ion channel receptors modulating, for instance, the synaptic glutamate and GABA receptor signaling [12–14]. GPCR-tyrosine receptor kinase (RTK) heteroreceptor complexes also exist [7, 15] in which the GPCR protomer can modulate the trophic function of the receptor tyrosine kinase (RTK) through allosteric modulation over the GPCR–RTK interface. Thus, GPCRs may modulate trophism and structural plasticity via allosteric modulation of RTK function in GPCR–RTK heteroreceptor complexes.

The contributors to this book will cover the large number of methodologies used to unravel the receptor–receptor interactions in heteroreceptor complexes in the CNS from the molecular to the network level.

The first three chapters deal with biochemical binding techniques, receptor autoradiography, and (^{35}S)GTP gammaS binding in autoradiography which were used already in the early work. The biochemical binding experiments in CNS membrane preparations in the early 1980s led to the concept of receptor–receptor interactions in the plasma membrane (Chap. 1). The autoradiography makes possible an improved regional analysis and the use of more intact brain membranes present in the brain sections. This may help maintain the intramembrane receptor–receptor interaction (Chaps. 2 and 3). Chapter 4 deals with the superfused synaptosome technique to understand the function of the receptor–receptor interactions at the presynaptic level. Chapter 5 gives the protocols for dissociating and maintaining primary neurons from the hippocampus in order to study the transactivation of the FGFR1 by muscarinic cholinergic receptors. Methods to determine neurite outgrowth and protein phosphorylation are also described. They are needed to determine the function of this RTK–GPCR interaction (Chap. 5). Protocols for electrophysiological approaches to determine GPCR–RTK interactions are found in Chap. 6. Patch-clamp protocols are provided to analyze the GPCR–RTK interaction.

At the network level, the protocols for the microdialysis technique are provided used to study the role of the receptor–receptor interactions in the brain circuits of the awake freely moving rat (Chap. 7). In Chap. 8, behavioral methods are given to determine the role of receptor–receptor interactions in fear and anxiety at the network level. The methods to use small interference RNA knockdown in rats to determine the function of receptor–receptor interactions in anxiety- and depression-like behaviors are provided in Chap. 9. Protocols for the use of double fluorescent knock-in mice are provided to investigate subcellular distribution of endogenous mu-delta opioid heteromers in primary hippocampal cultures from this mouse model (Chap. 10). Their coexistence in various brain regions can also be elegantly demonstrated.

Moving into biochemical methods Chap. 11 gives the protocols for coimmunoprecipitation (Co-IP) and protein affinity purification assays to determine D2R-associated protein complexes. Their D2R–protein interactions like D2R–Disc1 interactions play a significant role in regulating the D2R signaling. In Chap. 12, the protocols for BRET are introduced to determine receptor heterodimers in cellular models. Protocols are also given for assays to establish the existence of heterotrimers. These assays build on the use of a combination of BRET and bimolecular fluorescence complementation (BiFc) or of a sequential BRET-FRET (SRET) procedure. The restraints of the energy transfer assays are discussed and how they can be complemented by in situ proximity ligation assay (PLA). A more detailed analysis of a dimerization-induced BiFc is given in Chap. 13 introducing a high-resolution approach using confocal laser microscopy vs a low-resolution approach using automated cell imaging with a multi-mode plate reader. In Chap. 14, the protocols for flow cytometry-based FRET (fcFRET) are given. This technique will allow the detection and quantification of dynamic receptor–receptor interactions based on the analysis of the fc-FRET.

In Chap. 15, a novel strategy is given to study GPCR dimerization in living cells. It is a protein complementation assay involving a reconstitution of a luminescent protein, a so-called nanoluciferase. Complementary nanoluciferase fragments are fused to a receptor pair, which can combine to form a reporter if dimerization takes place. Another new method is provided in Chap. 16 to study GPCR dimerization based on proximity biotinylation. Its advantages and disadvantages versus other methods are discussed. In Chap. 17, a new method is presented for the conformational profiling of GPCRs using the fluorescein arsenical hairpin (FlAsH) binders as energy acceptors with the tetracysteine binding tag in different positions within the intracellular domains of the 5-HT2A receptor. This method uses a FlAsH-BRET approach with the *Renilla* luciferase fused to the C-terminal tail of the 5-HT2A receptor.

In Chap. 18, a detailed description of the GPCR-hetnet is given. The G protein-coupled receptor heterocomplex network database (GPCR-hetnet) is a database designed to store information on GPCR heteroreceptor complexes and their allosteric receptor–receptor interactions. It is an expert-authored and peer-reviewed database of well-documented GPCR–GPCR, GPCR–receptor tyrosine kinase, and GPCR–ionotropic receptor/ligand-gated ion channel interactions. This chapter describes in a basic protocol how to use, navigate, and browse through the GPCR-hetnet database to identify the clusters in which a receptor protomer of interest is involved, while further applicability is also described and introduced.

In Chap. 19, detailed protocols are given for the use of the in situ proximity ligation assay in the analysis of homo- and heteroreceptor complexes in the CNS. It involves the use of confocal laser microscopy and quantitation of the density of PLA-positive clusters per cell in the sampled fields of various brain regions. Protocols for demonstrating the brain functions of receptor oligomers are presented in Chap. 20 using receptor interface interacting peptides. The D1R-NMDAR heteromer is given as an example.

The book ends with a description of the methodology to study GPCR dimers and higher order oligomers at the single-molecule level through super-resolution level (Chap. 21). Photoactivated localization microscopy using photoactivatable dyes was employed to visualize the spatial organization of the homo-heteroreceptor complexes and perform quantitation of their composition in terms of monomers, dimers, trimers, etc.

Stockholm, Sweden   *Kjell Fuxe*
                                                          Dasiel O. Borroto-Escuela

References

1. Fuxe K, Agnati LF, Benfenati F, Celani M, Zini I, Zoli M, Mutt V (1983) Evidence for the existence of receptor–receptor interactions in the central nervous system. Studies on the regulation of monoamine receptors by neuropeptides. J Neural Transm Suppl 18:165–179

2. Fuxe K, Borroto-Escuela DO (2016) Heteroreceptor complexes and their allosteric receptor-receptor interactions as a novel biological principle for integration of communication in the CNS: targets for drug development. Neuropsychopharmacology 41(1):380–382. https://doi.org/10.1038/npp.2015.244

3. Borroto-Escuela DO, Wydra K, Pintsuk J, Narvaez M, Corrales F, Zaniewska M, Agnati LF, Franco R, Tanganelli S, Ferraro L, Filip M, Fuxe K (2016) Understanding the functional plasticity in neural networks of the basal ganglia in cocaine use disorder: a role for allosteric receptor-receptor interactions in A2A-D2 heteroreceptor complexes. Neural Plast 2016:4827268. https://doi.org/10.1155/2016/4827268

4. Borroto-Escuela DO, Narvaez M, Wydra K, Pintsuk J, Pinton L, Jimenez-Beristain A, Di Palma M, Jastrzebska J, Filip M, Fuxe K (2017) Cocaine self-administration specifically increases A2AR-D2R and D2R-sigma1R heteroreceptor complexes in the rat nucleus accumbens shell. Relevance for cocaine use disorder. Pharmacol Biochem Behav 155:24–31. https://doi.org/10.1016/j.pbb.2017.03.003

5. Borroto-Escuela DO, Li X, Tarakanov AO, Savelli D, Narvaez M, Shumilov K, Andrade-Talavera Y, Jimenez-Beristain A, Pomierny B, Diaz-Cabiale Z, Cuppini R, Ambrogini P, Lindskog M, Fuxe K (2017) Existence of brain 5-HT1A-5-HT2A Isoreceptor complexes with antagonistic allosteric receptor-receptor interactions regulating 5-HT1A Receptor recognition. ACS Omega 2(8):4779–4789. https://doi.org/10.1021/acsomega.7b00629

6. Borroto-Escuela DO, DuPont CM, Li X, Savelli D, Lattanzi D, Srivastava I, Narvaez M, Di Palma M, Barbieri E, Andrade-Talavera Y, Cuppini R, Odagaki Y, Palkovits M, Ambrogini P, Lindskog M, Fuxe K (2017) Disturbances in the FGFR1-5-HT1A heteroreceptor complexes in the Raphe-hippocampal 5-HT System develop in a genetic rat model of depression. Front Cell Neurosci 11:309. https://doi.org/10.3389/fncel.2017.00309

7. Borroto-Escuela DO, Tarakanov AO, Fuxe K (2016) FGFR1-5-HT1A heteroreceptor complexes: implications for understanding and treating major depression. Trends Neurosci 39(1):5–15. https://doi.org/10.1016/j.tins.2015.11.003

8. Borroto-Escuela DO, Wydra K, Ferraro L, Rivera A, Filip M, Fuxe K (2015) Role of D2-like heteroreceptor complexes in the effects of cocaine, morphine and hallucinogens. In: Preedy V (ed) Neuropathology of drug addictions and substance misuse, vol 1. Elsevier, London. pp 93–101. https://doi.org/10.15379/2409-3564.2015.02.01.5

9. Fuxe K, Borroto-Escuela DO, Tarakanov AO, Romero-Fernandez W, Ferraro L, Tanganelli S, Perez-Alea M, Di Palma M, Agnati LF (2014) Dopamine D2 heteroreceptor complexes and their receptor-receptor interactions in ventral striatum: novel targets for antipsychotic drugs. Prog Brain Res 211:113–139. https://doi.org/10.1016/B978-0-444-63425-2.00005-2

10. Fuxe K, Guidolin D, Agnati LF, Borroto-Escuela DO (2015) Dopamine heteroreceptor complexes as therapeutic targets in Parkinson's disease. Expert Opin Ther Targets 19(3):377–398. https://doi.org/10.1517/14728222.2014.981529

11. Borroto-Escuela DO, Brito I, Romero-Fernandez W, Di Palma M, Oflijan J, Skieterska K, Duchou J, Van Craenenbroeck K, Suarez-Boomgaard D, Rivera A, Guidolin D, Agnati LF, Fuxe K (2014) The G protein-coupled receptor heterodimer network (GPCR-HetNet) and its hub components. Int J Mol Sci 15(5):8570–8590. https://doi.org/10.3390/ijms15058570

12. Li S, Wong AH, Liu F (2014) Ligand-gated ion channel interacting proteins and their role in neuroprotection. Front Cell Neurosci 8:125. https://doi.org/10.3389/fncel.2014.00125

13. Wang M, Lee FJ, Liu F (2008) Dopamine receptor interacting proteins (DRIPs) of dopamine D1-like receptors in the central nervous system. Mol Cells 25(2):149–157

14. Lavine N, Ethier N, Oak JN, Pei L, Liu F, Trieu P, Rebois RV, Bouvier M, Hebert TE, Van Tol HH (2002) G protein-coupled receptors form stable complexes with inwardly rectifying potassium channels and adenylyl cyclase. J Biol Chem 277(48):46010–46019. https://doi.org/10.1074/jbc.M205035200

15. Flajolet M, Wang Z, Futter M, Shen W, Nuangchamnong N, Bendor J, Wallach I, Nairn AC, Surmeier DJ, Greengard P (2008) FGF acts as a co-transmitter through adenosine A(2A) receptor to regulate synaptic plasticity. Nat Neurosci 11(12):1402–1409. https://doi.org/10.1038/nn.2216

Contents

Contributors

LUIGI F. AGNATI • *Department of Neuroscience, Karolinska Institutet, Stockholm, Sweden*

DOUNGKAMOL ALONGKRONRUSMEE · *Department of Medicinal Chemistry and Molecular Pharmacology, Purdue University Institute for Integrative Neuroscience, Purdue University, West Lafayette, IN, USA*

PATRIZIA AMBROGINI • *Department of Biomolecular Science, Section of Physiology, University of Urbino, Urbino, Italy*

ANDRY ANDRIANARIVELO • *UPMC Univ Paris 06, INSERM, CNRS, Neurosciences Paris-Seine, Institut de Biologie Paris Seine (NPS-IBPS), Sorbonne Universités, Paris, France; UPMC Univ Paris 06, IPAL, CNRS, Sorbonnes Universités, Paris, France*

TIZIANA ANTONELLI • *Department of Medical Sciences, Section of Pharmacology, University of Ferrara, Ferrara, Italy*

SARAH BEGGIATO • *Department of Life Sciences and Biotechnology (SVEB), Section of Medicinal and Health Products, University of Ferrara, Ferrara, Italy*

NATALE BELLUARDO • *Department of Experimental Biomedicine and Clinical Neurosciences, University of Palermo, Palermo, Italy*

FABIO BENFENATI • *Center for Synaptic Neuroscience and Technology, The Italian Institute of Technology, Genoa, Italy; Laboratory of Molecular and Cellular Neurophysiology, University of Genova, Genoa, Italy*

KYLA BOURQUE • *Department of Pharmacology and Therapeutics, McGill University, Montreal, QC, Canada*

ISMEL BRITO • *Department of Neuroscience, Karolinska Institutet, Stockholm, Sweden; Observatorio Cubano de Neurociencias, Grupo Bohío-Estudio, Yaguajay, Cuba*

BARBARA CHRUŚCICKA • *APC Microbiome Ireland, University College Cork, Cork, Ireland; Department of Anatomy and Neuroscience, University College Cork, Cork, Ireland*

FRANCISCO CIRUELA • *Unitat de Farmacologia, Departament Patologia i Terapèutica Experimental, Facultat de Medicina, IDIBELL, Universitat de Barcelona, L'Hospitalet de Llobregat, Barcelona, Spain*

FIDEL CORRALES • *Facultad de Medicina, Instituto de Investigación Biomédica de Málaga, Universidad de Málaga, Málaga, Spain; Department of Neuroscience, Karolinska Institutet, Stockholm, Sweden; Observatorio Cubano de Neurociencias, Grupo Bohío-Estudio, Yaguajay, Cuba*

MINERVA CRESPO-RAMÍREZ • *Department of Neuroscience, Karolinska Institutet, Stockholm, Sweden*

RICCARDO CUPPINI • *Department of Biomolecular Science, Section of Physiology, University of Urbino, Urbino, Italy*

MAXWELL S. DENIES • *Laboratory of Cellular and Molecular Systems, Department of Mechanical Engineering, University of Michigan, Ann Arbor, MI, USA*

LYES DEROUICHE • *CNRS UPR3212, Institut des Neurosciences Cellulaires et Intégratives, Université de Strasbourg, Strasbourg, France*

DOMINIC DEVOST • *Department of Pharmacology and Therapeutics, McGill University, Montreal, QC, Canada*

VALENTINA DI LIBERTO • *Department of Experimental Biomedicine and Clinical Neurosciences, University of Palermo, Palermo, Italy*

MICHAEL DI PALMA • *Department of Biomolecular Science, Section of Physiology, University of Urbino, Urbino, Italy*

ZAIDA DÍAZ-CABIALE • *Facultad de Medicina, Instituto de Investigación Biomédica de Málaga, Universidad de Málaga, Málaga, Spain*

TIMOTHY G. DINAN • *APC Microbiome Ireland, University College Cork, Cork, Ireland; Department of Psychiatry and Neurobehavioural Science, University College Cork, Cork, Ireland*

CLÉMENTINE M. DRUELLE • *APC Microbiome Ireland, University College Cork, Cork, Ireland; Department of Anatomy and Neuroscience, University College Cork, Cork, Ireland*

DASIEL O. BORROTO-ESCUELA • *Department of Neuroscience, Karolinska Institutet, Stockholm, Sweden; Department of Biomolecular Science, Section of Physiology, University of Urbino, Urbino, Italy; Observatorio Cubano de Neurociencias, Grupo Bohio-Estudio, Yaguajay, Cuba*

SILVIA EUSEBI • *Department of Biomolecular Science, Section of Physiology, University of Urbino, Urbino, Italy*

LUCA FERRARO • *Department of Life Sciences and Biotechnology (SVEB), Section of Medicinal and Health Products, University of Ferrara, Ferrara, Italy*

MALGORZATA FILIP • *Laboratory of Drug Addiction Pharmacology, Institute of Pharmacology, Polish Academy of Sciences, Kraków, Poland*

ANTONIO FLORES-BURGESS • *Facultad de Medicina, Instituto de Investigación Biomédica de Málaga, Universidad de Málaga, Málaga, Spain*

RAFAEL FRANCO • *Centro de Investigación Biomédica en Red sobre Enfermedades Neurodegenerativas (CIBERNED), Instituto de Salud Carlos III, Madrid, Spain; Institute of Biomedicine of the University of Barcelona (IBUB), Barcelona, Spain; Departament de Bioquímica i Biologia Molecular, Facultat de Biologia, Universitat de Barcelona, Barcelona, Spain*

KRISTINA FRIEDLAND • *Molecular & Clinical Pharmacy, Friedrich-Alexander-Universität Erlangen, Erlangen, Germany*

KJELL FUXE • *Department of Neuroscience, Karolinska Institutet, Stockholm, Sweden; Department of Biomolecular Sciences, Section of Physiology, University of Urbino, Urbino, Italy*

BELÉN GAGO • *Facultad de Medicina, Instituto de Investigación Biomédica, Universidad de Málaga, Málaga, Spain*

AYLIN C. HANYALOGLU • *Department of Surgery and Cancer, Institute of Reproductive and Developmental Biology, Imperial College London, London, UK*

TERENCE E. HÉBERT • *Department of Pharmacology and Therapeutics, McGill University, Montreal, QC, Canada*

KIM C. JONAS • *Centre for Medical and Biomedical Education, St George's, University of London, London, UK*

JACE JONES-TABAH • *Department of Pharmacology and Therapeutics, McGill University, Montreal, QC, Canada*

DAVIDE LATTANZI • *Department of Biomolecular Science, Section of Physiology, University of Urbino, Urbino, Italy*

FRANKIE H. F. LEE • *Campbell Family Mental Health Research Institute, Centre for Addiction and Mental Health, Toronto, ON, Canada*

ALLEN P. LIU • *Laboratory of Cellular and Molecular Systems, Department of Mechanical Engineering, University of Michigan, Ann Arbor, MI, USA*

FANG LIU • *Campbell Research Institute, Centre for Addiction and Mental Health, University of Toronto, Toronto, ON, Canada; Department of Physiology, University of Toronto, Toronto, ON, Canada; Department of Psychiatry, University of Toronto, Toronto, ON, Canada*

EVA MARTÍNEZ-PINILLA • *Departamento de Morfología y Biología Celular, Facultad de Medicina, Instituto de Neurociencias del Principado de Asturias (INEUROPA), Universidad de Oviedo, Asturias, Spain; Instituto de Investigación Sanitaria del Principado de Asturias (ISPA), Universidad de Oviedo, Asturias, Spain*

DOMINIQUE MASSOTTE • *CNRS UPR3212, Institut des Neurosciences Cellulaires et Intégratives, Université de Strasbourg, Strasbourg, France*

CARMELO MILLÓN • *Facultad de Medicina, Instituto de Investigación Biomédica de Málaga, Universidad de Málaga, Málaga, Spain*

GIUSEPPA MUDÓ • *Department of Experimental Biomedicine and Clinical Neurosciences, University of Palermo, Palermo, Italy*

VICTORIA L. MURRAY • *Laboratory of Cellular and Molecular Systems, Department of Mechanical Engineering, University of Michigan, Ann Arbor, MI, USA*

JOSÉ ANGEL NARVÁEZ • *Facultad de Medicina, Instituto de Investigación Biomédica de Málaga, Universidad de Málaga, Málaga, Spain*

MANUEL NARVAEZ • *Facultad de Medicina, Instituto de Investigación Biomédica de Málaga, Universidad de Málaga, Málaga, Spain*

GEMMA NAVARRO • *CIBERNED, Centro de Investigación Biomédica en Red sobre Enfermedades Neurodegenerativas, Instituto de Salud Carlos III, Madrid, Spain; Institute of Biomedicine of the University of Barcelona (IBUB), Barcelona, Spain; Department of Biochemistry and Physiology, Faculty of Pharmacy, University of Barcelona, Barcelona, Spain*

STÉPHANE ORY • *CNRS UPR3212, Institut des Neurosciences Cellulaires et Intégratives, Université de Strasbourg, Strasbourg, France*

MIGUEL PÉREZ DE LA MORA • *División de Neurociencias, Instituto de Fisiología Celular, Universidad Nacional Autónoma de México, Mexico City, Mexico*

PAVEL POWLOWSKI • *Department of Pharmacology and Therapeutics, McGill University, Montreal, QC, Canada*

M. ÁNGELES REAL • *Facultad de Ciencias, Instituto de Investigación Biomédica, Universidad de Málaga, Málaga, Spain*

JOSÉ DEL CARMEN REJÓN-ORANTES • *Laboratorio Experimental de Farmacobiología, Facultad de Medicina, Universidad Autónoma de Chiapas, Chis, Mexico*

IRENE REYES-RESINA • *CIBERNED, Centro de Investigación Biomédica en Red sobre Enfermedades Neurodegenerativas, Instituto de Salud Carlos III, Madrid, Spain; Institute of Biomedicine of the University of Barcelona (IBUB), Barcelona, Spain; Department of Biochemistry and Molecular Biomedicine, Faculty of Biology, University of Barcelona, Barcelona, Spain*

ALICIA RIVERA • *Departamento de Biología Celular, Universidad de Málaga, Málaga, Spain; Facultad de Ciencias, Instituto de Investigación Biomédica, Universidad de Málaga, Málaga, Spain*

LUCIANA K. ROSSELLI-MURAI • *Laboratory of Cellular and Molecular Systems, Department of Mechanical Engineering, University of Michigan, Ann Arbor, MI, USA*

DEEPAK K. SAINI • *Department of Molecular Reproduction, Development and Genetics, Indian Institute of Science, Bangalore, India*

ESTEFANI SAINT-JOUR • *UPMC Univ Paris 06, INSERM, CNRS, Neurosciences Paris-Seine, Institut de Biologie Paris Seine (NPS-IBPS), Sorbonne Universités, Paris, France; UPMC Univ Paris 06, IPAL, CNRS, Sorbonnes Universités, Paris, France*

STEFANO SARTINI • *Department of Biomolecular Science, Section of Physiology, University of Urbino, Urbino, Italy*

DAVID SAVELLI • *Department of Biomolecular Science, Section of Physiology, University of Urbino, Urbino, Italy*

THORSTEN SCHAEFER • *Department of Neuroscience, Karolinska Institutet, Stockholm, Sweden; Molecular & Clinical Pharmacy, Friedrich-Alexander-Universität Erlangen, Erlangen, Germany*

HARRIËT SCHELLEKENS • *APC Microbiome Ireland, University College Cork, Cork, Ireland; Department of Anatomy and Neuroscience, University College Cork, Cork, Ireland*

KIRILL SHUMILOV • *Facultad de Ciencias, Instituto de Investigación Biomédica, Universidad de Málaga, Málaga, Spain; Departamento de Biología Celular, Universidad de Málaga, Málaga, Spain; Department of Neuroscience, Karolinska Institutet, Stockholm, Sweden*

KAMILA SKIETERSKA • *Laboratory of Eukaryotic Gene Expression and Signal Transduction (LEGEST), Ghent University, Ghent, Belgium*

RORY SLENO • *Department of Pharmacology and Therapeutics, McGill University, Montreal, QC, Canada*

ELISABETH M. STEEL • *Laboratory of Cellular and Molecular Systems, Department of Mechanical Engineering, University of Michigan, Ann Arbor, MI, USA*

CHRISTOPHE STOVE • *Laboratory for GPCR Expression and Signal Transduction (L-GEST)/Laboratory of Toxicology, Department of Bioanalysis, Ghent University, Ghent, Belgium*

PING SU • *Campbell Family Mental Health Research Institute, Centre for Addiction and Mental Health, Toronto, ON, Canada*

DIANA SUÁREZ-BOOMGAARD • *Institut de Recherche Interdisciplinaire en Biologie Humaine et Moléculaire, Université Libre de Bruxelles, Brussels, Belgium*

SERGIO TANGANELLI • *Department of Medical Sciences, Section of Pharmacology, University of Ferrara, Ferrara, Italy; Department of Life Sciences and Biotechnology (SVEB), University of Ferrara, Ferrara, Italy*

MARIA CRISTINA TOMASINI • *Department of Life Sciences and Biotechnology, Section of Medicinal and Health Products, University of Ferrara, Ferrara, Italy*

PIERRE TRIFILIEFF • *Nutrition and Integrative Neurobiology, INRA UMR-1286, Bordeaux, France; University of Bordeaux, Bordeaux, France*

ALEJANDRA VALDERRAMA-CARVAJAL • *Facultad de Ciencias, Instituto de Investigación Biomédica, Universidad de Málaga, Málaga, Spain*

ISMAEL VALLADOLID-ACEBES • *The Rolf Luft Research Center for Diabetes and Endocrinology, Karolinska Institutet, Karolinska University Hospital, Stockholm, Sweden*

KATHLEEN VAN CRAENENBROECK • *Laboratory for GPCR Expression and Signal Transduction (L-GEST)/Laboratory of Toxicology, Department of Bioanalysis, Ghent University, Ghent, Belgium; Laboratory of Eukaryotic Gene Expression and Signal Transduction (LEGEST), Ghent University, Ghent, Belgium*

RICHARD M. VAN RIJN • *Department of Medicinal Chemistry and Molecular Pharmacology, Purdue University Institute for Integrative Neuroscience, Purdue University, West Lafayette, IN, USA*

PETER VANHOUTTE • *UPMC Univ Paris 06, INSERM, CNRS, Neurosciences Paris-Seine, Institut de Biologie Paris Seine (NPS-IBPS), Sorbonne Universités, Paris, France; UPMC Univ Paris 06, IPAL, CNRS, Sorbonnes Universités, Paris, France; Neuronal Signaling and Gene Regulation, UMR-S 1130/UMR8246, Université Pierre et Marie Curie-Paris, Paris, France*

LAKSHMI VASUDEVAN • *Laboratory for GPCR Expression and Signal Transduction (L-GEST)/Laboratory of Toxicology, Department of Bioanalysis, Ghent University, Ghent, Belgium*

SHAUNA E. WALLACE FITZSIMONS • *APC Microbiome Ireland, University College Cork, Cork, Ireland; Department of Anatomy and Neuroscience, University College Cork, Cork, Ireland*

VAL J. WATTS • *Department of Medicinal Chemistry and Molecular Pharmacology, Purdue University Institute for Integrative Neuroscience, Purdue University, West Lafayette, IN, USA*

ELISE WOUTERS • *Laboratory for GPCR Expression and Signal Transduction (L-GEST)/ Laboratory of Toxicology, Department of Bioanalysis, Ghent University, Ghent, Belgium*

KAROLINA WYDRA • *Laboratory of Drug Addiction Pharmacology, Institute of Pharmacology, Polish Academy of Sciences, Kraków, Poland*

RAUNER ZALDIVAR-ORO • *Departamento de Bioquímica, Facultat de Biología, Universidad de la Habana, La Habana, Cuba*

MICHELE ZOLI • *Dipartimento di Scienze Biomediche, Metaboliche e Neuroscienze, Università degli studi di Modena e Reggio Emilia, Modena, Italy*

Chapter 1

Analysis and Quantification of GPCR Allosteric Receptor–Receptor Interactions Using Radioligand Binding Assays: The A2AR-D2R Heteroreceptor Complex Example

Dasiel O. Borroto-Escuela, Miguel Pérez de la Mora, Michele Zoli, Fabio Benfenati, Manuel Narvaez, Alicia Rivera, Zaida Díaz-Cabiale, Sarah Beggiato, Luca Ferraro, Sergio Tanganelli, Patrizia Ambrogini, Malgorzata Filip, Fang Liu, Rafael Franco, Luigi F. Agnati, and Kjell Fuxe

Abstract

There is a large body of biochemical and biophysical experimental evidences which establishes the existence of G protein-coupled receptors (GPCRs) as homo- and heteroreceptor complexes. The results indicate that there are allosteric interactions across the receptor–receptor interface of homo- and heteroreceptor complexes that modulate the binding properties of their receptor protomer components in terms of affinity and density, and thereby change their pharmacology. In the adenosine A2A-dopamine D2 heteroreceptor complexes (A2AR-D2R), the activation of the A2AR protomer by its standard receptor agonist CGS21680 causes a conformational change in the A2AR-D2R heteroreceptor complex. The allosteric wave passes over the receptor interface, invades the orthostatic dopamine binding site of the dopamine D2R protomer, and reduces the affinity of the high but not the low affinity D2R agonist binding site. In view of the complex nature of allosteric mechanisms, the detection, analysis, and quantification of the effects of this phenomenon rely on the use of competition radioligand binding assays to ensure proper demonstration of the high and low affinity D2R agonist binding sites. Outlined in this chapter is simple but useful experimental approaches for measuring the allosteric receptor–receptor interactions at GPCR heteroreceptor complexes. The readers will also find tips and discussion on the pitfalls of these assay and instructions for data analysis.

Key words Allosterism, Allosteric receptor–receptor interaction, Radioligand binding, G protein-coupled receptors, Heteroreceptor complexes, Dimerization

1 Introduction

GPCR homo- and heteroreceptor complexes are the subject of much current research [1–9]. There is a high interest in understanding the dynamics of the receptor–receptor and receptor–protein interactions in space and time and their integration in GPCR homo- and heteroreceptor complexes of the Central Nervous

Kjell Fuxe and Dasiel O. Borroto-Escuela (eds.), *Receptor-Receptor Interactions in the Central Nervous System*, Neuromethods, vol. 140, https://doi.org/10.1007/978-1-4939-8576-0_1, © Springer Science+Business Media, LLC, part of Springer Nature 2018

System (CNS) [10–17]. However, the concept of allosteric receptor–receptor interactions is not novel. Indeed, in 1980–1981 we introduced the concept that the monoamine and peptide signals in the CNS became integrated through direct peptide receptor–monoamine receptor interactions in the plasma membrane [18, 19]. This hypothesis was tested in membrane preparations of various CNS regions of the rat brain and we observed that neuropeptides could modulate the binding characteristics especially the affinity of the monoamine receptors in a receptor subtype specific way [18–20]. Thus, intramembrane receptor–receptor interactions did exist in addition to indirect actions via phosphorylation and changes in membrane potential. However, it took over 10 years before they began to have an impact in the GPCR receptor field.

Our results were in line with earlier findings by Lefkowitz, Limbird and colleagues in 1975 [21] showing negative cooperativity in β-adrenergic receptors, which could be explained by the existence of receptor homodimers leading to site–site interactions. In 1982, Birdsall [22] in an interesting opinion paper discusses several examples of the modulation of a receptor response in a cell by activation of a second receptor and highlights their importance as integrative and discriminative mechanisms, especially when the interactions are observed in membrane preparations. As a logical consequence for the indications of direct physical interactions between neuropeptide and monoamine receptors, the term heteromerization was introduced by us in 1993 to describe a specific interaction between different types of GPCRs [23] which sometimes can involve an adapter protein and sometimes require the assistance of scaffolding proteins to allow their interaction to occur. Even higher order heteromers called receptor mosaics were postulated [24]. The assembly of interacting GPCRs in homo- and heteroreceptor complexes leads to changes in the recognition, signaling, and trafficking of participating protomers via allosteric mechanisms [25–27] taking place intermolecularly. The term used for this phenomenon was receptor–receptor interactions [20, 28].

1.1 Allosteric Receptor–Receptor Interactions

Although allosterism encompasses many biologically relevant phenomena [29–31], the focus of this chapter is on the allosteric modulation which takes place via the receptor interface of the receptor protomers within a GPCR homo- or heteroreceptor complex (Fig. 1a–c). Allosteric receptor–receptor interactions occur when the binding of a ligand to an orthostatic or secondary (allosteric) site on a receptor alters or induces a conformational change in the partner receptor protomer within the same homo- or heteroreceptor complex. This allosteric action alters the ability of the partner receptor protomer ligand to bind to its orthosteric binding site which thus becomes modified (Fig. 1a–c). This modulation can be manifested as a change in affinity, and/or as a change in signaling efficacy. Allosteric potentiation of the orthosteric ligand binding of

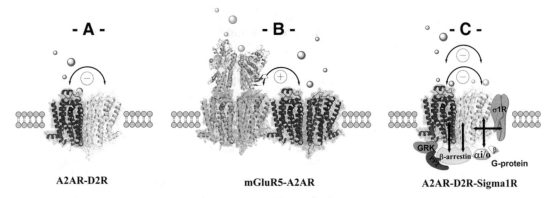

Fig. 1 Illustration of the GPCR-GPCR allosteric receptor–receptor interactions within a GPCR homo- or hetero-receptor complex (**a–c**). Allosteric receptor–receptor interactions occur when the binding of a ligand to an orthostatic or secondary (allosteric) site on a receptor alters or induces a conformational change in the partner receptor protomer within the same homo- or heteroreceptor complex. This modulation can be manifested as a change in affinity, and/or as a change in signaling efficacy. Allosteric potentiation of the orthosteric ligand binding of the partner receptor protomer is sometimes referred to as positive or agonistic allosteric modulation cooperativity, e.g., as observed in the mGluR5-A2AR heteroreceptor complex (panel **b**), whereas a reduction in binding affinity is usually termed negative cooperativity or antagonistic allosteric modulation, e.g., as observed in the A2AR-D2R and A2AR-D2R-Sigma1R heteroreceptor complexes (panel **a** and **c**)

the partner receptor protomer is sometimes referred to as positive cooperativity or agonistic allosteric modulation (Fig. 1b), whereas a reduction in binding affinity is usually termed negative cooperativity or antagonistic allosteric modulation (Fig. 1a). Furthermore, the allosteric receptor–receptor binding interactions are usually reciprocal in nature influencing the binding of one another.

Allosteric receptor–receptor interactions can be studied using either saturation/competition radioligand binding experiments [20, 32], biophysical (RT-FRET, Flash-BRET [33–35]) or functional assays (B-arrestin [36]). The former may be the most direct means for examining allosteric mechanisms that affect binding affinity. The protocols detailed in this chapter are therefore limited to the study of allosteric modulation in equilibrium binding assays. Therefore, they can be applied to a wide range of receptors in a variety of preparations, including purified and solubilized receptors, membrane preparations, whole cells, tissue slices, and even whole animals by selecting the appropriate radioligand and any other orthosteric agents known to be selective for the receptors of interest.

Overall, competition radioligand binding is a highly versatile method and easy to perform. It can even be automated, and the data obtained are usually tight and reproducible.

1.2 Examples of GPCR Allosteric Receptor–Receptor Interactions

There are a considerable number of reports on GPCR homo- and heteroreceptor complexes detected in radioligand binding studies [9, 37–49]. Many of these studies and their conclusions are supported by the use of other biochemical and biophysical approaches;

for example, co-immunoprecipitation, in situ proximity ligation assay, fluorescence and bioluminescence resonance energy transfer methods, and internalization assays.

However, a careful evaluation should be made before concluding that the allosteric receptor–receptor interaction findings provide evidence for the existence of a homo- or heteroreceptor complex. Furthermore, in several cases the reciprocal modulation of the receptor recognition of the two GPCR protomers in homo- and heteroreceptor complexes across the interface is either not detected due to failure to have e.g., radio ligands of high affinity and specificity or has not been investigated.

In addition, further optimization to the allosteric ternary complex model of the GPCR or those models suggested by other authors should be implemented. There is a need to consider other thermodynamic parameters, the multiple interactions of the receptor protomers with chaperon or scaffold proteins, the stoichiometry of the receptor complexes, the mobility of the heteroreceptor complexes in the cell membrane and their temporal and spatial distribution (compartmentalization).

Also the impact of the GPCR–G protein interactions and their allosteric mechanisms [50] should be considered in order to understand the role of the allosteric receptor–receptor interactions (Fig. 1c). The current view is that the receptor upon activation induces a reorganization of the G protein to enhance its dissociation of GDP and reach a stable conformation of Galpha free of nucleotides. Such a state of the G protein can then in turn help move the receptor into an active state of high stability represented by a "closed" conformation around the orthosteric agonist binding pocket. As a result, agonist dissociation is reduced and agonist binding increased [50]. It will be of high interest for the future to understand how the allosteric receptor–receptor interactions modulate the allosteric GPCR–G protein interactions, which could be of high relevance for their modulation of receptor–G protein coupling and thus for receptor signaling. The same holds true also for understanding how the allosteric receptor–receptor interactions can modulate the GRK–GPCR interactions [51].

2 Materials

1. Appropriate membrane preparation from transient transfected HEK293 cells or tissue dissected from the rat brain or human brain.
2. Polytron homogenizer (Brinkmann), chilled.
3. 50-ml round plastic centrifuge tubes (Nalgene).
4. Sorvall RC-5 centrifuge and SS-34 rotor.

5. Glass homogenizer and pestle (e.g., Potter-Elvehjem, VWR).

6. Additional reagents and equipment for determination of protein concentration (e.g., BCA protein assay, Pierce).

7. Incubation Tris-based buffer (50 mM Tris-HCl, 100 mM NaCl, 7 mM MgCl$_2$, 1 mM EDTA, 0.0% BSA, 1 mM DTT, pH 7.4) or incubation HEPES Assay buffer (11.92 g HEPES (50 mM), 0.51 g MgCl$_2$ (2.5 mM), 0.76 g EGTA (2 mM), dissolve with stirring in 900 ml ddH$_2$O, and adjust the pH at room temperature to 7.4 with 0.1 M NaOH, bring volume to 1 l with ddH$_2$O). For more details, see below.

8. Radioligand selective for D2R protomer (D2-likeR antagonist [^{3}H]-raclopride, 2 nM; specific activity 78.1 Ci/mmol, PerkinElmer Life Sciences, Stockholm, Sweden).

9. Unlabeled (non-radioactive) competitive (orthosteric) ligand for the D2-like receptor protomer (quinpirole).

10. Unlabeled (non-radioactive) ligand for determination of non-specific binding of D2R protomer ((+)-butaclamol, Sigma-Aldrich, Stockholm, Sweden).

11. Unlabeled test compound for the A2AR protomer (CGS 21680, A2A receptor agonist) (allosteric modulator of D2 receptor via receptor–receptor interactions).

12. Wash buffer (50 mM Tris-HCl, pH 7.4), ice cold.

13. Scintillation cocktail.

14. Shaking water bath.

15. Hydrophilic (LPB) Durapore ®Membrane, Flat-bottom 96-well filter plates (Millipore, Stockholm, Sweden).

16. MultiScreen™ Vacuum Manifold 96-well (Millipore Corp, Bedford, MA).

17. Scintillation counter and vials.

3 Protocol

3.1 Measurement of Allosteric Receptor–Receptor Interactions by Competition Radioligand Binding Experiments Under Equilibrium Conditions

The adenosine A2A-dopamine D2R heteroreceptor complex as an example.

Herein we present the protocol used to study and quantify the allosteric receptor–receptor interaction phenomena characteristic of the A2AR-D2R heteroreceptor complexes. A2AR-D2R heteroreceptor complexes with antagonistic allosteric receptor–receptor interactions are well-known to exist in the ventral and dorsal striatum [52–55]. Such antagonistic allosteric A2AR–D2R interactions have been demonstrated at the neurochemical levels [8, 11–13] and in the behavioral models, including cocaine reward and cocaine seeking in animals [56–61].

As mentioned above "competition" or "displacement" radioligand binding method is used to detect and quantify a certain type of allosteric receptor–receptor interactions. The use of this protocol allows us to measure the effects of an unlabeled test compound (CGS 21680, A2AR agonist), which binds specifically to the adenosine A2A receptor protomer, on the ability of increasing concentrations of quinpirole (D2R agonist) to compete for the binding of a fixed concentration of a radioligand [3H]-raclopride to the D2-like receptor (D2-like receptor radioligand antagonist). It is also possible to check the reverse effect, namely how the D2R agonist modulates the competition of an A2A receptor agonist with an A2A receptor radioligand for the A2A receptor agonist binding site.

The assay is run in a total volume of 250 μl in 96-well microtiter plates (e.g., flat-bottom, GF/C glass fiber 96-well filter plates or Hydrophilic (LPB) Durapore ®Membrane plates from Millipore). However, it should be possible to scale up the assay volume to 500 μl such that the assays can be in 12 × 75-mm borosilicate test tubes.

1. Prepare a fresh membrane preparation (for tissue membrane homogenate, see below Protocol 3.2). If membranes have been frozen, thaw them at room temperature or by placing them in warm water (~45 °C), then vortex to ensure suspension of the pellet and maintain on ice. Prolonged exposure (>15 min) to temperatures >4 °C can lead to reduced binding capacity, presumably as a result of receptor degradation. The use of protease inhibitors (e.g., 5 μg/ml benzamidine, 1 mM phenylmethylsulfonyl fluoride, 5 μg/ml leupeptin, 5 μg/ml soybean trypsin inhibitor, 10 mM $Na_4P_2O_7$, or a commercially available cocktail—e.g., cocktail of protease inhibitors from Roche Diagnostic) may help to reduce receptor degradation.

2. Determine protein concentration (e.g., BCA protein assay) and dilute the membrane preparation in incubation assay buffer (IB) to a protein concentration ten times greater than that desired in the final assay.

3. Just before use, prepare solutions of the radioligand (D2-likeR antagonist [3H]-raclopride, specific activity 78.1 Ci/mmol, PerkinElmer Life Sciences, Stockholm, Sweden) at a concentration ten times greater than the desired final assay concentration (e.g., 10 × Kd or in the current experiment 20 nM) in incubation assay buffer, 4 °C. The volume of the radioligand solution needed depends upon the number of wells in the assay. Calculation of the radioligand concentration in the stock container from the vendor could be performed using the *Quickcalcs* option of the GraphPad available at https://www.graphpad.com/quickcalcs/radcalcform.cfm. The final concentration of the radioligand should be as low as possible, usually between 0.1 and 1 × Kd. The best is to find and use a

minimal radioligand concentration that yields a reasonable signal-to-noise ratio.

4. Just before use, prepare solutions of quinpirole (the unlabeled competitive (orthosteric) ligand for D2R protomer) at 10× final assay concentrations in incubation assay buffer, 4 °C. Use at least five concentrations at half-log units above and below the expected IC_{50} value. In the current experiment, we use quinpirole concentrations from 0.3 nM to 3 mM range. To define nonspecific binding at the D2R protomer, a 10,000 × Kd concentration of the same unlabeled competitive (orthosteric) D2 agonist ligand (quinpirole) or a D2R antagonist (e.g., butaclamol) could be used. If the unlabeled orthosteric ligand is an agonist, then 1 mM GTP of a nonhydrolyzable GTP analogue should also be incorporated into the assay buffer to minimize the impact on the binding of any agonist-promoted receptor–G protein coupling.

5. Just before use, prepare dilutions or a single dose (usually three time the Kd value) of the unlabeled test compound for the A2AR protomer (CGS 21680, A2A receptor agonist 100 nM, allosteric modulator of D2 receptor via receptor–receptor interaction) at 10× final assay concentrations in incubation assay buffer, 4 °C.

6. To initiate the binding assay, pipette the appropriate amounts of incubation assay buffer, D2-like receptor radioligand ([³H]-raclopride, specific D2-likeR antagonist), unlabeled competitive (orthosteric) ligand (quinpirole, specific D2-likeR agonist), and the unlabeled test compound (CGS 21680, specific for A2AR) into the 96-well microtiter plates for a final volume of 250 μl.

   - *For total binding:* Combine 175 μl incubation assay buffer, 50 μl membrane homogenate, and 25 μl appropriate radioligand concentration ([³H]-raclopride).

   - *For wells containing the competitor:* Combine 150 μl incubation assay buffer, 50 μl membrane homogenate, 25 μl appropriate radioligand concentration ([³H]-raclopride), and 25 μl appropriate competitor concentration (quinpirole).

   - *For wells containing the competitor and the A2A receptor agonist, an allosteric modulator of D2 receptor via receptor–receptor interactions:* Combine 125 μl incubation assay buffer, 50 μl membrane homogenate, 25 μl appropriate radioligand concentration ([³H]-raclopride), 25 μl appropriate competitor concentration (quinpirole), and 25 μl appropriate A2A receptor agonist concentration (CGS 21680).

- *Nonspecific binding:* Combine 150 μl incubation assay buffer, 50 μl membrane homogenate, 25 μl appropriate radioligand concentration ([^{3}H]-raclopride), and 25 μl unlabeled competitor stock (quinpirole, 10,000 × Kd) or another competitor (10 μM butaclamol) at very high concentration, usually in the order of micromoles. Studies should be performed to determine if the test compound affects nonspecific binding, as allosteric modulators often affect the binding of the unlabeled competitive agent used to define nonspecific binding.

7. Shake the 96-well plates to ensure complete mixing of contents and incubate the plate for 90 min or longer at 37 °C to ensure binding equilibrium is attained. Because allosteric modulators of orthosteric ligand affinity can exert significant effects on the kinetics of radioligand binding, different incubation times should be tested to ascertain that the binding curve is generated under equilibrium conditions.

8. While the assay is being performed, soak a Hydrophilic (LPB) Durapore ®Membrane in Flat-bottom 96-well filter plates in 0.5% polyethyleneimine for 60 min, followed by 1 ml ice-cold homogenization buffer. This prechills the wells of the plate and serves to reduce dissociation of bound radioligand from the receptor, which can occur upon warming during separation of bound from free radioligand.

9. To terminate the binding assay reaction, the assay mixture incubated in a 96-well plate in step 6–7 is carefully transferred with a multichannel to the pre-soaked Hydrophilic (LPB) Durapore ®Membrane, Flat-bottom 96-well filter plates. And the wells content is filtered with a vacuum filtration manifold (e.g., MultiScreen™ Vacuum Manifold 96-well).

10. Wash the plate five times, each time with a 250 μl aliquot of ice-cold wash buffer.

11. Allow the filters on the plate to dry thoroughly and then place them in scintillation vials and add scintillation cocktail. The radioactivity bound to the membranes on the filter bottom of each well can be quantified by either adding scintillation cocktail to each well and counting with a plate counter (e.g., Packard TopCount), or by transferring the filter from each well, using a sharp forceps, to a 5-ml scintillation vial, adding scintillation cocktail and counting in a standard liquid scintillation counter. Allow the filters to become uniformly translucent prior to quantifying radioactivity in the scintillation counter.

12. Place 25 μl of each radioligand concentration directly into individual 5-ml scintillation vials. After quantifying this radioactivity and taking into account the specific activity of the

radioligand, the concentration of the radioligand in each condition can be accurately assessed.

13. Data Analysis. Data from the competition experiments is analyzed by nonlinear regression analysis using a commercial program GraphPad Prism 5.0 (GraphPad Software Inc., San Diego, CA) (Fig. 2). There are many commercial packages available to analyze receptor binding data. One available for Macintosh OS and Windows is the one used in this work (GraphPad Prism (GraphPad Software; http://www.graphpad.com)). This general graphing and data analysis software comes with an excellent tutorial for receptor binding data analysis and for nonlinear curve fitting in general. Procedures for analysis of competition binding experiments are somewhat similar to those for saturation binding. pKi_H and pKi_L values from several replications in each independent experiment are expressed as means±SEM. The effects of CGS 21680 on these values and on the percent proportion of D2R agonist binding sites in the high affinity state (RH) could be evaluated with paired Student's t-test and nonparametric Mann-Whitney U test, respectively.

3.2 Measurement in Tissue Membrane Homogenates

1. Rinse Polytron homogenizer with distilled deionized water and then with ice-cold homogenate wash buffer.

   Homogenate wash buffer: 50 mM Tris-HCl, pH 7.4, 7 mM $MgCl_2$, 1 mM EDTA and a cocktail of protease inhibitors (Roche Diagnostics, Mannheim, Germany) or 11.92 g HEPES (50 mM), 0.51 g $MgCl_2$ (2.5 mM), 0.76 g EGTA (2 mM), pH 7.4.

2. Place 10 ml homogenate wash buffer into a 50-ml round plastic centrifuge tube and keep on ice.

3. Decapitate a 225- to 300-g male Sprague-Dawley rat, remove the brain, and obtain the brain region of interest. Place tissue directly into the chilled homogenate wash buffer in the 50-ml centrifuge tube.

4. Chill the shaft of a Polytron homogenizer with ice-cold homogenate wash buffer, then homogenize tissue at setting 6–8 in three 15-s bursts separated by 15 s. Keep the centrifuge tube on ice while homogenizing because the Polytron generates a damaging amount of heat.

5. The polytron homogenized tissue is re-homogenized using a sonicator (Soniprep 150), setting at 5 in three 15-s bursts separated by 30 s.

6. The membranes are precipitated by centrifugation at 4 °C for 40 min at 40,000 × g (Thermo scientific, Sorvall Lynx 6000, Stockholm, Sweden). Discard supernatant and wash through rehomogenization in the same buffer once more.

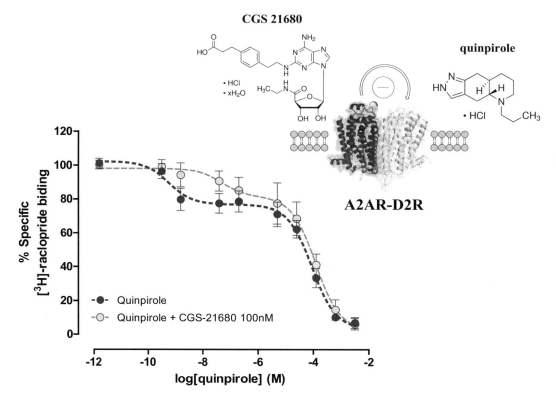

Fig. 2 Modulation by adenosine A2A agonist CGS 21680 (100 nM) of the D2R affinity in the high and low affinity states based on [³H]-raclopride/quinpirole displacement experiments in membrane preparations from the ventral striatum of rat brains. Thus, competition experiments involving dopamine D2-like receptor antagonist [³H]-raclopride binding versus increasing concentrations of quinpirole were performed in ventral striatum in the presence or absence of the adenosine A2A agonist CGS-21680 (100 nM) as indicated. Nonspecific binding was defined as the binding in the presence of 10 μM (+)-butaclamol. The curves are based on the means ± SEM of eight rats, each experiment was performed from one rat in triplicate. The binding values are given in percent of specific binding at the lowest concentration of quinpirole employed

7. Resuspend the pellet in 8 ml homogenate assay buffer and store up to 2 h on ice until use or store at −80 °C until required. Depending on the protein concentration being sought or the type of brain tissue examined, the volume of assay buffer used to resuspend the pellet will vary.

8. Set aside a small volume (10 μl) of the homogenate solution on ice and determine the protein concentration by means of BCA protein assay (Pierce, Sweden) using as a standard bovine serum albumin (BSA) when time allows. The BCA assay is easier to perform than the Lowry assay and has the advantage of not being affected by most types of buffer, eliminating the need for TCA precipitation. Pierce sells ready-to-use BCA kits.

Acknowledgments

The work was supported by the Swedish Medical Research Council (62X-00715-50-3) to K.F., by ParkinsonFonden 2015, 2016 and 2017, AFA Försäkring (130328) to K.F., and by Hjärnfonden (FO2016-0302) and Karolinska Institutet Forskningsstiftelser (2016–2017) to D.O.B-E. D.O.B-E. belongs to the "Academia de Biólogos Cubanos" group.

References

1. Borroto-Escuela DO, Narvaez M, Wydra K, Pintsuk J, Pinton L, Jimenez-Beristain A, Di Palma M, Jastrzebska J, Filip M, Fuxe K (2017) Cocaine self-administration specifically increases A2AR-D2R and D2R-sigma1R heteroreceptor complexes in the rat nucleus accumbens shell. Relevance for cocaine use disorder. Pharmacol Biochem Behav 155:24–31. https://doi.org/10.1016/j.pbb.2017.03.003

2. Borroto-Escuela DO, Narvaez M, Perez-Alea M, Tarakanov AO, Jimenez-Beristain A, Mudo G, Agnati LF, Ciruela F, Belluardo N, Fuxe K (2015) Evidence for the existence of FGFR1-5-HT1A heteroreceptor complexes in the midbrain raphe 5-HT system. Biochem Biophys Res Commun 456(1):489–493. https://doi.org/10.1016/j.bbrc.2014.11.112

3. Moreno JL, Muguruza C, Umali A, Mortillo S, Holloway T, Pilar-Cuellar F, Mocci G, Seto J, Callado LF, Neve RL, Milligan G, Sealfon SC, Lopez-Gimenez JF, Meana JJ, Benson DL, Gonzalez-Maeso J (2012) Identification of three residues essential for 5-hydroxytryptamine 2A-metabotropic glutamate 2 (5-HT2A. mGlu2) receptor heteromerization and its psychoactive behavioral function. J Biol Chem 287(53):44301–44319. https://doi.org/10.1074/jbc.M112.413161

4. Jonas KC, Huhtaniemi I, Hanyaloglu AC (2016) Single-molecule resolution of G protein-coupled receptor (GPCR) complexes. Methods Cell Biol 132:55–72. https://doi.org/10.1016/bs.mcb.2015.11.005

5. Han Y, Moreira IS, Urizar E, Weinstein H, Javitch JA (2009) Allosteric communication between protomers of dopamine class A GPCR dimers modulates activation. Nat Chem Biol 5(9):688–695. https://doi.org/10.1038/nchembio.199

6. Navarro G, Borroto-Escuela D, Angelats E, Etayo I, Reyes-Resina I, Pulido-Salgado M, Rodriguez-Perez AI, Canela EI, Saura J, Lanciego JL, Labandeira-Garcia JL, Saura CA, Fuxe K, Franco R (2018) Receptor-heteromer mediated regulation of endocannabinoid signaling in activated microglia. Role of CB1 and CB2 receptors and relevance for Alzheimer's disease and levodopa-induced dyskinesia. Brain Behav Immun 67:139–151. https://doi.org/10.1016/j.bbi.2017.08.015

7. Schellekens H, De Francesco PN, Kandil D, Theeuwes WF, McCarthy T, van Oeffelen WE, Perello M, Giblin L, Dinan TG, Cryan JF (2015) Ghrelin's orexigenic effect is modulated via a serotonin 2C receptor interaction. ACS Chem Neurosci 6(7):1186–1197. https://doi.org/10.1021/cn500318q

8. Zheng Y, Akgun E, Harikumar KG, Hopson J, Powers MD, Lunzer MM, Miller LJ, Portoghese PS (2009) Induced association of mu opioid (MOP) and type 2 cholecystokinin (CCK2) receptors by novel bivalent ligands. J Med Chem 52(2):247–258. https://doi.org/10.1021/jm800174p

9. Borroto-Escuela DO, Brito I, Romero-Fernandez W, Di Palma M, Oflijan J, Skieterska K, Duchou J, Van Craenenbroeck K, Suarez-Boomgaard D, Rivera A, Guidolin D, Agnati LF, Fuxe K (2014) The G protein-coupled receptor heterodimer network (GPCR-HetNet) and its hub components. Int J Mol Sci 15(5):8570–8590. https://doi.org/10.3390/ijms15058570

10. Borroto-Escuela DO, Fuxe K (2017) Diversity and bias through dopamine D2R heteroreceptor complexes. Curr Opin Pharmacol 32:16–22. https://doi.org/10.1016/j.coph.2016.10.004

11. Borroto-Escuela DO, Wydra K, Pintsuk J, Narvaez M, Corrales F, Zaniewska M, Agnati LF, Franco R, Tanganelli S, Ferraro L, Filip M, Fuxe K (2016) Understanding the functional plasticity in neural networks of the basal ganglia in cocaine use disorder: a role for allosteric receptor-receptor interactions in A2A-D2 heteroreceptor complexes. Neural Plast 2016:4827268. https://doi.org/10.1155/2016/4827268

12. Fuxe K, Borroto-Escuela DO (2016) Heteroreceptor complexes and their allosteric receptor-receptor interactions as a novel biological principle for integration of communication in the CNS: targets for drug development. Neuropsychopharmacology 41(1):380–382. https://doi.org/10.1038/npp.2015.244

13. Pintsuk J, Borroto-Escuela DO, Lai TK, Liu F, Fuxe K (2016) Alterations in ventral and dorsal striatal allosteric A2AR-D2R receptor-receptor interactions after amphetamine challenge: relevance for schizophrenia. Life Sci 167:92–97

14. Borroto-Escuela DO, Brito I, Di Palma M, Jiménez-Beristain A, Narvaez M, Corrales F, Pita-Rodríguez M, Sartini S, Ambrogini P, Lattanzi D, Cuppini R, Agnati LF, Fuxe K (2015) On the role of the balance of GPCR homo/heteroreceptor complexes in the brain. J Adv Neurosci Res 2:36–44

15. Fuxe K, Agnati LF, Borroto-Escuela DO (2014) The impact of receptor-receptor interactions in heteroreceptor complexes on brain plasticity. Expert Rev Neurother 14(7):719–721. https://doi.org/10.1586/14737175.2014.922878

16. Fuxe K, Borroto-Escuela D, Fisone G, Agnati LF, Tanganelli S (2014) Understanding the role of heteroreceptor complexes in the central nervous system. Curr Protein Pept Sci 15(7):647

17. Fuxe K, Borroto-Escuela DO, Ciruela F, Guidolin D, Agnati LF (2014) Receptor-receptor interactions in heteroreceptor complexes: a new principle in biology. Focus on their role in learning and memory. Neurosci Discov 2(1):6. https://doi.org/10.7243/2052-6946-2-6

18. Agnati LF, Fuxe K, Zini I, Lenzi P, Hokfelt T (1980) Aspects on receptor regulation and isoreceptor identification. Med Biol 58(4):182–187

19. Fuxe K, Agnati LF, Benfenati F, Cimmino M, Algeri S, Hokfelt T, Mutt V (1981) Modulation by cholecystokinins of 3H-spiroperidol binding in rat striatum: evidence for increased affinity and reduction in the number of binding sites. Acta Physiol Scand 113(4):567–569

20. Fuxe K, Agnati LF, Benfenati F, Celani M, Zini I, Zoli M, Mutt V (1983) Evidence for the existence of receptor--receptor interactions in the central nervous system. Studies on the regulation of monoamine receptors by neuropeptides. J Neural Transm Suppl 18:165–179

21. Limbird LE, Meyts PD, Lefkowitz RJ (1975) Beta-adrenergic receptors: evidence for negative cooperativity. Biochem Biophys Res Commun 64(4):1160–1168. doi: 0006-291X(75)90815-3 [pii]

22. Birdsall NJM (1982) Can different receptors interact directly with each other? Trends Neurosci 5:137–138

23. Zoli M, Agnati LF, Hedlund PB, Li XM, Ferre S, Fuxe K (1993) Receptor-receptor interactions as an integrative mechanism in nerve cells. Mol Neurobiol 7(3–4):293–334. https://doi.org/10.1007/BF02769180

24. Agnati LF, Fuxe K, Zoli M, Rondanini C, Ogren SO (1982) New vistas on synaptic plasticity: the receptor mosaic hypothesis of the engram. Med Biol 60(4):183–190

25. Koshland DE Jr, Nemethy G, Filmer D (1966) Comparison of experimental binding data and theoretical models in proteins containing subunits. Biochemistry 5(1):365–385

26. Monod J, Wyman J, Changeux JP (1965) On the nature of allosteric transitions: a plausible model. J Mol Biol 12:88–118

27. Tsai CJ, Del Sol A, Nussinov R (2009) Protein allostery, signal transmission and dynamics: a classification scheme of allosteric mechanisms. Mol BioSyst 5(3):207–216. https://doi.org/10.1039/b819720b

28. Fuxe K, Agnati LF (1987) Receptor-receptor interactions. A new intramembrane integrative mechanisms. McMillan Press, London

29. Birdsall NJ (2010) Class A GPCR heterodimers: evidence from binding studies. Trends Pharmacol Sci 31(11):499–508. https://doi.org/10.1016/j.tips.2010.08.003

30. Luttrell LM, Kenakin TP (2011) Refining efficacy: allosterism and bias in G protein-coupled receptor signaling. Methods Mol Biol 756:3–35. https://doi.org/10.1007/978-1-61779-160-4_1

31. Christopoulos A, Kenakin T (2002) G protein-coupled receptor allosterism and complexing. Pharmacol Rev 54(2):323–374

32. Borroto-Escuela DO, Marcellino D, Narvaez M, Flajolet M, Heintz N, Agnati L, Ciruela F, Fuxe K (2010) A serine point mutation in the adenosine A2AR C-terminal tail reduces receptor heteromerization and allosteric modulation of the dopamine D2R. Biochem Biophys Res Commun 394(1):222–227. https://doi.org/10.1016/j.bbrc.2010.02.168

33. Cottet M, Faklaris O, Falco A, Trinquet E, Pin JP, Mouillac B, Durroux T (2013) Fluorescent ligands to investigate GPCR binding properties and oligomerization. Biochem Soc Trans 41(1):148–153. https://doi.org/10.1042/BST20120237

34. Cottet M, Faklaris O, Maurel D, Scholler P, Doumazane E, Trinquet E, Pin JP, Durroux T (2012) BRET and Time-resolved FRET strategy to study GPCR oligomerization: from cell lines toward native tissues. Front Endocrinol 3:92. https://doi.org/10.3389/fendo.2012.00092

35. Comps-Agrar L, Maurel D, Rondard P, Pin JP, Trinquet E, Prezeau L (2011) Cell-surface protein-protein interaction analysis with time-resolved FRET and snap-tag technologies: application to G protein-coupled receptor oligomerization. Methods Mol Biol 756:201–214. https://doi.org/10.1007/978-1-61779-160-4_10

36. Borroto-Escuela DO, Romero-Fernandez W, Tarakanov AO, Ciruela F, Agnati LF, Fuxe K (2011) On the existence of a possible A2A-D2-beta-Arrestin2 complex: A2A agonist modulation of D2 agonist-induced beta-arrestin2 recruitment. J Mol Biol 406(5):687–699. https://doi.org/10.1016/j.jmb.2011.01.022

37. Albizu L, Holloway T, Gonzalez-Maeso J, Sealfon SC (2011) Functional crosstalk and heteromerization of serotonin 5-HT2A and dopamine D2 receptors. Neuropharmacology 61(4):770–777. https://doi.org/10.1016/j.neuropharm.2011.05.023

38. Borroto-Escuela DO, Romero-Fernandez W, Narvaez M, Oflijan J, Agnati LF, Fuxe K (2014) Hallucinogenic 5-HT2AR agonists LSD and DOI enhance dopamine D2R protomer recognition and signaling of D2-5-HT2A heteroreceptor complexes. Biochem Biophys Res Commun 443(1):278–284. https://doi.org/10.1016/j.bbrc.2013.11.104

39. Hill SJ, May LT, Kellam B, Woolard J (2014) Allosteric interactions at adenosine A(1) and A(3) receptors: new insights into the role of small molecules and receptor dimerization. Br J Pharmacol 171(5):1102–1113. https://doi.org/10.1111/bph.12345

40. Ferre S, von Euler G, Johansson B, Fredholm BB, Fuxe K (1991) Stimulation of high-affinity adenosine A2 receptors decreases the affinity of dopamine D2 receptors in rat striatal membranes. Proc Natl Acad Sci U S A 88(16):7238–7241

41. Barki-Harrington L, Luttrell LM, Rockman HA (2003) Dual inhibition of beta-adrenergic and angiotensin II receptors by a single antagonist: a functional role for receptor-receptor interaction in vivo. Circulation 108(13):1611–1618. https://doi.org/10.1161/01.CIR.0000092166.30360.78

42. Dasgupta S, Li XM, Jansson A, Finnman UB, Matsui T, Rinken A, Arenas E, Agnati LF, Fuxe K (1996) Regulation of dopamine D2 receptor affinity by cholecystokinin octapeptide in fibroblast cells cotransfected with human CCKB and D2L receptor cDNAs. Brain Res Mol Brain Res 36(2):292–299

43. Borroto-Escuela DO, Romero-Fernandez W, Tarakanov AO, Gomez-Soler M, Corrales F, Marcellino D, Narvaez M, Frankowska M, Flajolet M, Heintz N, Agnati LF, Ciruela F, Fuxe K (2010) Characterization of the A2AR-D2R interface: focus on the role of the C-terminal tail and the transmembrane helices. Biochem Biophys Res Commun 402(4):801–807. https://doi.org/10.1016/j.bbrc.2010.10.122

44. Perron A, Sharif N, Sarret P, Stroh T, Beaudet A (2007) NTS2 modulates the intracellular distribution and trafficking of NTS1 via heterodimerization. Biochem Biophys Res Commun 353(3):582–590. https://doi.org/10.1016/j.bbrc.2006.12.062

45. Koschatzky S, Tschammer N, Gmeiner P (2011) Cross-receptor interactions between dopamine D2L and neurotensin NTS1 receptors modulate binding affinities of dopaminergics. ACS Chem Neurosci 2(6):308–316. https://doi.org/10.1021/cn200020y

46. Pfeiffer M, Kirscht S, Stumm R, Koch T, Wu D, Laugsch M, Schroder H, Hollt V, Schulz S (2003) Heterodimerization of substance P and mu-opioid receptors regulates receptor trafficking and resensitization. J Biol Chem 278(51):51630–51637. https://doi.org/10.1074/jbc.M307095200

47. Pfeiffer M, Koch T, Schroder H, Laugsch M, Hollt V, Schulz S (2002) Heterodimerization of somatostatin and opioid receptors cross-modulates phosphorylation, internalization, and desensitization. J Biol Chem 277(22):19762–19772. https://doi.org/10.1074/jbc.M110373200

48. Romero-Fernandez W, Borroto-Escuela DO, Agnati LF, Fuxe K (2013) Evidence for the existence of dopamine D2-oxytocin receptor heteromers in the ventral and dorsal striatum with facilitatory receptor-receptor interactions. Mol Psychiatry 18(8):849–850. https://doi.org/10.1038/mp.2012.103

49. Borroto-Escuela DO, Li X, Tarakanov AO, Savelli D, Narvaez M, Shumilov K, Andrade-Talavera Y, Jimenez-Beristain A, Pomierny B, Diaz-Cabiale Z, Cuppini R, Ambrogini P, Lindskog M, Fuxe K (2017) Existence of brain 5-HT1A-5-HT2A isoreceptor complexes with antagonistic allosteric receptor-receptor interactions regulating 5-HT1A receptor recognition. ACS Omega 2(8):4779–4789. https://doi.org/10.1021/acsomega.7b00629

50. Mahoney JP, Sunahara RK (2016) Mechanistic insights into GPCR-G protein interactions. Curr Opin Struct Biol 41:247–254. https://doi.org/10.1016/j.sbi.2016.11.005

51. Komolov KE, Benovic JL (2017) G protein-coupled receptor kinases: past, present and future. Cell Signal 41:17–24. https://doi.org/10.1016/j.cellsig.2017.07.004

52. Fuxe K, Ferre S, Zoli M, Agnati LF (1998) Integrated events in central dopamine transmission as analyzed at multiple levels. Evidence for intramembrane adenosine A2A/dopamine D2 and adenosine A1/dopamine D1 receptor interactions in the basal ganglia. Brain Res Brain Res Rev 26(2–3):258–273

53. Trifilieff P, Rives ML, Urizar E, Piskorowski RA, Vishwasrao HD, Castrillon J, Schmauss C, Slattman M, Gullberg M, Javitch JA (2011) Detection of antigen interactions ex vivo by proximity ligation assay: endogenous dopamine D2-adenosine A2A receptor complexes in the striatum. BioTechniques 51(2):111–118. https://doi.org/10.2144/000113719

54. Borroto-Escuela DO, Romero-Fernandez W, Garriga P, Ciruela F, Narvaez M, Tarakanov AO, Palkovits M, Agnati LF, Fuxe K (2013) G protein-coupled receptor heterodimerization in the brain. Methods Enzymol 521:281–294. https://doi.org/10.1016/B978-0-12-391862-8.00015-6

55. Fuxe K, Borroto-Escuela DO, Romero-Fernandez W, Palkovits M, Tarakanov AO, Ciruela F, Agnati LF (2014) Moonlighting proteins and protein-protein interactions as neurotherapeutic targets in the G protein-coupled receptor field. Neuropsychopharmacology 39(1):131–155. https://doi.org/10.1038/npp.2013.242

56. Filip M, Frankowska M, Zaniewska M, Przegalinski E, Muller CE, Agnati L, Franco R, Roberts DC, Fuxe K (2006) Involvement of adenosine A2A and dopamine receptors in the locomotor and sensitizing effects of cocaine. Brain Res 1077(1):67–80. https://doi.org/10.1016/j.brainres.2006.01.038

57. Filip M, Zaniewska M, Frankowska M, Wydra K, Fuxe K (2012) The importance of the adenosine A(2A) receptor-dopamine D(2) receptor interaction in drug addiction. Curr Med Chem 19(3):317–355

58. Wydra K, Golembiowska K, Suder A, Kaminska K, Fuxe K, Filip M (2015) On the role of adenosine (A)(2)A receptors in cocaine-induced reward: a pharmacological and neurochemical analysis in rats. Psychopharmacology 232(2):421–435. https://doi.org/10.1007/s00213-014-3675-2

59. Wydra K, Golembiowska K, Zaniewska M, Kaminska K, Ferraro L, Fuxe K, Filip M (2013) Accumbal and pallidal dopamine, glutamate and GABA overflow during cocaine self-administration and its extinction in rats. Addict Biol 18(2):307–324. https://doi.org/10.1111/adb.12031

60. O'Neill CE, Hobson BD, Levis SC, Bachtell RK (2014) Persistent reduction of cocaine seeking by pharmacological manipulation of adenosine A1 and A 2A receptors during extinction training in rats. Psychopharmacology 231(16):3179–3188. https://doi.org/10.1007/s00213-014-3489-2

61. O'Neill CE, LeTendre ML, Bachtell RK (2012) Adenosine A2A receptors in the nucleus accumbens bi-directionally alter cocaine seeking in rats. Neuropsychopharmacology 37(5):1245–1256. https://doi.org/10.1038/npp.2011.312

Chapter 2

Analysis and Quantification of GPCR Heteroreceptor Complexes and Their Allosteric Receptor–Receptor Interactions Using Radioligand Binding Autoradiography

Manuel Narvaez, Fidel Corrales, Ismel Brito, Ismael Valladolid-Acebes, Kjell Fuxe, and Dasiel O. Borroto-Escuela

Abstract

G protein-coupled receptors (GPCRs) complexes and their allosteric receptor–receptor interactions represent a new fundamental principle in molecular medicine for integration of transmitter signals in the plasma membrane. The allosteric receptor–receptor interactions in heteroreceptor complexes give diversity, specificity, and bias to the receptor protomers due to conformational changes in discrete domains leading to changes in receptor protomer function and their pharmacology. Therefore, a novel understanding of the molecular basis of central nervous system diseases should consider this phenomena and new strategies for mental and neurodegenerative disorders treatment should target heteroreceptor complexes based on a new pharmacology with combined treatment, multi-targeted drugs and heterobivalent drugs. In this chapter, it is described a technique to visualize the majority of G protein-coupled receptor allosteric receptor–receptor interactions in sections of frozen brain tissue using receptor autoradiography. The basic procedure involves incubating slide-mounted tissue sections with radioligands, washing and drying of the sections with specifically bound ligands under conditions that preserve ligand binding, and visualizing and quantifying the binding sites in the tissues. Protocols for brain extraction and sectioning, radioligand exposure, autoradiogram generation, and data quantification are provided, as are the optimal incubation conditions for the autoradiographic visualization of allosteric receptor–receptor interactions using agonist and antagonist radioligands.

Key words Receptor, Autoradiography, Radioligand binding, Brain cryosections, Brain tissue, Allosteric receptor–receptor interaction, Heteroreceptor complexes

1 Introduction

Receptor autoradiography is the localization of radioactive ligands bound to specific receptors in tissue sections. The distribution of the radioligand in the tissue is mapped when the energy emitted from the radioactive molecules collides with nuclear emulsion or film opposed to the tissue section. Thus, the autoradiograms generated provided a detailed localization of specific receptor or other molecules of interest. Autoradiography is considered a very sensi-

Kjell Fuxe and Dasiel O. Borroto-Escuela (eds.), *Receptor-Receptor Interactions in the Central Nervous System*, Neuromethods, vol. 140, https://doi.org/10.1007/978-1-4939-8576-0_2, © Springer Science+Business Media, LLC, part of Springer Nature 2018

tive approach which allows the detection of low levels of receptors in specific zones [1].

1.1 History

The first autoradiography was obtained accidently around 1867 when Niepce de St. Victor observes blackening produced on emulsions of silver chloride and iodide by uranium salts. The incident was noticed by Henri Becquerel in 1869, when he observed that opaque paper placed between the uranium nitrate and the emulsion experienced the same darkening effect [2]. In 1924, the first biological experiment involving autoradiography traces the distribution of polonium in biological specimens [3]. The modern technique was first used in the 1940s, and autoradiograms prepared by exposing lantern slides were used to trace ^{131}I in thyroid sections. The resolution of these autoradiographs was improved by painting slide emulsion directly onto the tissue sections, and the stripping of emulsion from slides led to the development of stripping film [1, 3]. In the late 1950s, ^{131}I-Labeled trifluoroiodomethane was used to measure rat cerebral blood flow, despite the high volatility of the tracer compound, and the low resolution of the resulting images. When the tracer was replaced with ^{14}C-labeled antipyrine, permeability across the blood–brain barrier was reduced [1–3]. In the late 1970s, ^{14}C-Labeled iodoantipyrine yields accurate measurements of cerebral blood flow, and becomes the reference diffusible tracer for this experiment [2].

Important step forward on the use of receptor autoradiography took place during the 1980s, when quantitative receptor autoradiography was used in studies on the distribution of transmitter receptors in the central nervous system [4–6]. In 1993 were shown autoradiographic evidences for receptor–receptor interaction in the rat brain [7]. Nowadays, autoradiography is still becoming a part of the tool box to study how agonist activation of one receptor can alter the binding characteristics of another receptor in distinct brain regions. Therefore, it helps to understand the contribution of receptor–receptor interactions to the brain functions studied [8].

1.2 Components of Autoradiography

Autoradiography consists of the following parts: the biological samples, the radionuclide that is used as a tracer to label the sample, and the detecting medium which is a film in traditional method.

- Specimen/sample. There are many types of specimen that are used in autoradiography either from biological organisms, chromatography, or electrophoresis. The prepared samples should be very thin (few μm) whenever the main interest is high resolution. The samples can be acquired from the entire organism (animals) or just take slices of tissue.

- Radionuclides. The main considerations that need to be taken into account when choosing the radioisotopes to perform autoradiography are the characteristics of the radiation emitted from the biological aspect. Radioactive decay produces emitted particles which are capable of ionizing matter. Of particular interest in autoradiography are beta particles, positrons, and alpha particles. For receptor autoradiography are important principles governing general receptor binding ligands studies. An appropriate radioligand with high specific activity, having saturable binding and pharmacological specificity for the receptor of interest, must be available. Many ligands bind to more than one site; in that case, the radioligand may still be used if one of the sites can be masked by a cold ligand. A ligand also may be radiolabeled antibody specific for the protein of interest [9]. Two commonly used radioisotopes are ^{125}I and ^{3}H. Tritium produces low-energy β particles that do not penetrate further than 5 μm in tissue. The thickness of the sections above 5 μm is not crucial since only the labeling on the surface of the section will be recorded on the film. The resolution is high but the exposure times needed are longer. With ^{125}I-ligands the high-energy γ rays they emit penetrate the tissue much further from the source, reducing the resolution of the autoradiograms. The advantage of the higher energy emitting isotopes is that the exposure times are from hours to days.

- Film. The film for autoradiography consists of three main layers: a flexible base, the photosensitive emulsion, and a protective supercoat (Fig. 1a). The base is 200 μm thick and is made of polyester or triacetate. The supercoat is 1–10 μm thick and is made of non-photosensitive gelatin. The photographic emulsion is usually 10–30 μm thick, and is composed of silver halide grains (AgI, AgBr, AgCl) dispersed within gelatin. The grains are 1 μm or greater in diameter; the larger grains show better sensitivity, but display low resolution. Conversely, smaller grains will produce the reverse effect. The latent image formation is due to the oxidization of the bromine atoms, causing the release of free electrons. The free electrons travel trough the crystal until it becomes trapped in one of the imperfections in the lattice. The electrons attract the silver ions, which reduce to silver atom and form clumps called latent image centers. When a single silver halide grain contains a threshold number of latent image centers, the grain becomes developable. The developable grains form the dark areas of the image when the emulsion is developed (Fig. 1b) [3]. The process of development transforms the latent image into a visible image by reducing silver ions into twisted strands of metallic silver. The reducing agent hydroquinone is used to react with silver ions to form silver atoms and quinone. The reaction is catalyzed by the latent image centers and can be accelerated by the agents metol or phenidone. Development is carried out

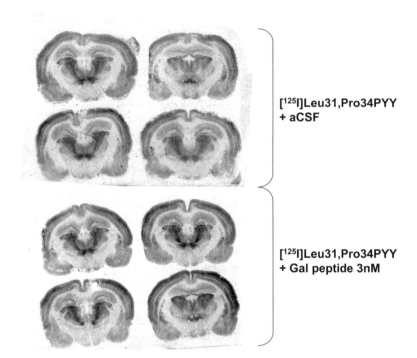

$[^{125}I]$Leu31,Pro34PYY + aCSF

$[^{125}I]$Leu31,Pro34PYY + Gal peptide 3nM

Fig. 1 Representative autoradiograms of [125I]-[Leu31,Pro34]PYY binding in the hippocampal dentate gyrus (Bregma—3.5 mm) showing a higher NPYY1R agonist binding following the icv administration of Galanin 3 nmol

until nearly all the silver contained within developable grains is reduced. Overdevelopment causes fogging, in which silver from grains not containing latent image centers is incorporated into strands. After development, the image is fixed with a thiosulfate solution, which forms complexes with undeveloped silver halide and hardens the gelatin within the photographic emulsion. The image is then ready for qualitative or quantitative analysis.

2 Materials

1. Cryostate (HM550, Microm International).
2. Gelatin-coated slides.
3. Appropriate radioligand.
4. Coplin jars.
5. Krebs-Ringer phosphate buffer (KRP): 5 mM KH_2PO_4, 1 mM $MgSO_4$, 1 mM $CaCl_2$, 136 mM NaCl, 4.7 mM KCl.
6. KRP buffer supplemented with 0.1% Bovine Serum Albumin and 0.05% bacitracin.

7. Autoradiography Imaging Film. Below are given the sensitivities of Amersham films for receptor studies:

8. Hyperfilm MP: ^{3}H, ^{125}I, ^{35}S, ^{14}C.

9. BioMax MR: ^{35}S, ^{14}C.

10. Hyperfilm b-max: ^{125}I, ^{35}S, ^{14}C.

11. Hyperfilm 3H: ^{3}H, ^{125}I.

12. X-ray cassettes.

13. Kodak D-19 Developer (Kodak, Rochester, NY).

14. Kodak Rapid fix (Kodak, Rochester, NY).

3   Method and Protocol

3.1   Brain Removal

For receptor autoradiography the tissue is usually not fixed at all, since most receptors are no longer recognized by their natural ligands with tissue treated with a fixative as formaldehyde. Animals-treated are killed by decapitation without anesthesia and their brains are rapidly removed from the skull. Best tissue quality is preserved through shock-freezing. Strongly recommended it is not to freeze the tissue by simply putting it into a −20 °C refrigerator. The method of choice is to prepare a freezing mixture with dry ice and organic solvent (mostly isopentane) kept at −45 to −40 °C for several minutes. If you wait too long, it will reach −78 °C (the sublimation temperature of CO_2), so temperature must be supervised with a thermometer. To avoid freezing artifacts in the middle of the tissue, at least one dimension must be below 10 mm. For a rat brain, a 100 ml beaker will be sufficient, while for a human brain slice, a 1000 ml jar will be necessary. Then the brain should be transferred to crushed dry ice to prevent expansion and/or cracking of some part of it. For keeping receptor stability during days or weeks, it is recommended to store brains at −80 °C.

3.2   Cutting of Sections

To allow the preparation of thin slices, the tissue is kept frozen in a cryostat. The frozen tissue is transferred to the cryostat chamber (kept at −20 °C) and mounted with a cryo-gel, mostly "Tissue-Tec® O.C.T. compound" (from optimal cutting temperature) to metal holders that can be fixed to the microtome. The mounted tissue is moved across the knife, leaving on it the semi-thin sections (10–15 μm). We have used different thickness to study neuropeptides and transmitter receptor–receptor interactions (10 μm for dorsal raphe nucleus [10–12] or 14 μm for nucleus of the solitary tract [13], arcuate nucleus [14], amygdala [15], and hippocampus [8, 11, 12]).

Thickness of tissue affects on resolution, since emission sources that are distant from the photographic film (as in thick slices) may expose a broader area of the emulsion than sources that are close. In this way, resolution is increased as tissue thickness decreases. The frozen section is taken up with a coated glass slide by "thaw mounting," so the tissue is transformed from the frozen to the unfrozen state. Coated glass slides are available (e.g., coated with poly-lysine). A warm bar moved slowly across the back of the slide produces a controlled wave of melting. The tissue sections are allowed to dry on the coated slide (usually several per slide) at low temperature, but without freezing (to avoid freezing artifacts) [16]. After drying, the slides are stored at −20 °C.

Multiple slides with almost identical tissue sections (i.e., from the same neuroanatomical level) must be obtained. At least will be obtained one slide for total and another for nonspecific binding (specific binding = total binding − nonspecific binding).

3.3 Preincubation

The beneficial effects of preincubation in buffer, which can increase the amount of specific binding, often are attributed to the removal of competing endogenous ligands from the tissue [17]. Temperature, ionic content, and length of preincubation should be optimized empirically. We usually preincubate sections for 1 h at room temperature in a Krebs-Ringer phosphate buffer (KRP) at pH 7.4 [8].

3.4 Incubation

The coating keeps the tissue slice during the incubation to the glass, and the radioligand diffuses freely to its binding sites. The radioligand penetrates the slice and binds to receptors.

Tissue sections are incubated in buffer with radioactive ligand concentration near the K_d (i.e., total binding). The K_d and B_{max} should be calculated from the tissue sections using densitometry or by counting the radioactivity, or based on kinetics of binding to tissue homogenates.

The section is brought in contact with the radioligand either by immersing the slide into a bath or by covering the section with a droplet containing the radioligand.

The tissue should always be incubated in a neutral buffered solution. Fragile tissues may benefit from the inclusion of bovine serum albumin (0.1–1%) and/or protease inhibitors, which also protect peptide ligands.

For visualizing receptor distributions, the length of the incubation should be determined by the optimal signal-to-noise ratio.

Dilution of radioligand depends on number of days under storage to calculate percent remaining. Then stock concentration it is necessary (based on original isotope dilution and activity) to dilute the stock to the final concentration.

We incubate slides with tissue for 2 h in KRP buffer supplemented with 0.1% BSA and 0.05% bacitracin [8].

Nonspecific binding is determined by incubation with cold and radiolabeled ligands together.

3.5 Washing and Drying After Incubation

Specific receptor binding generally dissociates at a temperature-dependent rate. By contrast, nonspecific binding is not temperature dependent and is eliminated by the largest possible number of short-time immersions into cold buffers [16].

To avoid diffusion of reversibly bound ligand, tissue on the slides should be dried as quickly and completely as possible.

In our laboratory, we wash in ice-cold KRP (washing buffer) four times during 2 min. After that, slides are washed (Dip) in distilled water twice (0–4 °C) to remove salts and quickly dry in a stream of cold air [8].

3.6 Generating Autoradiograms

Different types of autoradiographic films are commercially available, each with specific characteristics. Choice depends on the isotope used, levels of sensitivity, or the resolution desired. There are double-sided films coated with emulsion on both faces and monoface films with emulsion on one single side. Double-sided films have greater sensitivity than monoface ones, but provide lower resolution.

Films should be handled under darkroom conditions using the recommended safe light. Tritium sensitive film must be handled carefully since it has no anti-scratch coating. The slides are mounted onto a sheet to prevent shifting, the sheet of slides is put into a cassette, and the film is placed on top. All the slides and standards must be at the same thickness. The length of exposure of the autoradiograms is based on the estimated amount of radioactivity in the tissue.

The film is developed typically in D-19 (Kodak, Rochester, NY) for 2 min, rinsed, and fixed in Kodak Rapid fix for 4 min. The film is rinsed and allowed to dry.

In our protocol, sections are placed in X-ray cassettes and exposed against Hyperfilms (Kodak Biomax MR film, Kodak, Rochester, NY) for 6 days together with ^{125}I microscales (Amersham International) as reference standards [8].

3.7 Analysis of Autoradiograms

The grey levels of the film/screen are evaluated by comparison with brain-mash containing known amounts of radioactivity. Alternatively, calibrated plastic strips are commercially available [18]. Optical density (OD) it is calculated as: $OD = \log(I_o/I)$, where I_o is the intensity of light before passing object, and I is the intensity of light after passing object.

A set of standards must be used with each piece of film since the optical density will depend on several factors, including the film development. Imaging systems are available for quantitating film autoradiograms, as image J system (NIH, USA). A standard curve is made by plotting the radioactivity per area (dpm/mm^2) of the standards versus the optical density of the film. Plots of film density (log of opacity) versus the log of exposure are called characteristic curves, or Hurter–Driffield curves.

The binding may be quantitated as dpm/mg of tissue. Nonspecific binding of an adjacent section can be subtracted from the autoradiograph.

Acknowledgments

The work was supported by the Swedish Medical Research Council (62X-00715-50-3) to K.F., by ParkinsonFonden 2015, 2016 and 2017, AFA Försäkring (130328) to K.F., and by Hjärnfonden (FO2016-0302) and Karolinska Institutet Forskningsstiftelser (2016–2017) to D.O.B-E. D.O.B-E. belongs to the "Academia de Biólogos Cubanos" group.

References

1. Marjorie A (1998) Receptor localization. Laboratory methods and procedures. Wiley, New York

2. Baker JR (1989) Autoradiography: a comprehensive overview. Oxford University Press, Oxford

3. Rogers AW (1979) Techniques of autoradiography. Elsevier/North-Holland Biomedical Press, Amsterdam

4. Fuxe K, Agnati LF, Benfenati F, Andersson K, Camurri M, Zoli M (1983) Evidence for the existence of a dopamine receptor of the D-1 type in the rat median eminence. Neurosci Lett 43(2–3):185–190

5. Fuxe K, Agnati LF, Benfenati F, Cimmino M, Algeri S, Hokfelt T, Mutt V (1981) Modulation by cholecystokinins of 3H-spiroperidol binding in rat striatum: evidence for increased affinity and reduction in the number of binding sites. Acta Physiol Scand 113(4):567–569

6. Fuxe K, Agnati LF, Benfenati F, Celani M, Zini I, Zoli M, Mutt V (1983) Evidence for the existence of receptor–receptor interactions in the central nervous system. Studies on the regulation of monoamine receptors by neuropeptides. J Neural Transm Suppl 18:165–179

7. Fior DR, Hedlund PB, Fuxe K (1993) Autoradiographic evidence for a bradykinin/ angiotensin II receptor-receptor interaction in the rat brain. Neurosci Lett 163(1):58–62

8. Narvaez M, Borroto-Escuela DO, Millon C, Gago B, Flores-Burgess A, Santin L, Fuxe K, Narvaez JA, Diaz-Cabiale Z (2016) Galanin receptor 2-neuropeptide Y Y1 receptor interactions in the dentate gyrus are related with antidepressant-like effects. Brain Struct Funct 221(8):4129–4139. https://doi.org/10.1007/ s00429-015-1153-1

9. Watson JT, Adkins-Regan E, Whiting P, Lindstrom JM, Podleski TR (1988) Autoradiographic localization of nicotinic acetylcholine receptors in the brain of the zebra finch (Poephila guttata). J Comp Neurol 274(2):255–264. https://doi.org/10.1002/ cne.902740209

10. Diaz-Cabiale Z, Parrado C, Narvaez M, Puigcerver A, Millon C, Santin L, Fuxe K, Narvaez JA (2011) Galanin receptor/neuropeptide Y receptor interactions in the dorsal raphe nucleus of the rat. Neuropharmacology 61(1–2):80–86. https://doi.org/10.1016/j. neuropharm.2011.03.002

11. Millon C, Flores-Burgess A, Narvaez M, Borroto-Escuela DO, Santin L, Gago B, Narvaez JA, Fuxe K, Diaz-Cabiale Z (2016) Galanin (1-15) enhances the antidepressant effects of the 5-HT1A receptor agonist 8-OH-DPAT: involvement of the raphe-hippocampal 5-HT neuron system. Brain Struct Funct 221(9):4491–4504. https://doi. org/10.1007/s00429-015-1180-y

12. Flores-Burgess A, Millon C, Gago B, Narvaez M, Borroto-Escuela DO, Mengod G, Narvaez JA, Fuxe K, Santin L, Diaz-Cabiale Z (2017) Galanin (1-15) enhancement of the behavioral effects of Fluoxetine in the forced swimming test gives a new therapeutic strategy against depression. Neuropharmacology 118:233–241. https://doi. org/10.1016/j.neuropharm.2017.03.010

13. Diaz-Cabiale Z, Parrado C, Rivera A, de la Calle A, Agnati L, Fuxe K, Narvaez JA (2006) Galanin-neuropeptide Y (NPY) interactions in central cardiovascular control: involvement of the NPY Y receptor subtype. Eur J Neurosci 24(2):499–508. https://doi. org/10.1111/j.1460-9568.2006.04937.x

14. Parrado C, Diaz-Cabiale Z, Garcia-Coronel M, Agnati LF, Covenas R, Fuxe K, Narvaez JA (2007) Region specific galanin receptor/neuropeptide Y Y1 receptor interactions in the tel- and diencephalon of the rat. Relevance for food consumption. Neuropharmacology 52(2):684–692. https://doi.org/10.1016/j.neuropharm.2006.09.010

15. Narvaez M, Millon C, Borroto-Escuela D, Flores-Burgess A, Santin L, Parrado C, Gago B, Puigcerver A, Fuxe K, Narvaez JA, Diaz-Cabiale Z (2015) Galanin receptor 2-neuropeptide Y Y1 receptor interactions in the amygdala lead to increased anxiolytic actions. Brain Struct Funct 220(4):2289–2301. https://doi.org/10.1007/s00429-014-0788-7

16. Herkenham M, Pert CB (1982) Light microscopic localization of brain opiate receptors: a general autoradiographic method which preserves tissue quality. J Neurosci 2(8):1129–1149

17. Pasternak GW, Wilson HA, Snyder SH (1975) Differential effects of protein-modifying reagents on receptor binding of opiate agonists and antagonists. Mol Pharmacol 11(3):340–351

18. Rainbow TC, Biegon A, Berck DJ (1984) Quantitative receptor autoradiography with tritium-labeled ligands: comparison of biochemical and densitometric measurements. J Neurosci Methods 11(4):231–241

Chapter 3

On the Study of D₄R-MOR Receptor–Receptor Interaction in the Rat Caudate Putamen: Relevance on Morphine Addiction

Alicia Rivera, Alejandra Valderrama-Carvajal, Diana Suárez-Boomgaard, Kirill Shumilov, M. Ángeles Real, Kjell Fuxe, and Belén Gago

Abstract

Receptor–receptor interactions that occur in G protein-coupled receptors (GPCRs) oligomers can be explored using three different functional approaches as starting point: (1) quantitative receptor autoradiography (saturation assay); (2) agonist-stimulated [^{35}S]GTPγS binding in autoradiography; and (3) immunohistochemistry. Together, they allow to explore functional changes in receptors signaling transduction, i.e., receptor recognition, G protein activation, and downstream signaling cascades. Here, we describe these three selected methods that have been successfully employed in the study of the functional interaction of dopamine D₄ (D₄R) and μ opioid (MOR) receptors in the rat caudate putamen in the context of morphine addiction.

Key words G protein-coupled receptors, Receptor–receptor interaction, Quantitative receptor autoradiography, [^{35}S]GTPγS binding in autoradiography, Immunohistochemistry

1 Introduction

Evidences accumulated over the past two decades confirm the existence of a supramolecular organization of the G protein-coupled receptors (GPCRs) to form homo- and heterodimers, as well as higher order multimer complexes in the central nervous system (CNS) [1–3]. The assembly of GPCRs in oligomers leads to allosteric receptor–receptor interaction mechanisms, which may induce changes in the agonist recognition, signaling and trafficking of the participating protomers [2]. Thus, GPCR oligomers have emerged as integrative signaling centers, but also as a source of diversity in receptor responsiveness [4]. To date, a large number of GPCR oligomers has been identified in the CNS [5, 6]. The study of the functional relevance of these oligomers, their role in several neurological disorders (e.g., Parkinson's disease, schizophrenia,

Kjell Fuxe and Dasiel O. Borroto-Escuela (eds.), *Receptor-Receptor Interactions in the Central Nervous System*, Neuromethods, vol. 140, https://doi.org/10.1007/978-1-4939-8576-0_3, © Springer Science+Business Media, LLC, part of Springer Nature 2018

depression, or drug addiction), and their potentiality as target for novel drug treatments represents one of the main challenges of the molecular medicine [4, 7].

In previous works we have postulated the existence of a putative heteroreceptor complex composed by the dopamine D_4 (D_4R) and the μ opioid receptor (MOR) in the rat caudate putamen (dorsal striatum) [8], more specifically in the striosomal compartment. This hypothetical D_4R-MOR heteroreceptor was initially proposed on the bases of the coexistence of these two receptors in the striosomes [9] and the ability of the D_4R to modulate MOR expression in this striatal compartment [10]. Afterwards, preliminary results obtained with biochemical (co-immunoprecipitation) and biophysical (bioluminescence resonance energy transfer, BRET) techniques have confirmed the existence of a D_4R-MOR heteroreceptor in living cells, and more recently with an antibody-based method (in situ proximity ligation assay, PLA) in native tissue (unpublished results). At the functional level, D_4R counteracts molecular, cellular, and behavioral actions induced by morphine, without any effect on the analgesic properties of this drug [11]. These counteractive effects seem to occur through the proposed D_4R-MOR heteroreceptor and the capacity of the D_4R to act as an allosteric modulator of MOR, affecting both MOR recognition and signaling [11–13].

Nowadays, morphine and other opioids are the first-line choice for the management of chronic pain in patients. However, their improperly and widespread use has resulted in a severe public health problem with an epidemic of opioid overdose deaths and addictions [14, 15]. Thus, the D_4R-MOR heteroreceptor emerges as a potential pharmacological target to prevent the adverse effects of morphine in the treatment of pain.

1.1 Methods to Study Receptor–Receptor Interaction in the Central Nervous System

A wide set of methods based on biophysical, biochemical, and structural approaches is available for the study and characterization of GPCR oligomers [16, 17]. In the last years, biophysical techniques such as resonance energy transfer (BRET and FRET), bimolecular fluorescence complementation (BiFC), or a combination of both have taken advantage in the demonstration of GPCR oligomers in living cells [18]. More recently, in situ PLA have been successfully used to demonstrate the molecular existence of GPCR oligomers in their native environment [19]. A large number of methods cannot be used to clearly demonstrate the physical proximity between GPCRs, that is indicative of their oligomerization, but they are applied to explore its physiological significance. In fact, the collection of these indirect functional data has been critical for the development of the novel concept of receptor–receptor interaction in the paradigm of GPCR multimer complexes [8]. Of especial interest is that many of these functional approaches allow an examination of GPCR oligomers function in native tissue.

In this chapter, we propose three different functional approaches as a useful starting point to elucidate the physiological role of GPCR oligomers: (1) quantitative receptor autoradiography; (2) agonist-stimulated [^{35}S]GTPγS binding in autoradiography; and (3) immunohistochemistry. Together, these approaches allow to explore receptor–receptor interaction in three critical points of the receptor signaling transduction process, i.e., receptor recognition, G protein activation, and downstream signaling cascades. We have successfully used these three methods to study the D$_4$R-MOR receptor–receptor interaction in the rat caudate putamen in the context of morphine addiction [11–13, 20].

2   Materials

2.1   Animals

For the study of GPCRs interaction in the central nervous system, we use Sprague-Dawley rats (Charles River, Barcelona, Spain) weighing 220–240 g. Animals were maintained on a standard 12 h light/dark cycle, constant room temperature (20 ± 2 °C), and relative humidity (45 ± 5%). Food pellets and tap water were available ad libitum.

2.1.1   Drug Administration

1. Needles (25G) and 3 ml syringes.
2. Osmotic pump 2ML1 from Alzet® (Cupertino, CA, USA).
3. Morphine sulfate and PD168,077 (D$_4$R agonist, Cat. # 1065, Tocris Bioscience, Avonmouth, UK).
4. Sterile 0.9% NaCl solution.

2.1.2   Subcutaneous Osmotic Pump Implantation

1. Sodium ketamine and medetomidine.
2. Topical antiseptic, usually povidone-iodine 10% solution (Betadine).
3. Surgical instruments and suture material.

2.2   Quantitative μ Opioid Receptor Autoradiography

2.2.1   Tissue Preparation

1. Guillotine and skull cutting forceps.
2. Isopentane.
3. Dry ice.
4. Cryostat (Microm HM 550, Microm Laborgerate S.L., Barcelona, Spain).
5. Silane-coated slides (see **Note 1**).

2.2.2   Autoradiography in Saturation Binding Assay

1. [^{3}H]DAMGO (MOR agonist; specific activity ~50 Ci/nmol, Cat. # NET-902, PerkinElmer, Waltham, MA, USA) stored at −20 °C.
2. Assay buffer: 5% bovine serum albumin (BSA) in Tris-HCl buffer, pH 7.4.

3. Tris-HCl buffer: 50 mM Tris-HCl, pH 7.4.

4. Naloxone (MOR antagonist; Cat. # N7758, Sigma-Aldrich, St. Louis, MO, USA) stored at −20 °C.

5. Coplin jars and humidity chamber.

6. Tritium-sensitive films (BioMax MR Film, Kodak, Rochester, NY, USA) stored at 4 °C.

7. Prefabricated ^{3}H-labeled polymer standard strips (RPA506, GE Healthcare, Piscataway, NJ, USA).

8. Autoradiography cassettes, cards, and adhesive tape.

9. LX 24 X-ray developer (Kodak) and AL-4 X-ray fixer (Kodak).

10. Three trays and print tongs.

11. Darkroom with safelight for processing films.

12. Scintillation fluid (PerkinElmer) and liquid scintillation counter (Beckman LS6500, Beckman Coulter, Brea, CA, USA).

2.2.3   Image Analysis of Autoradiograms

1. Film scanner.

2. Computer-analyzing system ImageJ (NIH, Bethesda, MD, USA) (https://imagej.nih.gov/ij/) or other suitable software.

3. GraphPad Prism (GraphPad Software, Inc., La Jolla, CA, USA) or other curve-fitting software.

2.3   μ Opioid Receptor-Stimulated [^{35}S]GTPγS Binding in Autoradiography

The same material described in Sect. 2.2.1.

2.3.1   Tissue Preparation

2.3.2   [^{35}S]GTPγS Binding in Autoradiography

1. Assay buffer: 50 mM Tris-HCl, 3 mM MgCl$_2$, 2 μM EGTA, 100 μM NaCl, pH 7.4.

2. Tris-HCl buffer: 50 mM Tris-HCl, pH 7.4.

3. [^{35}S]GTPγS (~1250 Ci/nmol; Cat. # NEG030, PerkinElmer) stored at −20 °C.

4. GDP (Cat. # G7127, Sigma-Aldrich) stored at −20 °C.

5. DAMGO (Cat. # 1171, Tocris Bioscience) stored at −20 °C.

6. Coplin jars and humidity chamber.

7. Autoradiographic [^{14}C] microscale (GE Healthcare).

8. ^{35}S-sensitive films (BioMax MR film, Kodak) stored at 4 °C.

9. Autoradiography cassettes, cards, and double adhesive tape.

10. LX 24 X-ray developer (Kodak) and AL-4 X-ray fixer (Kodak).

11. Three trays and print tongs.

12. Darkroom with safelight for processing films.

2.3.3 Image Analysis of Autoradiograms	The same material described in Sect. 2.2.3.

2.4 Immunohisto-chemistry

2.4.1 Tissue Preparation

1. Sodium pentobarbital.
2. Perfusion pump (Dinko D-25VXi, 30 rpm; Dinko Instruments, Barcelona, Spain) equipped with 5 mm tubing.
3. Dissection tools, syringes (3 ml), and needles (25G).
4. PBS: 0.1 M phosphate-buffered saline, pH 7.4.
5. 4% paraformaldehyde (w/v) in PBS (*see* **Note 2**).
6. Cryoprotectant solution: 30% sucrose and 0.1% sodium azide in 0.1 M PBS, pH 7.4.
7. Multiwell culture dishes in 12 well format to store free floating sections.

2.4.2 Immunohisto-chemistry for FosB/ΔFosB

1. Multiwell culture dishes in 6 and 12 well format to be used as incubation chambers.
2. Glass Pasteur pipettes with hook tip (*see* **Note 3**).
3. Laboratory shaker.
4. PBS: 0.1 M phosphate-buffered saline, pH 7.4.
5. Hydrogen peroxide 30%, store at 4 °C.
6. PBS-TX buffer: 0.2% Triton X-100 in PBS.
7. Primary antibody to target ΔFosB. The protocol is illustrated with the polyclonal FosB/ΔFosB antisera raised in rabbit (Cat. # sc-7203, Santa Cruz Biotechnology, Santa Cruz, CA, USA), store at 4 °C (*see* **Note 4**).
8. Secondary antibody, in this chapter illustrated with biotinylated goat anti-rabbit IgG (Cat. # BA-1000, Vector Laboratories, Burlingame, CA, USA), store at −20 °C.
9. Horseradish peroxidase-conjugated streptavidin (Cat. # E2886, Sigma-Aldrich), store at 4 °C.
10. 3-3′-Diaminobenzidine tetrahydrochloride hydrate (DAB) (Cat. # D5637, Sigma-Aldrich) and nickel ammonium sulfate (Cat. # A1827, Sigma-Aldrich).
11. Reagent for dehydrating and clearing sections: series of increasing ethanol concentrations (50°, 70°, 96°, and 100°) and xylene.
12. Mounting medium for coverslips, e.g., DPX mountant (Merck Millipore, Burlington, MA, USA).
13. Silane-coated slides (*see* **Note 1**) and coverslips.

2.4.3 Image Analysis of Immunostained Sections

1. Digital camera coupled to a light microscope.
2. Computer-analyzing system ImageJ (NIH) or other suitable software.

3 Methods

3.1 Drugs Administration

Chronic administration of drugs was performed by an osmotic pump subcutaneously implanted, which allows the continuous dosing of rats [21]. The use of osmotic pumps is a reliable method for the controlled in vivo drugs delivery, which minimizes unwanted experimental variables, suppresses the need for nighttime or weekend dosing, and reduces handling and stress to animals. In this section, we describe the use of the osmotic pump 2ML1 (Alzet®) with a reservoir volume of 2 ml, rate of release of 10 μl/h, and 7 days of delivery.

3.1.1 Filling and Priming Osmotic Pumps

1. Weigh the empty pump (reservoir and flow moderator).

2. Fill up slowly the reservoir of the osmotic pump with the drug solution (10 mg/kg/day of morphine and/or 1 mg/kg/day of PD168,077) using a syringe attached to a 25G needle (*see* **Note 5**). Be sure that syringe and attached needle are free of air bubbles.

3. Carefully insert the flow moderator (*see* **Note 6**).

4. Weigh the filled pump. The weight difference between the empty and filled pump gives the net weight of the loaded solution. In the case of most dilute aqueous solutions, the weight in milligrams is equivalent to the same as the volume in microliters. The fill volume should be more than 90% of the reservoir volume of the osmotic pump.

5. It is recommended an in vitro priming of the osmotic pump prior to the in vivo implantation. For that, place the filled pump in sterile 0.9% NaCl at 37 °C for at least 6–8 h.

3.1.2 Subcutaneous Osmotic Pump Implantation

Osmotic pumps can be subcutaneously or intraperitoneally implanted, so that the content released from the pump reaches local capillaries resulting in systemic administration. In this section, we present the protocol for the subcutaneous implantation of the osmotic pump in the back of the animals, which has consistently provided the best systemic drug infusion in our hands.

1. Deeply anesthetize the rat with sodium ketamine (75 mg/kg, i.p.) and medetomidine (0.5 mg/kg, i.p.) and check the loss of pedal reflex (toe pinch).

2. Shave and disinfect the skin over the implantation site with a topical antiseptic, as povidone-iodine 10% solution (Betadine).

3. Make an incision in the skin between the shoulder blades.

4. Insert a hemostat into the incision and expand the subcutaneous tissue to create a pocket for the pump by opening and closing the jaws of the hemostat (*see* **Note 7**).

5. Insert the filled pump into the pocket and close the skin incision with sutures. Apply a topical antiseptic and give appropriate postoperative care.

3.2 Quantitative μ Opioid Receptor Autoradiography

Quantitative receptor autoradiography is a widely used technique to demonstrate receptor distribution and density within a tissue, and the most sensitive technique to measure receptor affinity [22]. Since the binding parameters of a receptor may change in response to multiple factors (e.g., disease state or drug exposure) [23], this technique is a powerful tool to investigate the pharmacological properties of transmitter receptors and its possible pathophysiological relevance [24]. Particularly for the GPCRs in the central nervous system, quantitative receptor autoradiography has also been used to provide consistent evidences of receptor–receptor interaction, in both normal and pathological conditions [11, 23, 25, 26].

This chapter describes the application of quantitative receptor autoradiography in saturation assays in brain tissue from rats that were chronically treated with morphine and/or the D_4R agonist PD168,077 [12]. For this purpose, coronal brain sections were incubated with increasing amounts of the MOR radioligand [³H] DAMGO and the radioactivity in each of them was determined by image analysis. The plot of data on a saturation isotherm allowed the determination of the dissociation constant (K_d) and the number of binding sites (B_{max}). The K_d is a measure of the affinity of the ligand for its binding sites, whereas B_{max} represent receptor density.

As the majority of radioligands can also bind at low levels to other constituents of the tissue, the non-specific binding (NSB) should be also determined. For this purpose, a number of sections (one for each radioligand concentration being tested) are incubated in presence of a high concentration of a competing ligand. The specific binding is then calculated as the difference between total binding and NSB. Naloxone is typically used for determination of NSB in MOR autoradiography assays.

3.2.1 Tissue Preparation for Autoradiography

1. Sacrifice awake rats (n = 4–6 per experimental group) by swift decapitation using a guillotine. This procedure can only be done by experienced personnel.

2. Remove the brain rapidly from the skull.

3. Freeze the brain by immersion in dry ice-cooled isopentane (−30 °C) and store at −80 °C until sectioning (*see* **Note 8**).

4. Cut consecutive 14 μm thick brain sections using a cryostat (*see* **Note 9**). Thaw-mount a section from every rat belonging to the same experimental group onto the same silane-coated slide and store at −80 °C until use. As standard, 20 sections are required for each saturation curve, 10 for the determination of total binding, and 10 for NSB. Thus, the total would be 20 slides per experimental group, with 4–6 sections each. Additionally, three sections are collected into microcentrifuge tubes to determine protein concentration by the Bradford method [27].

3.2.2 Autoradiography in Saturation Binding Assay

1. Dilute the stock solution of [³H]DAMGO (140 nM) to get the highest concentration to 14 nM by adding the suitable amount of assay buffer (*see* **Note 10**).

2. Using this highest concentration, prepare a serial dilution to obtain a total of ten concentrations over the range of 14–0.36 nM. Use a 50 μl aliquot from each of the serial dilutions to determine the amount (disintegrations per minute, dpm) of radioligand (*see* **Note 11**). These values are necessary for further analysis of the data. Half of the volume of each dilution is used to determine total binding (total solution). Add naloxone (10 μM) to the other half of the volume of each dilution and use it to define non-specific binding (NSB solution).

3. Remove endogenous opioids by preincubating the sections for 30 min at RT with the assay buffer. This step can be done in coplin jars.

4. In a humidity chamber, incubate the sections for 1 h at RT with 200 μl of each of total or NSB solutions. This time of incubation is enough to reach binding equilibrium.

5. Transfer the slides to racks and sequentially wash for 5 min, each in ice-cold Tris-HCl buffer (two times) and distiller water (one time). Dry in a stream of cold air.

6. Mount the slides onto a card with double adhesive tape, together with a slide bearing prefabricated ³H-labeled polymer standard strip (GE Healthcare), and place in an autoradiography cassette.

7. In a darkroom with an appropriate safelight, appose a tritium-sensitive film over the slides and clamp shut the autoradiographic cassette. Expose for 6–8 weeks at 4 °C.

8. In the darkroom with a safelight, process the film, by placing it for 5 min at 20 °C in a tank containing LX 24 X-ray developer (Kodak), followed by 30 s in a tank with deionized water and finally for 10 min in a tank with AL-4 X-ray fixer (Kodak) (*see* **Note 12**).

3.2.3 Image Analysis of Autoradiograms

1. Scan the film to obtain gray images of every autoradiographic tissue section. Digitized images should be in TIFF format.

2. Calibrate the image analyzer (ImageJ) to obtain optical density values from gray levels.

3. Delineate with the cursor the regions of interest within the images from both total binding and NSB sections and measure the optical densities.

4. Construct a standard curve from the ³H-labeled polymer standard strips in order to correlate optical densities to known amounts of radioactivity. For that, measure the optical density of every band of the strip and correct with the optical density

of the background. Enter the known amount of radioactivity (nCi/mg) for each band that is provided by the supplier. Generate the logarithmic plot of optical density vs. radioactivity to give a linear relationship and interpolate optical density values obtained from brain sections.

5. The specific activity of the radioligand and the amount of proteins determined in a brain section (see Sect. 3.2.1, step 4) can be used to convert nCi/mg into fmol/mg of protein.

6. Run a suitable computer program (e.g., GraphPad Prism) and enter values (in fmol/mg protein) for total binding and NSB from the image analysis for each concentration of the radioligand.

7. Select a nonlinear regression analysis and choose the model to be fitted (one-site, two-site, etc.). The program calculates specific binding subtracting the NSB from the total binding, plots the data on a saturation isotherm, and estimates the K_d and B_{max} values.

3.3 μ Opioid Receptor-Stimulated [^{35}S]GTPγS Binding in Autoradiography

Agonist-stimulated [^{35}S]GTPγS binding autoradiography is a simple and sensitive assay to localize and measure the level of G protein activation following agonist binding to a GPCR [24, 28] (*see* **Note 13**). Thus, this assay provides much important information since it analyzes a functional consequence of receptor occupancy at one of the earliest receptor-mediated events, with the advantage that is not subjected to amplification or other modulation that may occur further downstream of the receptor.

This method is based on the heterotrimeric G proteins (Gα, Gβ, Gγ subunits) activation/inactivation cycle. At resting state, Gα is bound to Gβ/Gγ dimer and guanosine diphosphate (GDP). The activation of a GPCR by an agonist promotes the exchange of GDP for GTP on the Gα. Binding of GTP causes dissociation of G proteins into Gα and Gβ/Gγ dimer that in turn activate or inactivate multiple downstream effectors. The intrinsic GTPase activity of Gα leads to conversion of bound GTP into GDP and hence its reassociation with Gβ/Gγ dimer and therefore the inactivation of the G protein. When the poorly hydrolyzed [^{35}S]GTPγS is used instead of GTP, the half-life of the GTPγS-bound Gα is increased, allowing the analyses of the bound GTP analog.

This chapter describes the application of MOR agonist-stimulated [^{35}S]GTPγS binding autoradiography in brain tissue from rats that were chronically treated with morphine and/or the D$_4$R agonist PD168,077 [12].

3.3.1 Tissue Preparation

Prepare tissue as has been described above (see Sect. 3.2.1).

3.3.2 [^{35}S]GTPγS Assay in Autoradiography

1. Wash the sections for 15 min at RT with assay buffer. This step can be done in coplin jars.

2. In a humidity chamber, preincubate the sections for 15 min at RT with 2 μM GDP in assay buffer (*see* **Note 14**).

3. Discard the previous buffer and incubate the sections with 0.04 nM [^{35}S]GTPγS and 0.2 μM GDP in assay buffer with (stimulated) or without (basal) 3 μM DAMGO (*see* **Note 15**). Incubate for 40 min at RT.

4. Transfer the slides to racks and sequentially wash twice in cold 50 mM Tris-HCl buffer and distiller water. Dry in a stream of cold air.

5. Mount the slides onto a card with double adhesive tape, together with a slide bearing ^{14}C microscale (GE Healthcare), and place in an autoradiography cassette.

6. In a darkroom with an appropriate safelight, appose a ^{35}S-sensitive film (BioMax MR Film, Kodak) over the slides. Expose the sections for 48 h at RT.

7. Process the film in a darkroom with a safelight as it has been described above (see Sect. 3.2.2, step 8).

3.3.3 Image Analysis of Autoradiograms

1. Scan the film and obtain TIFF format gray images.

2. Calibrate the image analyzer (ImageJ) to obtain optical density values from gray levels.

3. Delineate with the cursor the regions of interest within the images from both stimulated and basal sections and measure the optical densities.

4. Correct the optical density values by subtraction of the background measure.

5. Convert optical density values into nCi/g using a standard curve constructed from the measure of optical density values of a ^{14}C microscale.

6. Set at "0" the [^{35}S]GTPγS binding values of basal sections (non-stimulated) and determine the percent of [^{35}S]GTPγS binding of agonist-stimulated sections.

3.4 Immunohisto chemistry

The activation of GPCRs and the subsequent G protein-mediated downstream signaling cascades promote the regulation of a variety of transcription factors, which in turn modulate other genes expression [29]. In this sense, the studies of the last decades regarding transcription factors expression have significantly contributed to the knowledge of the physiological role of the GPCRs and its relevance in pathological conditions. The use of the immunohistochemical technique takes advantages as it is an easy, standardized, and suitable approach when combined with image analysis to evaluate changes in transcription factors expression upon GPCR stimulation. Furthermore, a wide variety of antibodies against different transcription factors is available nowadays.

This chapter describes the use of the immunohistochemical techniques for the detection of FosB/ΔFosB in brain tissue from rats that were chronically treated with morphine and/or the D_4R agonist PD168,077 [13]. FosB and the truncated form ΔFosB are members of the Fos family of transcription factors. It has been extensively documented the accumulation of ΔFosB in the caudate putamen after repeated exposure to many types of drugs of abuse [30]. This ΔFosB response seems to represent a crucial mechanism by which drugs of abuse induce long-lasting and stable changes in the brain, contributing to the physiological and behavioral addiction phenotype [31].

3.4.1 Tissue Preparation

1. Anesthetize the rat with sodium pentobarbital (60 mg/kg, i.p.) and check the loss of pedal reflex (toe pinch).

2. Cut open the rat below the diaphragm and the rib cage on the lateral edges to expose the heart.

3. Insert and clamp a needle into the left ventricle, then cut the right atrium to allow flow.

4. Perfuse transcardially with cold PBS for 5 min to remove blood from vessels (*see* **Note 16**) followed by 400 ml of cold 4% paraformaldehyde in PBS for 30–40 min (*see* **Note 17**).

5. Remove the brain and postfix by immersion in the same fixative at 4 °C for 2 h.

6. Transfer the brain to a cryoprotectant solution until sink to the bottom of the jar, usually after 72 h. Then, freeze the brain rapidly in dry ice.

7. Cut the brain with a freezing microtome to obtain free-floating 30 μm thick sections. Store the sections in 0.02% sodium azide in PBS at 4 °C until use for immunohistochemistry (*see* **Note 18**).

3.4.2 Immunohisto chemistry for ΔFosB

The free-floating sections should be incubated in multiwall culture dishes (6 or 12 well format) with gentle shaking during the whole immunohistochemical staining process. It is recommended to use 6 well format for the washes with PBS and 12 well format for the incubation with antibodies. The glass Pasteur pipettes with hook tip (*see* **Note 3**) are used to transfer the sections from one well to another.

1. Quench the endogenous peroxidase activity of the tissue by incubating it with 3% hydrogen peroxide in PBS for 15 min.

2. Wash sections with PBS, three times for 10 min each.

3. Incubate with the polyclonal ΔFosB antibody diluted in PBS-TX with 0.1% sodium azide for 24 h at room temperature (RT) (*see* **Note 19**).

4. Wash the sections, as in step 2 above, and incubate for 1 h at RT with the appropriate secondary antisera (biotinylated goat anti-rabbit IgG at 1:500) diluted in PBS-TX with 0.1% sodium azide.

5. Wash the sections again, as in step 2 above, and incubate for 1 h at RT with horseradish peroxidase-conjugated streptavidin (1:2000 in PBS-TX) (*see* **Note 20**).

6. Develop the peroxidase reaction with a fresh solution of 0.05% DAB (*see* **Note 21**), 0.08% nickel ammonium sulfate (*see* **Note 22**), and 0.01% H_2O_2 in PBS. First, incubate for 10 min with DAB and nickel ammonium sulfate, and then add the H_2O_2. Monitor sections for development of the reaction product.

7. Stop the peroxidase reaction by transferring the sections into fresh PBS.

8. Mount the immunostained sections onto silane-coated slides, air-dry, dehydrate in a series of increasing ethanol concentrations (50°, 70°, 96°, and 100°), and clear with xylene.

9. Coverslip with a mounting medium such as DPX.

3.4.3 Image analysis of Δ FosB Immunoreactivity

1. Use a digital camera coupled to the light microscope to acquire microphotographs covering the areas of interest from the immunostained sections. Digitized images should be in grayscale and TIFF format.

2. Using a computer analyzing system, set a threshold range to detect immunopositive nuclei apart from the background. This will convert the image to binary.

3. Set the minimum size and the maximum pixel area size to exclude anything that is not a nucleus in the image.

4. Apply automatic particle counting to obtain the number of immunopositive nuclei per area.

4   Notes

1. To prepare silane-coated slides for tissue adhesion, clean slides with acetic acid:ethanol (1:100) and dry. Immerse slides in dry acetone for 1 min followed by a freshly prepared 2% solution of silane (3-aminopropyl triethoxysilane; A3648, Sigma-Aldrich) in dry acetone for 1 min. Wash twice in distilled water and dry 24 h at 37 °C. Store silane-coated slides at RT. A coplin jar reserved exclusively for silane-coating should be used. Make all the procedure in a fume hood.

2. To prepare 1 l of 4% paraformaldehyde, heat 500 ml of distiller water, add 40 g of paraformaldehyde, and mix inside the fume hood. Add several drops of NaOH until clear. Add 500 ml of

0.2 M PBS and filter the solution. Use within the following 24 h.

3. Heat the tip of the pipette and bent to form a smooth and blunt hook.

4. This antibody recognized all FosB members, i.e., FosB and its truncated ΔFosB forms [32].

5. Rapid filling of osmotic pumps should be avoided as it can produce air bubbles into the reservoir.

6. The insertion of the flow moderator normally displaces a small amount of the solution from the filled pump.

7. The pocket should be 1 cm longer than the pump to allow some free movement. A pocket too large should be avoided as this will favor the pump turn around or slip down on the flank of the rat.

8. Avoid keeping the brains at −80 °C for a long time, this is only a tip in case that the number of brains to cut is high and it is impossible to get all of them sectioned in the same day.

9. Cryostat temperature should range between −15 and −23 °C. If the tissue is too cold the sections will curl and if it is too warm the sections will stick to the knife.

10. GraphPad QuickCalcs (https://www.graphpad.com/quick-calcs/radcalcform/) is an useful free online resource for radio-activity calculations.

11. Dilute 50 μl of radioligand in 4 ml of scintillation fluid and use a scintillation counting for dpm determination.

12. Old developer and fixer will negatively affect the image quality of processed autoradiograph films. It is recommended to change developer and fixer every month even if they are infrequently used.

13. The agonist-stimulated [^{35}S]GTPγS assay works best with Gi-coupled GPCRs.

14. The assay is usually run in the presence of high concentration of GDP in order to obtain optimal agonist-stimulated [^{35}S]GTPγS binding and reduce background. GDP acts by filling empty nucleotide binding sites of the Gα subunit and hence reduces the basal levels of [^{35}S]GTPγS binding. GDP also suppresses binding of [^{35}S]GTPγS to non-heterotrimeric G protein targets.

15. DAMGO is a synthetic peptide with high selectivity for the μ opioid receptor.

16. Bleaching of the liver is an indication that washing of blood vessels with PBS is suitable.

17. Tremor of the extremities resulting from the aldehyde-crosslinking of nerves and muscle is an indication that fixation is taking place.

18. Under these conditions, sections may be stored for several months at 4 °C.

19. Proper ΔfosB antibody dilution is previously determined by testing different concentrations.

20. Avoid using sodium azide in the horseradish peroxidase-conjugated streptavidin dilution buffer, as the enzyme is deactivated.

21. DAB is carcinogenic. Work with appropriate safety systems to avoid all contact, especially breathing in the powder, when handling it. Prepare a 10× DAB stock solution in 0.05 M Tris buffer (pH 7.6) and store at −20 °C in an appropriate volume for one reaction (aliquots of 500 μl).

22. Nickel ammonium sulfate is used to intensify the immunohistochemical signal, yielding a dark purplish color.

Acknowledgments

This work has been supported by CTS161 (Junta de Andalucía, Spain).

References

1. Fuxe K, Borroto-Escuela D (2016) Volume transmission and receptor-receptor interactions in heteroreceptor complexes: understanding the role of new concepts for brain communication. Neural Regen Res 11:1220

2. Fuxe K, Borroto-Escuela DO, Marcellino D et al (2012) GPCR heteromers and their allosteric receptor-receptor interactions. Curr Med Chem 19:356–363

3. Ferré S, Casadó V, Devi LA et al (2014) G protein-coupled receptor oligomerization revisited: functional and pharmacological perspectives. Pharmacol Rev 66:413–434

4. Ciruela F, Vallano A, Arnau JM et al (2010) G protein-coupled receptor oligomerization for what? J Recept Signal Transduct Res 30:322–330

5. Khelashvili G, Dorff K, Shan J et al (2010) GPCR-OKB: the G protein coupled receptor oligomer knowledge base. Bioinformatics 26:1804–1805

6. Borroto-Escuela DO, Brito I, Romero-Fernandez W et al (2014) The G protein-coupled receptor heterodimer network (GPCR-HetNet) and its hub components. Int J Mol Sci 15:8570–8590

7. Guidolin D, Agnati LF, Marcoli M et al (2015) G-protein-coupled receptor type A heteromers as an emerging therapeutic target. Expert Opin Ther Targets 19:265–283

8. Fuxe K, Marcellino D, Rivera A et al (2008) Receptor-receptor interactions within receptor mosaics. Impact on neuropsychopharmacology. Brain Res Rev 58:415–452

9. Rivera A, Cuéllar B, Girón FJ et al (2002) Dopamine D4 receptors are heterogeneously distributed in the striosomes/matrix compartments of the striatum. J Neurochem 80:219–229

10. Gago B, Fuxe K, Agnati L et al (2007) Dopamine D4 receptor activation decreases the expression of μ-opioid receptors in the rat striatum. J Comp Neurol 502:358–366

11. Rivera A, Gago B, Suárez-Boomgaard D et al (2017) Dopamine D4 receptor stimulation prevents nigrostriatal dopamine pathway activation by morphine: relevance for drug addiction. Addict Biol 22:1232–1245

12. Suárez-Boomgaard D, Gago B, Valderrama-Carvajal A et al (2014) Dopamine D_4 receptor counteracts morphine-induced changes in μ opioid receptor signaling in the striosomes of the rat caudate putamen. Int J Mol Sci 15:1481–1498

13. Gago B, Suárez-Boomgaard D, Fuxe K et al (2011) Effect of acute and continuous morphine treatment on transcription factor expression in subregions of the rat caudate putamen. Marked modulation by D4 receptor activation. Brain Res 1407:47–61

14. Volkow ND, McLellan AT (2016) Opioid abuse in chronic pain—misconceptions and mitigation strategies. N Engl J Med 374:1253–1263

15. Novak SP, Håkansson A, Martinez-Raga J et al (2016) Nonmedical use of prescription drugs in the European Union. BMC Psychiatry 16:274

16. Guo H, An S, Ward R et al (2017) Methods used to study the oligomeric structure of G-protein-coupled receptors. Biosci Rep 37:pii: BSR20160547

17. Szidonya L, Cserzo M, Hunyady L (2008) Dimerization and oligomerization of G-protein-coupled receptors: debated structures with established and emerging functions. J Endocrinol 196:435–453

18. Fuxe K, Marcellino D, Guidolin D et al (2008) Heterodimers and receptor mosaics of different types of G-protein-coupled receptors. Physiology (Bethesda) 23:322–332

19. Borroto-Escuela DO, Hagman B, Woolfenden M et al (2016) In situ proximity ligation assay to study and understand the distribution and balance of GPCR homo- and heteroreceptor complexes in the brain. In: Luján R, Ciruela F (eds) Receptor and ion channel detection in the brain. Neuromethods, vol 110. Humana, New York

20. Gago B, Fuxe K, Brené S et al (2013) Early modulation by the dopamine D4 receptor of morphine-induced changes in the opioid peptide systems in the rat caudate putamen. J Neurosci Res 91:1533–1540

21. Herrlich S, Spieth S, Messner S et al (2012) Osmotic micropumps for drug delivery. Adv Drug Deliv Rev 64:1617–1627

22. Stumpf WE (2005) Drug localization and targeting with receptor microscopic autoradiography. J Pharmacol Toxicol Methods 51:25–40

23. Tena-Campos M, Ramon E, Rivera D et al (2014) G-protein-coupled receptors oligomerization: emerging signaling units and new opportunities for drug design. Curr Protein Pept Sci 15:648–658

24. Zhang R, Xie X (2012) Tools for GPCR drug discovery. Acta Pharmacol Sin 33:372–384

25. Flores-Burgess A, Millón C, Gago B et al (2017) Galanin (1-15) enhancement of the behavioral effects of Fluoxetine in the forced swimming test gives a new therapeutic strategy against depression. Neuropharmacology 118:233–241

26. Llorente-Ovejero A, Manuel I, Giralt MT et al (2017) Increase in cortical endocannabinoid signaling in a rat model of basal forebrain cholinergic dysfunction. Neuroscience 362:206–218

27. Bradford MM (1976) A rapid and sensitive method for the quantitation of microgram quantities of protein utilizing the principle of protein-dye binding. Anal Biochem 72:248–254

28. Harrison C, Traynor JR (2003) The [35S] GTPgammaS binding assay: approaches and applications in pharmacology. Life Sci 74:489–508

29. Ho MKC, Su Y, Yeung WWS et al (2009) Regulation of transcription factors by heterotrimeric G proteins. Curr Mol Pharmacol 2:19–31

30. Nestler EJ, Barrot M, Self DW (2001) DeltaFosB: a sustained molecular switch for addiction. Proc Natl Acad Sci U S A 98:11042–11046

31. Ruffle JK (2014) Molecular neurobiology of addiction: what's all the (Δ)FosB about? Am J Drug Alcohol Abuse 40:428–437

32. Grande C, Zhu H, Martin AB et al (2004) Chronic treatment with atypical neuroleptics induces striosomal FosB/ΔFosB expression in rats. Biol Psychiatry 55:457–463

Chapter 4

Use of Superfused Synaptosomes to Understand the Role of Receptor–Receptor Interactions as Integrative Mechanisms in Nerve Terminals from Selected Brain Region

Sarah Beggiato, Sergio Tanganelli, Tiziana Antonelli,
Maria Cristina Tomasini, Kjell Fuxe, Dasiel O. Borroto-Escuela,
and Luca Ferraro

Abstract

Synaptosomes are sealed presynaptic nerve terminals obtained by homogenizing selected brain regions under iso-osmotic conditions. This preparation has been extensively used to study the mechanism of neurotransmitter release in vitro because they preserve the biochemical, morphological, and electrophysiological properties of the synapse. This allows an unequivocal interpretation of results obtained under accurately specified experimental conditions. By using superfused synaptosome technique, it is possible to individuate the co-localization of different receptors on the same nerve endings and the presence of synergistic/antagonistic interactions between them. The latter opportunity renders superfused synaptosome technique particularly suitable to understand the functional role of receptor–receptor interactions at the presynaptic level.

This chapter is mainly focused on a general description of the technique and on its application to investigate the functional relevance of receptor–receptor interactions in regulating neurotransmitter release from selected brain region nerve terminals.

Key words Receptor heteromers, Nerve terminals, Synaptic transmission, Spontaneous and evoked release, D2R-σ1R-N-type calcium channel complex

1 Introduction

It is now well accepted that release modulation through presynaptic receptors is a general mechanism that involves all the identified transmitters. When referred to a receptor, the term "presynaptic" indicates its anatomical localization on the nerve terminal (i.e., presynaptic axon bouton) and, except for this, presynaptic receptors are equal to postsynaptic ones. The advent of the era of

Kjell Fuxe and Dasiel O. Borroto-Escuela (eds.), *Receptor-Receptor Interactions in the Central Nervous System*, Neuromethods, vol. 140, https://doi.org/10.1007/978-1-4939-8576-0_4, © Springer Science+Business Media, LLC, part of Springer Nature 2018

presynaptic receptors prompted the development of experimental procedures suitable to separate nerve terminals in order to differentiate presynaptic from postsynaptic neurotransmitter release mechanisms. The procedure to pinch off nerve terminals from selected brain regions by gentle homogenization, and to successively purify these structures from the other subcellular fragments and glial cells by density-gradient centrifugation, was originally developed by Gray and Whittaker [1–3]. Rapidly, it became clear that these so-called synaptosomes, even though isolated from their physiological environment, contain all the components necessary to store, release, and retain neurotransmitters, thus maintaining the whole stimulus-secretion coupling mechanisms [4]. After about a decade, a simple apparatus for evaluating the release of neurotransmitters from synaptosomes was proposed. The idea was to plate standard synaptosomes as very thin layers on microporous filters and to up-down superfuse them with a physiological solution [5, 6]. Superfusion synaptosome technique quickly proved particularly useful in the identification and characterization of receptors that are localized on central nervous system (CNS) axon terminals and involved in transmitter release regulation. Using this technique, hundreds of researchers from many laboratories have demonstrated that several ligands modulate transmitter release by acting at their respective receptors present on the nerve terminals. Actually, superfused synaptosomes still represent the preparation of choice in studies aimed at the identification and characterization of release-regulating presynaptic receptors. In particular, this technique can provide relevant and specific information on: (1) the indisputable localization of native receptors on the nerve terminal releasing a specific neurotransmitter; (2) the characterization of receptor transduction mechanisms; (3) the pharmacological characterization of the mechanism of action of novel compounds acting on presynaptic receptors; (4) the impact of genetic modifications on presynaptic neurotransmitter release; (5) the existence of metabotropic receptors adopting a constitutively activated conformation and the functional activity of positive and negative allosteric modulators acting at both ionotropic and metabotropic receptors; (6) the possible co-localization of different receptors on the same nerve endings and the presence of a synergistic/antagonistic interactions between them [7]. The latter opportunity renders superfused synaptosome technique particularly suitable to understand the role of receptor–receptor interactions as integrative mechanisms in nerve terminals from selected brain region.

The main principles and methodological details of superfused synaptosome technique as well as its advantages and limitations have been largely described elsewhere [3, 5, 6, 8–10] and we refer to those reviews for detailed descriptions. This chapter is mainly focused on a general description of the technique and on its application to investigate the functional relevance of receptor–receptor interactions in regulating integrative mechanisms in nerve terminals from selected brain region.

2   Synaptosome Overview

Synaptosomes (i.e., isolated nerve endings) are sealed particles that contain small, clear vesicles and sometimes larger dense-cored vesicles, indicating their presynaptic origin. From a morphological point of view, synaptosomes look like small (1–2 μm in diameter) rounded pockets. Occasionally, larger dense-cored vesicles can be found, which presumably store modulatory transmitters such as neuropeptides or catecholamines. Furthermore, more frequently, larger nonspherical membranes, possibly representing endoplasmic reticulum or endosomes, are present in a synaptosome preparation. The plasma membrane of synaptosomes regularly presents fragments of electron-dense membranes attached, which represent the postsynaptic density to which the terminal is functionally coupled [9].

Synaptosomes retain on the neuronal membrane structures present in the nerve ending in vivo (i.e., receptors, carriers, etc.) and hold all the components necessary to store, release, and retain neurotransmitters, along with viable mitochondria, enabling production of ATP and active energy metabolism. Furthermore, synaptosomes retain the Ca^{2+}-buffering capacity, as evidenced by their ability to maintain resting internal Ca^{2+} concentrations of 100–200 nM in the presence of 2 mM extracellular Ca^{2+} [11]. Finally, synaptosomes also maintain a normal membrane potential (which is regulated by a Na^+/K^+-ATPase) and express functional uptake carriers and ion channels in their plasma membranes [5, 9, 12].

These subcellular particles of functionally active nerve tissue can be easily prepared by means of a specific homogenization of selected brain regions [2]. Subsequent homogenate centrifugations in a density gradient (i.e., Percoll gradient) allow the purification of synaptosomes from contaminating particles originating from other cells, including astrocytes [13, 14]. This aspect reduces eventual sources of bias for downstream analyses due to the presence of neuronal/glial network, thus rendering synaptosomes advantageous with respect to other in vitro experimental models, such as primary neuronal cultures or tissue slices. Another particular advantage of synaptosomes is that they are directly prepared from neurons that were formed, developed, and differentiated in their natural environment, the living brain, allowing a reliable analysis of the events that presynaptically occurs in CNS [12].

In the 1970s, the group of Raiteri introduced the superfusion technique in which synaptosomes are layered in a monolayer on microporous filters and up-down superfused with a physiological solution (Fig. 1a), thus strongly facilitating the studies of neurotransmitter release from nerve endings [5]. In this context, one important problem to solve is, for example, to distinguish between drugs able to directly affect release and drugs that could do it indirectly, i.e., by altering transmitter reuptake and/or metabolism. Superfused synaptosomes allow this, as released neurotransmitters are removed by the superfusion medium quickly enough to escape reuptake and/or metabolism (Fig. 1b). Furthermore, during synaptosome superfusion the instantaneous removal of endogenous compounds strongly reduces their concentration at nerve terminals and, as a consequence, the possibility that they can act on adjacent particles, thus inducing indirect effects (Fig. 1b). Moreover, the monolayer renders impossible the action of endogenous molecules on synaptosomes located below those particles from which they are released. Under these conditions, the targets on the synaptosomal membranes such as receptors and/or carriers

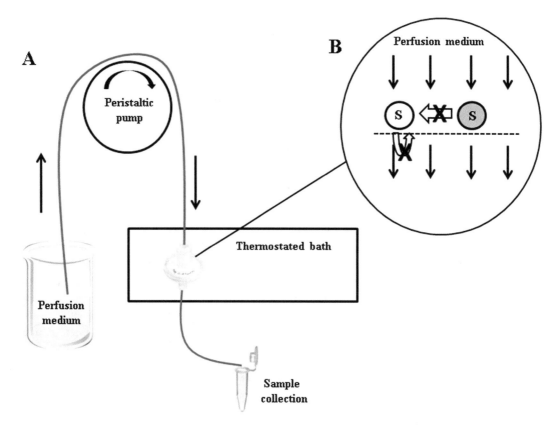

Fig. 1 Schematic representation of the principles of superfused synaptosome technique. Panel **a**: Experimental setup. Synaptosomes are monolayered in a microporous nylon filter and continuously up-down superfused with a physiological solution. A peristaltic pump maintains a constant perfusion flow rate during the experiment. Treatments are performed by adding the investigated compounds to the perfusion medium. Panel **b**: specific features of the superfusion technique. The neurotransmitter released by a monolayered synaptosome (S) preparation is rapidly removed by the perfusion medium. In this way, neurotransmitter reuptake and indirect effects are practically excluded

can be selectively activated/inhibited by ligands present in the perfusion medium. Therefore, any effects on the release of a neurotransmitter may be attributed exclusively to an action on targets present on the nerve ending in selected synaptosomal subpopulations [7, 12]. As in other techniques used to measure neurotransmitter release, the perfusion of a low-Ca^{2+} medium allows distinguishing exocytotic from non-exocytotic effects [15].

Notably, the synaptosomal suspension originating from a selected brain region is composed of particles that are isolated from different subfamilies of nerve terminals. However, the selective labeling with radioactive tracers mimicking endogenous transmitters as well as the monitoring of the release of endogenous molecules (e.g., glutamate, GABA, catecholamines and neuropeptides) permits to isolate a given subfamily of nerve terminals from the entire synaptosomal populations, impeding artifacts that may originate from preparation heterogeneity [7]. Thus, the heterogeneity of synaptosomal preparations, generally a major drawback under some circumstances, has relatively little relevance when a given target present on one family of nerve endings is investigated.

By using superfused synaptosomes, it is possible to measure either spontaneous or evoked neurotransmitter release. In fact, following the application of diverse depolarizing stimuli (e.g., high K^+ concentrations, veratridine, and 4-aminopyridine) multiple neurotransmitters (including amino acids, peptides, and catecholamines) are released into the extracellular medium as a consequence of Ca^{2+} entering via high voltage-sensitive Ca^{2+} channels and exocytosis of docked vesicles. 4-Aminopyridine-induced neurotransmitter release from synaptosomes is sensitive either to the Na^+ channel blocker tetrodotoxin (TTX) or to a low-Ca^{2+} medium. This contrasts with the exocytosis evoked by high K^+ that is unchanged by TTX. In fact, although elevation of the extracellular K^+ concentration is an efficient means of depolarizing synaptosomes, the immediate clamping of the membrane potential in the depolarized state bypasses presynaptic Na^+-channel activation, the physiological trigger for action-potential-dependent Ca^{2+} entry and subsequent transmitter release. Finally, the neurotransmitter-transporter-mediated release evoked by veratridine is highly TTX-sensitive and does not require activation of Ca^{2+} channels [16].

It is worth noting that the use of superfused synaptosomes made it possible to discover that some neurotransmitters, i.e., those for which a reuptake carrier exists on the presynaptic terminal, can be released not only by vesicular exocytosis but, under some conditions, through the carrier working in the inside-out direction.

Actually, perfused synaptosome technique can be also used for evaluating receptor desensitization and trafficking, alterations in receptor subunits structures, ex-vivo assessment of receptor function after chronic treatment with drugs, chronic changes in the exposure to environmental toxins or genetic modifications during development or aging. In addition, by means of synaptosomal preparations it is also possible to study presynaptic receptors or neurotransmitter release from human tissue [17–20]. Finally, the possibility to prepare synaptosomes from frozen tissues or to cryopreserve them represents a further undoubtable advantage of this procedure [21–23].

3   Experimental Details

3.1   Synaptosome Preparation

Several articles describing different procedures for the preparation of synaptosomes from brain tissue have been published [14, 15, 24–26]. A general protocol, suitable to prepare synaptosomes for superfusion experiments, is described below [27, 28]:

1. The animal is sacrificed under anesthesia, the brain is rapidly removed, and the selected brain region is quickly placed on an ice-cold plate; assuming the use of 12-parallel superfusion chambers, both striata of one rat are necessary.

2. The selected brain region is rapidly placed in a glass Teflon tissue grinder containing ice-cold 0.32 M sucrose, buffered to pH 7.4 with TRIS (final concentration 0.01 M) and previously carbogenated (95% O_2/5% CO_2; 2–3 min); 5–10 volumes of buffer are generally used (about 3–7 ml).

3. The tissue is homogenized by means of 15–17 up-and-down gentle strokes.

4. The homogenate is centrifuged at $1000 \times g$ for 5 min (4 °C), to remove nuclei and debris; the supernatant is then gently stratified on a discontinuous Percoll gradient (6%, 10%, and 20% v/v in Tris-buffered sucrose) and centrifuged at $33,500 \times g$ for 5 min (4 °C).

5. The layer between 10% and 20% Percoll (synaptosomal fraction) is then collected and washed by centrifugation.

6. The synaptosomal pellet was resuspended in 5–6 ml of physiological medium (standard medium), gassed with 95% O_2 and 5% CO_2; for example, a solution having the following composition (mM) might be used: NaCl, 140; KCl, 3; $MgSO_4$, 1.2; $CaCl_2$, 1.2; NaH_2PO_4, 1.2; $NaHCO_3$, 5; HEPES, 10 mM, glucose, 10; pH 7.2–7.4. We also usually use a standard Krebs' solution (mM: NaCl 118.5, KCl 4.7, $CaCl_2$ 1.2, KH_2PO_4 1.2, $MgSO_4$ 1.2, $NaHCO_3$ 25, glucose 10).

7. An aliquot (50–100 μl) of the solution is stored at −80 °C to determine synaptosomal protein contents according to Bradford [29], while the remaining synaptosomal preparation is used for release experiments.

3.2 Labeled Neurotransmitter Incubation

When necessary, especially due to low analytical method sensitivity, the selective labeling of synaptosomes with tritiated tracers mimicking endogenous transmitters could be used. In this case, selective inhibitors of neurotransmitter metabolism or degradation could be added to the perfusion medium [e.g., ascorbic acid (0.05 mM) and disodium EDTA (0.03 mM) to prevent tritiated-dopamine (DA) degradation; aminooxyacetic acid (0.1 mM) to prevent tritiated-GABA degradation] [28, 30]. To do this:

1. Place synaptosomal suspension in a vial and maintain it slightly carbogenated at 37 °C.

2. Add tritiated neurotransmitter and wait for 15–30 min.

This step is not necessary when endogenous neurotransmitters are analyzed. In this case, wait for 20–30 min before starting with the perfusion experiments.

3.3 Superfusion Experiment

The release of labeled or endogenous transmitters from synaptosomes is measured in a continuous superfusion system as initially described by Raiteri et al. [5].

1. The lag time between the introduction of a substance into the perfusate medium and its collection should be measured before, and considered during, the experiment.

2. Identical aliquots of synaptosomal suspension are distributed on microporous filters (0.5 ml/filter), placed at the bottom of a set of parallel superfusion chambers maintained at 37 °C, and perfused with aerated (95% O_2/5% CO_2) physiological solution (0.3 ml/min). Some authors suggest to use aliquots of the synaptosomal suspension containing approximately 0.24 mg of protein; it is useful to prime the filters with 1 ml of physiological solution and to perfuse them for at least 10 min before synaptosome distribution. Synaptosomes are monolayered by moderate vacuum applied to the bottom of the superfusion chambers or, alternatively, by placing them into the filters with a syringe and by exerting a moderate pressure. Be sure that air does not remain entrapped into the filter.

3. Synaptosomes are superfused with the medium (37 °C; continuously carbogenated) for 20–30 min to obtain a stable outflow of neurotransmitter (equilibration period). The perfusion medium was replaced every 10 min by fresh medium continuously aerated and maintained at 37 °C at the top of the superfusion chambers (when a peristaltic pump is used to maintain the constant perfusion flow rate, this is not necessary). When the ionic composition of the standard medium is changed, the medium remnant at the bottom of the superfusion chambers is previously removed by moderate vacuum

through the tubes connected to the vacuum pump (Fig. 1) without altering the synaptosome monolayer.

4. Collect consecutive perfusate samples according to experimental necessity (e.g., 5 min samples from 30th to 80th min from the onset of the superfusion period).

5. To study the effects of treatments on spontaneous neurotransmitter release: After the collection of three basal samples, the compound under investigation is added to the perfusion medium for a limited time period or till the end of the experiment. Control synaptosomes, perfused with the physiological solution for the entire experimental period, are also assayed in parallel; to study the effect of antagonist and agonist drugs in combination, antagonists are generally added at 5–10 min before agonists to permit the antagonist drug to reach a state of equilibrium with the receptor population present in the preparation. Generally, the antagonist effect, by itself, on the neurotransmitter release is tested, although by the methodology used, no direct effect on the neurotransmitter release is expected.

6. To study K^+-evoked neurotransmitter release: After the collection of three basal samples, the standard perfusion medium is replaced by 15–30 mM K^+ (substituting for an equimolar concentration of NaCl)-containing medium for 60–90 s. After this period, synaptosomes are perfused with the original medium. When required, tested compounds are added, alone or in combination, concomitantly with the depolarizing stimulus.

7. At the end of the superfusion period, neurotransmitter levels in the perfusate samples are measured by liquid scintillation (labeled neurotransmitter; the radioactivity in the filters containing the synaptosomes is also measured by completely covering them with scintillant liquid) or by means of specific analytical methods (endogenous neurotransmitters).

3.4 Examples of Data Evaluation

Below are reported two examples of data evaluation. Other approaches are possible.

1. The amount of radioactivity released into each fraction is generally expressed as a percentage of the total synaptosomal tritium present at the start of the respective collection period (fractional release). K^+-evoked $[^3H]$-neurotransmitter overflow is calculated as net extra-outflow (NER) by subtracting the basal $[^3H]$-neurotransmitter (determined by interpolation of the outflow measured 5 min before and 10 min after the onset of the stimulation) from the total tritium released during the 90 s stimulation and subsequent 8.5 min washout (i.e., two samples). This difference is calculated as a percentage of the total tissue tritium content at the onset of stimulation (frac-

tional rate net extra-outflow). The effect of treatments is expressed as the percent ratio of the depolarization-evoked neurotransmitter overflow calculated in the presence of the drug versus that obtained in control group, i.e., K$^+$-evoked [^{3}H]-neurotransmitter release measured in untreated synaptosomes, always assayed in parallel [28, 31].

2. Spontaneous neurotransmitter levels in each sample are expressed in nmol/min/g of protein. The effects of K$^+$ stimulation on endogenous neurotransmitter levels during the third fraction are expressed as percentage changes of basal values, as calculated by the means of the two fractions collected prior to treatment [28].

4   Superfused Synaptosome to Study Receptor–Receptor Interactions

The study of synaptosome functions has been one of the cornerstones of neurochemistry research, being instrumental to the understanding of the molecular machinery of neurotransmission and synaptic protein–protein interaction networks, including receptor–receptor interactions.

It is well established that the allosteric receptor–receptor interactions in heteroreceptor complexes give diversity, specificity, and bias to the receptor protomers due to conformational changes in discrete domains leading to changes in receptor protomer function and their pharmacology [32]. Heteroreceptor complexes and their receptor–receptor interactions represent a new fundamental principle in molecular medicine for integration of transmitter signals in the plasma membrane. Thus, the understanding of functional consequences of heteromer recruitment under either physiological or pathological conditions could strongly facilitate the development of new therapeutic strategies aimed at targeting heteroreceptor complexes with combined treatment, multi-targeted drugs and heterobivalent drugs. In this context, it is important to be able to take advantage of multiple techniques for use in cellular models, brain tissue and in vivo studies to understand the functional role of heteromers in discrete brain circuits. Among the available methods, superfused synaptosomes represent the preparation of choice to understand the role of receptor–receptor interactions as integrative mechanisms in nerve terminals (i.e., at presynaptic level) from selected brain region. Once the existence of a specific receptor heteromer has been demonstrated or postulated at presynaptic level, its functional relevance into the brain can be assessed by evaluating how ligands of one receptor protomer modify the changes in neurotransmitter levels induced by an agonist/antagonist of another receptor protomer. In fact, when in superfused synaptosomes the addition to the superfusion medium of an agonist selective for one receptor modifies the effect on release induced by a ligand for another receptor, it is possible to conclude that (1) the two receptors coexist on the same axon terminal; (2) there may be a direct intramembrane cross-talk between the two receptors. The typical methodological advantages of using superfused synaptosomes, instead of other in vitro technique, are particularly relevant when this preparation is used to investigate on the possible existence of receptor–receptor interactions. In fact, by using synaptosomes: (1) it is possible to exclude indirect effects as possible mechanisms underlying the observed effects; (2) by measuring different neurotransmitters in the same preparation, it is possible to discriminate the existence and the absence of an investigated receptor heteromer on different nerve terminal subfamilies; (3) it is possible to evaluate the functional role of receptor heteromers, and its modification, after chronic treatment with drugs, in genetically modified animals and in animal model of pathologies. A particular aspect in which superfused synaptosomes will probably be of great help in the near future is related to the mechanisms of neurotransmitter release. In fact, exocytosis and endocytosis are increasingly emerging as very complex processes in which several presynaptic actors are involved. Several proteins (e.g., synaptobrevins, syntaxins, SNAP-25, synaptotagmins, synapsins, and others) participate in this complex scenario and might be implicated into the mechanisms by which receptor heteromers regulate neurotransmitter release. Finally, there are unique insights to be gained by directly examining receptor heteromer functions in human synaptosomes, which cannot be substituted by animal models [17–20]. For all these reasons, superfused synaptosome technique has been used to investigate the functional neurochemical aspects of presynaptic receptor–receptor interaction [12, 28, 33–35].

5   Superfused Synaptosome to Study How Cocaine Modulates Allosteric D2-σ1 Receptor–Receptor Interactions on Dopamine and Glutamate Nerve Terminals from Rat Striatum

In order to highlight the general aspects to consider when superfused synaptosomes are used to investigate on receptor–receptor interactions, the procedure we recently followed to evaluate the possible effects of cocaine on sigma1 receptor (σ_1R)/DA D2 receptor (D2R) interactions is described [28].

5.1   Rationale

Previous studies demonstrated that the activation of σ_1Rs mediated different aspects of cocaine abuse [36, 37]. These receptors are highly localized in the CNS, including the dorsal striatum and the nucleus accumbens [38] and it has been proven that σ_1Rs can physically interact with D_2Rs in mouse striatum. By possibly interacting with this heteromer, cocaine demonstrated to modulate downstream signaling in both cultured cells and in mouse striatum [39]. Thus, it was postulated that the previously demonstrated allosteric action of cocaine at D2Rs [40, 41] might be mediated by the σ_1R protomer at the presynaptic D2R-σ_1R heteroreceptor complex. In order to verify this hypothesis, we tested the effects of nanomolar cocaine concentrations, possibly not blocking the DA transporter (DAT) activity, on striatal D2R-σ_1R heteroreceptor complexes. Furthermore, the possible role of σ_1Rs in the cocaine-provoked amplification of quinpirole-induced reduction of K^+-evoked [^{3}H]-DA and glutamate release from rat striatal synaptosomes was also investigated [28].

5.2 Experimental Procedure

With the aim to confirm that D2Rs negatively modulate [^{3}H]-DA and glutamate release from rat striatal nerve terminals, we used the selective D2R agonist quinpirole, added to the perfusion medium at 10 nM, 100 nM, and 1 μM concentrations. We initially observed that at all concentrations tested, quinpirole failed to significantly affect either [^{3}H]-DA or glutamate levels. On the contrary, quin-pirole (100 nM and 1 μM) caused a concentration-dependent reduction of the K^+ (15 mM; 90 s)-evoked [^{3}H]-DA release and glutamate from rat striatal synaptosomes, while it was ineffective at the lower 10 nM concentration. To confirm the involvement of D2Rs in these effects, we pretreated synaptosomes with the D2R antagonist (S)-$(-)$-sulpiride (100 nM; by itself ineffective), which fully counteracted the quinpirole-induced reduction in [^{3}H]-DA or glutamate release.

The previous demonstration that the σ_1R agonist PRE084 induced a significant decrease in the ability of D2Rs to signal through Gi [39], prompted us to investigate the possible functional consequences of this biochemical findings, by measuring [^{3}H]-DA or glutamate release from striatal synaptosomes. As we used the selective σ_1R antagonist BD1063 to investigate the presynaptic D2R-σ_1R heteroreceptor complex, an increase in D2R-mediated control of [^{3}H]-DA and glutamate release was expected. Thus, we found that in the presence of BD1063 (100 nM; by itself ineffective), quinpirole at a concentration by itself ineffective (10 nM) significantly decreased K^+-evoked [^{3}H]-DA, but not glutamate, release from rat striatal synaptosomes (Fig. 2). This effect was not observed when quinpirole was given with lower BD1063 concentrations (1 and 10 nM). Furthermore, BD1063 (100 nM) significantly amplified the quinpirole (100 nM)-induced reduction of K^+-evoked [^{3}H]-DA, but not glutamate, release from rat striatal synaptosomes. Based on these and previous findings, it has been postulated that both the D2Rs and possibly the σ_1R modulate a voltage-dependent N-type calcium channels in the striatal DA, but not glutamate, terminals. Successively, we demonstrated that cocaine (already at 1 nM) significantly enhanced the quinpirole (100 nM)-induced decrease of K^+-evoked [^{3}H]-DA release from

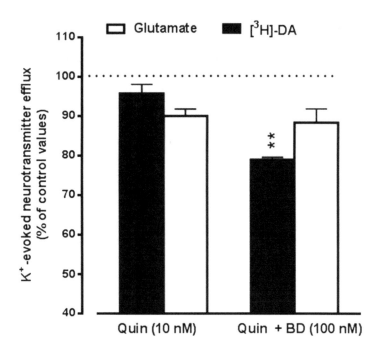

Fig. 2 Effects of the σ₁ receptor antagonist BD1063 (BD; 100 nM) on the quin-pirole (Quin; 10 nM)-induced modification of K⁺-evoked [³H]-DA and glutamate release from rat striatal synaptosomes. The drugs were added alone or in combination concomitantly with the depolarizing stimulus (15 mM K⁺, 90 s). The same volume of vehicle (Krebs' solution) was combined to the depolarizing stimulus in the control groups. The effect of the treatments on K⁺-evoked [³H]-DA and glutamate release is expressed as percent of control values. Each treatment bar represents the mean ± SEM of 6–8 determinations run in duplicate. **$p < 0.01$ significantly different from the respective control and other groups, according to ANOVA Tukey's Multiple Comparison post-test (modified from [28])

rat striatal synaptosomes. This result supported previous findings demonstrating that cocaine at nanomolar concentrations can enhance Gi/o-mediated D2R signaling by direct actions on the D2 heteroreceptor complexes. Interestingly, the σ₁R receptor antagonist BD1063, at nM concentrations, counteracted the enhanced cocaine–quinpirole interactions on DA terminals (Fig. 3). Thus, it seems possible that this effect is mediated by the altered allosteric receptor–receptor interactions induced by cocaine targeting the σ₁R protomer in the postulated D2R-σ₁R-N-type calcium channel complex. Similar events were found to take place in the K⁺-evoked glutamate release experiments under the influence of cocaine but only at 10 nM, thus validating the preferential role of these effects on striatal DA terminals. The possible relevance of this finding is underlined by the evidence that antagonism of σ₁R receptors attenuates cocaine-induced convulsions, acute locomotor stimulation, locomotor sensitization as well as cocaine-induced conditioned place preference. Thus, by using superfused

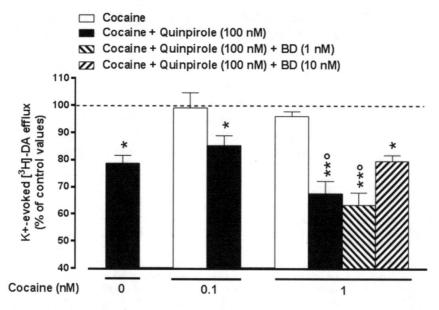

Fig. 3 Effects of the σ_1 receptor antagonist BD1063 (BD; 1 and 10 nM) on cocaine-provoked amplification of quinpirole (Quin; 100 nM)-induced reduction of K+-evoked [³H]-DA release from rat striatal synaptosomes. The drugs were added alone or in combination concomitantly with the depolarizing stimulus (15 mM K+, 90 s). The same volume of vehicle (Krebs' solution) was combined to the depolarizing stimulus in the control groups. The effect of the treatments on K+-evoked [³H]-DA release is expressed as percent of control values. Each treatment bar represents the mean ± SEM of 6–8 determinations run in duplicate. *$p < 0.05$, **$p < 0.01$ significantly different from the control and cocaine alone groups; °$p < 0.05$ significantly different from the Quin alone group, according to ANOVA followed by Tukey's Multiple Comparison post-test (modified from [28])

synaptosomes, it has been possible to provide a molecular mechanism, involving a possible receptor heteromer, underlying these behavioral effects.

6 Conclusions

Superfused synaptosomes have been utilized in studies of neurotransmitter release during the last 40–45 years. This simple and versatile "ex-vivo" technique allows clarifying several aspects of synaptic neurochemistry including neurotransmitter transport, receptor localization, receptor heterogeneity, and mechanisms of neurotransmitter exocytosis–endocytosis. As receptor–receptor interactions take place likewise also at presynaptic level, superfused synaptosomes can relevantly contribute to the understanding of the functional relevance of these integrative mechanisms associated with the plasma membrane. When combined with other neurochemical techniques such as tissue slices or in vivo microdialysis, synaptosomes offer the possibility to discriminate pre- and postsynaptic receptor heteromer-mediated regulation of neurotransmitter release. Furthermore, this technique also allows the possibility to establish whether receptor heteromer functions are modified during the development of a specific pathology, thus suggesting these complexes as emerging therapeutic targets [42]. Finally, the possibility to perform experiments in synaptosomes prepared from human tissue could strongly contribute to clarify the role of receptor heteromers in various human developmental, traumatic, and degenerative brain disorders. Thus, resuming the charming title of a published review [6], it is possible to conclude that "synaptosomes are still viable after 45 years of superfusion."

References

1. Gray EG, Whittaker VP (1960) The isolation of synaptic vesicles from the central nervous system. J Physiol 153:35–37

2. Gray EG, Whittaker VP (1962) The isolation of nerve endings from brain: an electron microscopic study of cell fragments derived by homogenisation and centrifugation. J Anat 96:82–83

3. Whittaker VP (1993) Thirty years of synaptosome research. J Neurocytol 22:735–742

4. Breukel AIM, Besselsen E, Ghijsen WEJM (1997) Synaptosomes: a model system to study release of multiple classes of transmitters. Methods Mol Biol 72:33–47

5. Raiteri M, Angelini F, Levi G (1974) A simple apparatus for studying the release of neurotransmitter from synaptosomes. Eur J Pharmacol 25:411–414

6. Raiteri L, Raiteri M (2000) Synaptosomes still viable after 25 years of superfusion. Neurochem Res 25:1265–1274

7. Pittaluga A (2016) Presynaptic release-regulating mGlu1 receptors in central nervous system. Front Pharmacol 7:295

8. Middlemiss DN (1988) Receptor-mediated control of neurotransmitter release from brain slices or synaptosomes. Trends Pharmacol Sci 9:83–84

9. Ghijsen WE, Leenders AG, Lopes da Silva FH (2003) Regulation of vesicle traffic and neurotransmitter release in isolated nerve terminals. Neurochem Res 28:1443–1452

10. Kamat PK, Kalani A, Tyagi N (2014) Method and validation of synaptosomal preparation for isolation of synaptic membrane proteins from rat brain. MethodsX 1:102–107

11. Fontana G, Blaustein MP (1993) Calcium buffering and free $Ca2+$ in rat brain synaptosomes. J Neurochem 60:843–850

12. Marchi M, Grilli M, Pittaluga AM (2015) Nicotinic modulation of glutamate receptor function at nerve terminal level: a fine-tuning of synaptic signals. Front Pharmacol 29:6–89

13. Nagy A, Delgado-Escueta AV (1984) Rapid preparation of synaptosomes from mammalian brain using nontoxic isosmotic gradient material (Percoll). J Neurochem 43:1114–1123

14. Dunkley PR, Jarvie PE, Robinson PJ (2008) A rapid Percoll gradient procedure for preparation of synaptosomes. Nat Protoc 3:1718–1728

15. Garcia-Sanz A, Badia A, Clos MV (2001) Superfusion of synaptosomes to study presynaptic mechanisms involved in neurotransmitter release from rat brain. Brain Res Brain Res Protoc 7:94–102

16. Galván E, Sitges M (2004) Characterization of the participation of sodium channels on the rise in Na+-induced by 4-aminopyridine (4-AP) in synaptosomes. Neurochem Res 29:347–355

17. Haberland N, Hetey L (1987) Studies in postmortem dopamine uptake. II Alterations of the synaptosomal catecholamine uptake in postmortem brain regions in schizophrenia. J Neural Transm 68:303–313

18. Ferraro L, Tanganelli S, Caló G et al (1993) Noradrenergic modulation of gamma-aminobutyric acid outflow from the human cerebral cortex. Brain Res 629:103–108

19. Postupna NO, Keene CD, Latimer C et al (2014) Flow cytometry analysis of synaptosomes from post-mortem human brain reveals changes specific to Lewy body and Alzheimer's disease. Lab Investig 94:1161–1172

20. Jhou JF, Tai HC (2017) The study of postmortem human synaptosomes for understanding Alzheimer's disease and other neurological disorders: a review. Neurol Ther 6(Suppl 1):57–68

21. Dodd PR, Hardy JA, Baig EB, Kidd AM, Bird ED, Watson WE, Johnston GA (1986) Optimization of freezing, storage, and thawing conditions for the preparation of metabolically

active synaptosomes from frozen rat and human brain. Neurochem Pathol 4:177–198

22. Valtier D, Dement WC, Mignot E (1992) Monoaminergic uptake in synaptosomes prepared from frozen brain tissue samples of normal and narcoleptic canines. Brain Res 588:115–119

23. Gleitz J, Beile A, Wilffert B, Tegtmeier F (1993) Cryopreservation of freshly isolated synaptosomes prepared from the cerebral cortex of rats. J Neurosci Methods 47:191–197

24. Dunkley PR, Jarvie PE, Heath JW et al (1986) A rapid method for isolation of synaptosomes on Percoll gradient. Brain Res 372:115–129

25. López-Pérez MJ (1994) Preparation of synaptosomes and mitochondria from mammalian brain. Methods Enzymol 228:403–411

26. Evans GJ (2015) Subcellular fractionation of the brain: preparation of synaptosomes and synaptic vesicles. Cold Spring Harb Protoc 2015:462–466

27. Musante V, Summa M, Cunha RA et al (2011) Pre-synaptic glycine GlyT1 transporter–NMDA receptor interaction: relevance to NMDA autoreceptor activation in the presence of Mg2+ ions. J Neurochem 117:516–527

28. Beggiato S, Borelli AC, Borroto-Escuela D et al (2017) Cocaine modulates allosteric D2-σ1 receptor-receptor interactions on dopamine and glutamate nerve terminals from rat striatum. Cell Signal 40:116–124

29. Bradford MM (1976) A rapid and sensitive method for the quantitation of microgram quantities of protein utilizing the principle of protein-dye binding. Anal Biochem 72:248–254

30. Chiodi V, Uchigashima M, Beggiato S et al (2012) Unbalance of CB1 receptors expressed in GABAergic and glutamatergic neurons in a transgenic mouse model of Huntington's disease. Neurobiol Dis 45:983–991

31. Ferraro L, Tanganelli S, Fuxe K et al (2001) Modafinil does not affect serotonin efflux from rat frontal cortex synaptosomes: comparison with known serotonergic drugs. Brain Res 894:307–310

32. Fuxe K, Borroto-Escuela DO (2016) Heteroreceptor complexes and their allosteric receptor-receptor interactions as a novel biological principle for integration of communication in the CNS: targets for drug development. Neuropsychopharmacology 41:380–382

33. Grilli M, Summa M, Salamone A et al (2012) In vitro exposure to nicotine induces endocytosis of presynaptic AMPA receptors modulating dopamine release in rat nucleus accumbens nerve terminals. Neuropharmacology 63:916–926

34. Chiodi V, Ferrante A, Ferraro L et al (2016) Striatal adenosine-cannabinoid receptor interactions in rats over-expressing adenosine A2A receptors. J Neurochem 136:907–917

35. Morató X, Luján R, López-Cano M et al (2017) The Parkinson's disease-associated GPR37 receptor interacts with striatal adenosine A2A receptor controlling its cell surface expression and function in vivo. Sci Rep 7:9452

36. Maurice T, Martin-Fardon R, Romieu P et al (2002) Sigma(1) receptor antagonists represent a new strategy against cocaine addiction and toxicity. Neurosci Biobehav Rev 26:499–527

37. Matsumoto RR, Nguyen L, Kaushal N et al (2014) Sigma (σ) receptors as potential therapeutic targets to mitigate psychostimulant effects. Adv Pharmacol 69(2014):323–386

38. Alonso G, Phan V, Guillemain I et al (2000) Immunocytochemical localization of the sigma(1) receptor in the adult rat central nervous system. Neuroscience 97:155–170

39. Navarro G, Moreno E, Bonaventura J et al (2013) Cocaine inhibits dopamine D2 receptor signaling via sigma-1-D2 receptor heteromers. PLoS One 8:e61245

40. Genedani S, Carone C, Guidolin D et al (2010) Differential sensitivity of A2A and especially D2 receptor trafficking to cocaine compared with lipid rafts in cotransfected CHO cell lines. Novel actions of cocaine independent of the DA transporter. J Mol Neurosci 4:347–357

41. Ferraro L, Frankowska M, Marcellino D et al (2012) A novel mechanism of cocaine to enhance dopamine d2-like receptor mediated neurochemical and behavioral effects. An in vivo and in vitro study. Neuropsychopharmacology 37:856–1866

42. Guidolin D, Agnati LF, Marcoli M et al (2015) G-protein-coupled receptor type A heteromers as an emerging therapeutic target. Expert Opin Ther Targets 19:265–283

Chapter 5

Detection of Fibroblast Growth Factor Receptor 1 (FGFR1) Transactivation by Muscarinic Acetylcholine Receptors (mAChRs) in Primary Neuronal Hippocampal Cultures Through Use of Biochemical and Morphological Approaches

Valentina Di Liberto, Giuseppa Mudó, Dasiel O. Borroto-Escuela, Kjell Fuxe, and Natale Belluardo

Abstract

In addition to their canonical intracellular signals involved in the regulation of neuronal plasticity, G-protein coupled receptors can also rapidly transactivate tyrosine kinase receptors and their downstream intracellular signaling in absence of specific ligands. Here we describe our protocol for dissociating and maintaining hippocampal primary neurons in high- and low-density culture, followed by a description of methods employed to evaluate neurite outgrowth and protein phosphorylation associated with fibroblast growth factor receptor 1 transactivation by muscarinic acetylcholine receptors. Our goal was to provide the reader with detailed protocols of the abovementioned techniques and to highlight advantages and limitations of the used approaches as compared to other valid alternatives.

Key words Primary neuronal culture, Hippocampus, Fibroblast growth factor receptor, Transactivation, Western blotting, Tyrosine kinase receptor, Muscarinic acetylcholine receptor, Receptor–receptor interactions, Phosphorylation, Neurite growth

1 Introduction

G-protein coupled receptors (GPCRs) can stimulate the signaling activity of tyrosine kinase receptors (TKRs). In this way, the broad diversity of GPCRs is connected with the potent signaling capacities of TKRs. This cross-communication between receptors, also called transactivation, controls multiple physiological processes [1]. Transactivation of TKRs by GPCRs can occur through ligand-dependent and ligand-independent mechanisms. In the first case, GPCR-mediated TKR transactivation depends on the activation of membrane-bound matrix metalloproteases, which in turn cleaves the TKR ligand precursor to form the

Kjell Fuxe and Dasiel O. Borroto-Escuela (eds.), *Receptor-Receptor Interactions in the Central Nervous System*, Neuromethods, vol. 140, https://doi.org/10.1007/978-1-4939-8576-0_5, © Springer Science+Business Media, LLC, part of Springer Nature 2018

active ligand causing the subsequent activation of the TKR. The ligand-independent transactivation is usually mediated by the physical association between the two receptors (heteromerization), which leads to the TKR phosphorylation and involves the recruitment of intracellular kinases [1]. As an example, GABA-B receptors can transactivate insulin growth factor-1 receptors leading to survival signaling in primary neurons [2]. Instead serotonin receptor 1A (5-HT1A) is able to transactivate fibroblast growth factor receptor 1 (FGFR1) in the rat hippocampus with enhancement of hippocampal plasticity [3].

This chapter describes a detailed protocol for primary neuronal hippocampal cultures from E18 rat embryos and some practical applications based on biochemical and morphological analysis aimed to detect and analyze FGFR1 transactivation by muscarinic acetylcholine receptors (mAChRs) activation and the related functional effects [4].

2   Materials

In Fig. 1 is shown a scheme of workflow with main steps of protocol for preparing neuronal hippocampal primary cultures.

2.1   Preparation of Poly-lysinated Plates and Coverslips

In order to promote neuronal adhesion to glass substrate, coverslips need to be treated with acid to make the surface rough enough for cell attachment. To this end, German glass coverslips (12 mm diameter) were incubated overnight on a shaker with concentrated nitric acid (70% wt/wt), rinsed four times with sterile water, sterilized in EtOH 70%, and dried in the laminar flow hood. Immediately before use, poly-D-lysine (molecular weight 70–150 kDa; Sigma-Aldrich, cat. no. P6407) was dissolved at 0.5 mg/ml in sterile 150 mM sodium borate buffer (pH 8.5), which enhances poly-lysine adsorption to cell culture surface, and filter-sterilized. Coverslips, placed in 24-well plates, and 12-well plates were coated with poly-lysine solution and incubated for 2 h in a 37 °C, 5% CO_2 humidified incubator. Coverslips and plates were rinsed twice with sterile water, dried under a laminar flow hood, and stored at 4 °C for up to 1 week. According to our experience, plates and coverslips not coated/cleaned properly or stored too long before use cause cells detachment or clumping of cell bodies, neurite fasciculation, and impairment of neuron survival.

2.2   Hippocampus Dissection

Hippocampus was dissected from brain of E18-E19 Wistar rat embryos. Cultures were prepared from late-stage embryos because at that time the generation of pyramidal neurons is essentially complete, while the major gliogenesis occurs either later in gestation or postnatally [5]. Timed pregnancies were obtained by daily checking vaginal washings for sperm and were confirmed by manual

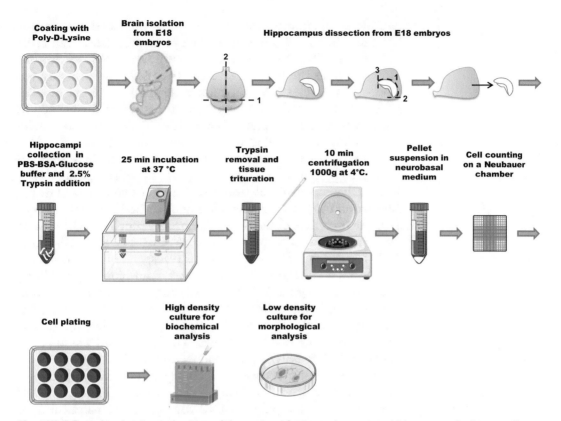

Fig. 1 Workflow showing the main steps of the protocol for preparing neuronal hippocampal primary cultures. Figure designed using http://smart.servier.com/

palpation. At the appropriate stage of gestation, the pregnant rat was euthanized, the uterus was dissected out and placed in a sterile 100-mm dish containing cold PBS 1×. Working in a laminar flow hood, over an ice support, the embryos were removed from the uterus, the brains dissected out and placed in a 30-mm dish containing cold PBS 1×. Hippocampal tissue was dissected under a dissecting microscope as follows: the cerebellum and brain stem were cut away, brains were incised along midline and separated into two hemispheres. To avoid fibroblast-like or epithelial-like cell contamination of neuronal culture, meninges were removed and the hippocampus, a curved structure that starts in the distal part of the hemisphere and bends ventrally, was separated from the cortex by a cut parallel to the hippocampal fissure, along the convex outer side, and by transverse cuts at its rostral and caudal ends.

2.3 Culture of Primary Hippocampal Neurons and Pharmacological Treatment

Hippocampi from 10 to 14 embryos were collected and transferred in a 15-ml sterile tube containing 4.5 ml of cold PBS (1×)-BSA (0.1%)-D-Glucose (0.18%) buffer. After the addition of 2.5% trypsin (0.5 ml, Gibco), hippocampal tissue was incubated for 25 min in a water bath at 37 °C. The diluted concentration of trypsin used in this protocol does allow longer times of enzyme incubation to increase individual cellular disassociation. On the other hand, prolonged time of trypsin action can cause cell damage. Trypsin solution was gently removed, leaving the hippocampi at the bottom of the tube, and trypsin activity was stopped by adding 2.5 ml of horse serum and 2.5 ml of PBS-BSA-Glucose buffer. The hippocampal tissue was dissociated by repeatedly pipetting up and down with a Pasteur pipette, avoiding foaming by expelling the suspension against the wall of the tube, until no big clumps of tissue remained. This step has proven to be crucial in consistent neuronal culture, since excessive mechanical manipulation can cause massive neuronal death. After trituration, the tube was left undisturbed for 5 min in such a way that chunks of tissue left can precipitate. The cell solution was transferred to another tube and centrifuged for 10 min at $1000 \times g$ at 4 °C. The cell pellet was resuspended in plating neurobasal medium (Gibco) containing 1% antibiotics (penicillin/streptomycin, Sigma-Aldrich P4333) and L-Glutamine (0.5 mM, Gibco), and living cells were counted on a Neubauer chamber with trypan blue. This step is helpful in evaluating the quality and health of the dissociated cells. The mean yield was approximately 5×10^5 cells/hippocampus, and 90% of the cells were viable. 7×10^5 living cells/well were plated on a 12-well plate, as high-density culture for biochemical studies (Fig. 2a), while 3×10^4 cells were plated on 12 mm glass coverslips, as low-density culture for morphological studies (Fig. 3a). After 2 h, once the adhesion is complete, floating dead cells and non-adherent cells were removed by replacing the medium with maintenance neurobasal medium containing 2% B27 (Gibco), L-Glutamine (0.5 mM), and 1% antibiotics. The day after plating, cytosine arabinoside 10 μM (AraC, Sigma-Aldrich C1768) was added to prevent proliferation of dividing cells. Forty-eight hours later, in order to avoid AraC toxicity toward neurons [6], the medium was carefully replaced with fresh medium. Neurons were maintained in culture with only half of the medium changed every fourth day in order to improve long-term neurons survival, which is strictly dependent on soluble factors released by neurons themselves in the medium. Neurons were cultured for 48 h for analysis of neurite outgrowth or for 12 days for evaluation of FGFR1 phosphorylation. In the second case, in order to minimize acute influence of growth factors and signaling molecules present in the medium on FGFR1 phosphorylation, before stimulation with muscarinic acetylcholine receptor agonist Oxotremorine-M (Oxo-M) (O100, Sigma-Aldrich) neurons were "starved" for 1 h by replacing full maintenance medium with neurobasal medium containing only antibiotics. Effects of Oxo treatment on FGFR1 phosphorylation were evaluated by western blotting.

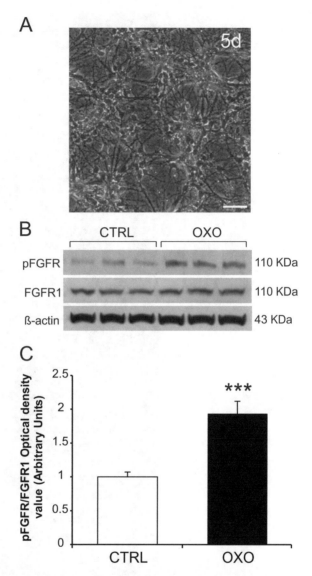

Fig. 2 High-density hippocampal primary neuronal culture for biochemical investigations of FGFR phosphorylation. Dissociated hippocampal neurons from E18 rat brain were plated at a density of 7×10^5 cells/well on a 12-well plate. After 12 days in culture, neurons were treated for 5 min with Oxotremorine 100 nM and FGFR phosphorylation was measured in cell extracts by western blotting. (**a**) Representative picture of hippocampal neuronal culture 5 days (5d) after plating, showing the established complex neurites network. Scale bar 50 μm. (**b**) Representative western blotting bands of phosphorylated FGFR (pFGR), FGFR1 and β-actin in untreated neurons (CTRL) and neurons treated with Oxotremorine (Oxo) 100 nM for 5 min. (**c**) Histogram showing the quantification of pFGFR signal normalized with FGFR1 total levels in CTRL and Oxo-treated neurons. β-Actin was used as internal loading control. The data were evaluated by t-test analysis. ***$p < 0.005$

62   Valentina Di Liberto et al.

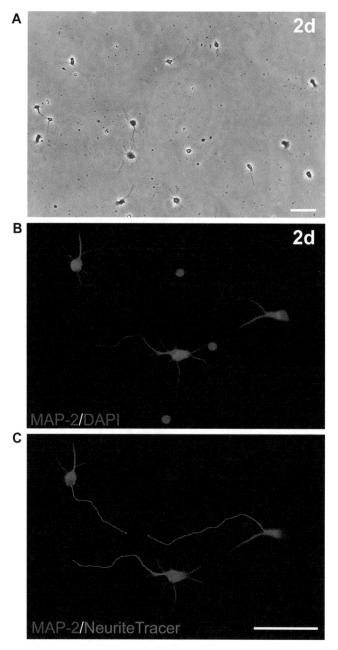

Fig. 3 Low-density hippocampal primary neuronal culture for morphological investigations of neurites outgrowth. Dissociated hippocampal neurons from E18 rat brain were plated at a density of 3×10^4 cells on 12 mm coverslips. Twenty-four hours after plating, cells were treated with Oxotremorine for 24 h, fixed with 4% formaldehyde solution, and processed for MAP-2 immunocytochemistry. (**a**) Representative picture of hippocampal neuronal culture 48 h (2d) after plating, where single neurites are clearly distinguishable. Scale bar 50 μm. (**b**) Representative picture of hippocampal neurons stained with MAP-2 (red) to label neuronal perikarya and neurites. Cell nuclei were counterstained with DAPI (blue). (**c**) Neuritis labeling (purple) using Simple Neurite Tracer plugin of ImageJ 1.51n software®

3   Methods

3.1   Protein Extraction and Western Blotting

After 12 days in culture, neurons, seeded in 12-well plate, were washed with 1 ml ice cold PBS 1× and scraped in 70 μl cold radio-immunoprecipitation assay (RIPA) buffer (Tris 50 mM, pH 7.4, NaCl 150 mM, Triton 1%, SDS 0.1%), supplemented with a protease inhibitor cocktail (P8340, Sigma-Aldrich, 5 μl/ml) and phosphatase inhibitor cocktail (P5726, Sigma-Aldrich, 10 μl/ml). RIPA buffer is a strong lysis buffer useful for whole cell extracts, membrane-bound and nuclear proteins since it contains detergents (SDS and Triton) helpful in isolating proteins from membranes. In addition to chemical lysis, a physical extraction/solubilization procedure was applied to extract our proteins of interest: samples were sonicated (30 pulsations/min), while keeping the suspension on ice to prevent excessive heating. Importantly, as soon as lysis occurs, proteolysis, dephosphorylation, and denaturation begin. These events can be significantly slowed down if samples are kept on ice or at 4 °C at all times and appropriate inhibitors are added fresh to the lysis buffer. After sonication, the samples were incubated 15 min in ice, quantified by the Lowry method [7], and stored at −80 °C. For western blotting, protein samples (60 μg per lane) were diluted 1:1 with Laemmli sample buffer and heated to dry plate for 5 min at 100 °C. Samples and molecular weight marker (161–0375, BIO-RAD) were run on polyacrylamide 7% gel for detection of phosphorylated-FGFR (pFGFR) and transferred onto nitrocellulose membrane (RPN303E, GE Healthcare Life Science) using Biorad Trans-Blot Turbo Transfer System. We have used Acrylamide/Bis-Acrylamide Solution with 37.5:1 ratio, which allows the separation and the migration of high molecular weight FGFR proteins. The membranes were incubated for 1 h in blocking buffer (TBS 1×, Tween-20 0.1%, nonfat dry milk 5%), washed once with TBS/Tween (TBS/T), and incubated with gentle shaking overnight at −4 °C with specific antibody in primary antibody dilution buffer: TBS 1×, Tween-20 0.1%, BSA 5%. For detection of pFGFR, mouse anti-pFGFR (Tyr653/654) antibody (55H2, Cell Signaling, 1:1000) was used. The day after, membranes were washed three times for 10 min with TBS/T and incubated for 1 h RT with goat anti-mouse IgG (SC-2005, Santa Cruz Biotechnology) or goat anti-rabbit IgG (SC-2004, Santa Cruz Biotechnology) horseradish peroxidase-conjugated (1:7000). After three washings with TBS-T, immuno-complexes were visualized with chemiluminescence reagent (RPN2236, GE Healthcare Life Science) according to the manufacturer's instructions. The Hyperfilms (28,906,837, GE Healthcare Life Science) were developed using developer and fixer (1,900,984 and 1,902,485, Kodak GBX). Each film was exposed for at least three different times to evaluate saturation problems.

FGFR1 protein (Swiss-Prot Acc P11362) comprises 822 amino acids, which form a 100–135 kDa glycoprotein. We detected a band of approximately 110 kDa, which corresponds to the glycosylated/phosphorylated form. In addition, we confirmed the specificity of the obtained band by the observation that membrane Calf intestinal alkaline phosphatase treatment after blotting abolishes the reactivity of the antibody.

Since the measured levels of a phosphoprotein may change with treatment or through gel loading errors, the membranes were stripped at 37 °C for 30 min in buffer pH 2.2 containing glycine 200 mM, SDS 3 mM, and Tween-20 1%, washed twice with TBS/T, and re-probed with goat anti-FGFR1 (SC-31169, Santa Cruz Biotechnology, 1:500) and anti β-actin (SC-32258, Santa Cruz Biotechnology, 1:5000). In this way, phosphorylated FGFR fraction relative to the total fraction can be determined and β-actin can serve as an internal loading control. Densitometric evaluation of bands was performed by measuring optical density after background subtraction using ImageJ 1.51n software® (Rasband, W.S., ImageJ, U. S. National Institutes of Health, Bethesda, Maryland, USA, https://imagej.nih.gov/ij/, 1997–2017). We report here (Fig. 2b, c), as a general example, a representative image and the relative quantification of pFGFR levels in primary hippocampal culture following 5 min treatment with Oxo (100 nM).

3.2 Neuronal Growth Analysis

To evaluate neuronal functional effects of FGFR1 transactivation by mAChRs, we analyzed neurite outgrowth, by examining neurite length and density. Cells were grown at density of 3×10^4 on poly-D-lysine/boric acid-coated glass coverslips. Twenty-four hours after plating, cells were treated with Oxo for 24 h and fixed with formaldehyde 4% for 15 min RT (Fig. 3a). Cells were washed twice with PBS, pre-incubated in blocking solution (BSA 5 mg/ml and Triton 0.1% in PBS 1×) for 15 min, and incubated overnight with anti-MAP-2 antibody 1:4000 (M4403, Sigma-Aldrich) diluted in blocking solution. MAP-2 is a neuron-specific protein that promotes assembly and stability of the microtubule network [8]. Antibodies to MAP-2 are therefore excellent markers on neuronal cells, their perikarya and neuronal processes. The day after, cells were washed twice with PBS and incubated for 1 h with a rhodamine-conjugated anti-mouse IgG Cy3 antibody 1:150 (115-165-003, Jackson ImmunoResearch). After two washing in PBS, DAPI staining was performed to label cell nuclei (Fig. 3b). The coverslips were mounted on slides in buffered 70% glycerol and examined, using the 50× objective, under a fluorescence microscope (DMRBE, Leica Microsystems GmbH), equipped with digital video camera (Spot-RT Slider, Diagnostic Instruments). Number of neurites/cell and primary neurite length were measured in a sample of at least 300 cells per each condition of five independent experiments, using Simple Neurite Tracer plugin of ImageJ 1.51n software® which labels and measures neurites in neuronal cultures (Fig. 3c) and identifies and counts neuronal nuclei [9]. The results obtained by NeuriteTracer were further validated by comparison to manual tracing.

4   Notes

4.1   Neuronal Culture Features

Primary culture of neurons is a helpful and commonly used model for addressing a wide range of questions in molecular and cellular neurobiology and for elucidating the functional role of several factors and mechanisms on affecting neuronal network features, which evolve over time in parallel with neuronal network differentiation and maturation. The most important advantage in using primary neuronal cultures rather than continuous tumor-derived cell lines from the central nervous system (SH-SY5Y, NT2, PC12) is indeed the possibility to follow the process of neuron maturation. Neurons in vitro gradually mature passing through defined stages of differentiation [10], making synapses [11–13] and showing morphological and biochemical features that recapitulate the physiological properties of neuronal cells in vivo [14, 15]. As disadvantage, unlike cell lines that provide unlimited supplies of homogeneous cells, the preparation and culture of neuronal primary cells is more challenging. Primary neuronal cultures are not immortal, they do not proliferate, are less easy to transfect and start to degenerate at some point in culture, and therefore the number of cells and the time window available for experiments is limited. Furthermore, brain tissue is made up of different types of cells, making the separation of neurons from glial cells or the enhancement of neural survival necessary to obtain a pure neuronal culture. Finally, primary neuronal cultures are characterized by intrinsic variability, which can affect reproducibility of experimental data.

Among the regions of the brain used to extract primary nerve cells, hippocampus is one of the main sources of neurons. The hippocampus is a defined structure that can be easily isolated from the whole brain; it contains a relatively simple cell population, mostly pyramidal neurons [16] which express many of their key phenotypic features in culture, like development of neurites and formation of synaptically connected networks [5].

For the analysis of neuronal growth, we have established a low-density culture system in which individual cells and their processes could be easily distinguished in vitro (Fig. 3a, b). One of the major factors determining survival of neurons and neurite outgrowth in culture is cell density, since it affects neuronal morphology and synaptic density, due to variations in cell-to-cell contact and the global concentration of extrinsic factors [17, 18]. Higher cell densities result in better survival rates and improve neuronal differentiation [5]. Indeed, most procedures, which have been described for culturing neurons, employ high cell densities in order to enhance long-term cell survival. This inevitably leads to extensive cell aggregation, where the processes from individual cells are hardly distinguishable. Alternatively, neurons can be grown in presence of hippocampal explants or over an astrocyte feeder layer [5, 19, 20]. The first option causes inhomogeneous culture conditions in the well, since cell survival is strictly dependent on the proximity to the explant. The second option is also not well suited for morphological investigations because the neurons and glia are intermixed and the effects of drugs on neuronal growth could be affected by factors secreted by glial cells. It has been also shown that fibroblast growth factor-2 (FGF-2) induces survival of hippocampal neurons and promotes proliferation of neuronal progenitors from embryonic tissue [21]. However, for the purpose of investigating FGFR1 transactivation by mAChRs, the addition of FGF-2 to the medium would have affected the results.

We choose for our experiments a low-density neuronal culture (160 cells/mm^2), which permits a reasonable survival and neurite outgrowth avoiding extensive cell aggregation (Fig. 3a, b). We observed 50% viability at 48 h after plating. The viability was still unchanged after 72 h. However, we have not checked at later time points since we were interested in evaluating Oxo early action on the initial neurite outgrowth and not its long-term effects on neuronal morphology and survival. With regard to this aspect, many studies suggest a progressive decrease in long-term cell survival starting already 72 h after plating a low-density neuronal culture [18].

Usually, the initiation of neurite extension occurs within the first 24 h in culture. We have observed that, when the initial cell density is 160 cells/mm^2, 25% of the applied cells (usually larger, elliptical or pyramidal in shape) extend neurites. This percentage is low but still comparable to other investigations using co-culture methods [22], likely thanks to the use of poly-lysine as coating substrate and to the support of neurobasal medium supplemented with B27 and L-glutamine [23].

The rest of the cells instead, small and round when plated, do not develop neurites. According to literature, the two cell populations are to be considered different in their stage of development. In detail, the majority of nerve cells which survive and develop processes after dissociation are post-mitotic cells which have already completed their last round of DNA synthesis 12–48 h prior to dissociation. On the contrary, earlier post-mitotic cells which at the time of dissociation have already started to develop their definitive neuritis or cells still in division or immediately post-mitotic, hardly survive [18].

Based on our experience, supported by an extensive database, we below shortly describe maturation steps of primary neurons in culture. About 1 h after plating, cells are attached to the substrate, show a symmetric appearance, and extend a lamella all around the cell body before starting extending several minor neurites. We could observe that the development of neuronal polarity occurs during the first 2 days in culture with an outgrowing axon and several dendrites clearly visible. Indeed, after plating, the cell neurites may become axons or dendrites in the first hours, while within 48 hours only one neurite acquires axonal characteristics, growing quickly, and the remaining neurites become dendrites, growing more slowly. For example, if the original axon is cut, neurons can develop a new axon from another neurite [24–26]. At 5–8 days, neurons show higher number of outgrowing and branching dendrites and start synapses formation (Fig. 2a). Within the next days, neuronal dendrites form spines and extend the network of synaptic connections. After 12 days in culture, neuronal cells morphology and synaptic transmission are fully developed [22, 27]. For morphological analysis, we fixed our cells 48 h after plating, thus catching the impact of FGFR1 transactivation on neurite formation and axon elongation. For biochemical analysis we have used hippocampal neurons at 12 days in high-density culture (1900 cells/mm^2), when functional synaptic transmission is fully developed [22, 27].

4.2 Analysis of TKR Phosphorylation

TKRs control several cellular responses, including the regulation of cell growth, differentiation, migration, metabolism, and survival [28]. TKR activation is strictly dependent on phosphorylation, a key post-translational modification that regulates protein activity [29].

A classical method of directly measuring protein phosphorylation involves the incubation of whole cells with radiolabeled ^{32}P-orthophosphate, extraction and separation of proteins by SDS-PAGE, and exposure to film. However, this method requires multi-hour incubations and the use of radioisotopes [30]. Alternatively, two-dimensional gel electrophoresis can be used, since phosphorylation alters the mobility and isoelectric point of the protein, causing a shift in protein migration [31]. Mass spectrometry allows the identification of novel phosphorylated proteins and sites of phosphorylation, but it is often not used since it is biased toward particular phosphorylated sites, it is in general not quantitative, unless special applications (e.g., inductively coupled plasma mass spectrometry) are used [32, 33] and require expensive instrumentation. A valid alternative method is based on the development of specific antibodies for phosphorylated protein [34], that can be used either in western blotting or ELISA [30]. However, the main issue in using phospho-specific antibodies is that successful detection is dependent on the specificity and affinity of the antibody for the phosphoprotein of interest. Furthermore, the site of phosphorylation of interest has to be known in advance, thus limiting the analysis to only well-characterized phosphorylations, and antibodies made for every single phosphosite and phosphoprotein are not always available. Alternatively, good generic phosphotyrosine antibodies can be used, preceded by immunoprecipitation of the protein of interest.

To detect pFGFR1, we have used an antibody which targets FGFRs only when phosphorylated at tyrosines 653/654 essential for the catalytic activity of FGFR and signaling [35]. The antibody used in the present work does not distinguish among members of FGFR family. However, we can support the hypothesis that FGFR1 gives the major contribution since all the experiments have been performed in hippocampal neuronal cultures, which express mainly FGFR1, whereas FGFR2 and FGFR3 are expressed in astroglial cells [36] http://www.sciencedirect.com/science/article/pii/S0304416516304020 - bb0040. Moreover, heterocomplexes between mAChRs and FGFR1, responsible for the observed transactivation mechanism, have been detected [4].

The use of phospho-specific antibodies is strictly dependent on the protein amount in the sample. For example, if the phosphorylated form of the protein is only a small fraction, higher amount of proteins need to be loaded. Alternatively, the protein of interest can be captured by immunoprecipitation and probed with generic anti-phosphotyrosine antibodies on a western blotting [37]. We have used this strategy to detect pFGFR1 in the rat hippocampus, where the heterogeneous cell population does not allow an adequate concentration of the protein of interest [3].

5   Conclusions

The ability to produce in vitro cultures of neuronal cells has been a key step to advancing our understanding of the nervous system functions. Culture of hippocampal neurons was initially designed to study cell maturation steps but today this model has become one of the most powerful tools to study neuronal properties and to answer different molecular and cellular questions. For example, synaptogenesis, synaptic properties, and network functioning can be electrophysiologically investigated. Neuronal ultrastructure and molecular dynamics of single cells can be studied by time-lapse imaging techniques. In our study receptor functions and their pharmacology, including the mechanisms of action of new drugs, were biochemically tested. However, recently three-dimensional neuronal cultures, based on the generation of organoids/spheroids or bio-scaffolding systems, have been developed, which make experiments with neuronal cultures more similar to in vivo cellular responses, and can overcome some of the limitations of standard 2D culture systems [38].

References

1. Di Liberto V, Mudo G, Fuxe K, Belluardo N (2014) Interactions between cholinergic and fibroblast growth factor receptors in brain trophism and plasticity. Curr Protein Pept Sci 15(7):691–702

2. Tu H, Xu C, Zhang W, Liu Q, Rondard P, Pin JP, Liu J (2010) GABAB receptor activation protects neurons from apoptosis via IGF-1 receptor transactivation. J Neurosci 30(2):749–759. https://doi.org/10.1523/JNEUROSCI.2343-09.2010

3. Borroto-Escuela DO, Romero-Fernandez W, Mudo G, Perez-Alea M, Ciruela F, Tarakanov AO, Narvaez M, Di Liberto V, Agnati LF, Belluardo N, Fuxe K (2012) Fibroblast growth factor receptor 1- 5-hydroxytryptamine 1A heteroreceptor complexes and their enhancement of hippocampal plasticity. Biol Psychiatry 71(1):84–91. https://doi.org/10.1016/j.biopsych.2011.09.012

4. Di Liberto V, Borroto-Escuela DO, Frinchi M, Verdi V, Fuxe K, Belluardo N, Mudo G (2017) Existence of muscarinic acetylcholine receptor (mAChR) and fibroblast growth factor receptor (FGFR) heteroreceptor complexes and their enhancement of neurite outgrowth in neural hippocampal cultures. Biochim Biophys Acta 1861(2):235–245. https://doi.org/10.1016/j.bbagen.2016.10.026

5. Kaech S, Banker G (2006) Culturing hippocampal neurons. Nat Protoc 1(5):2406–2415. https://doi.org/10.1038/nprot.2006.356

6. Wallace TL, Johnson EM Jr (1989) Cytosine arabinoside kills postmitotic neurons: evidence that deoxycytidine may have a role in neuronal survival that is independent of DNA synthesis. J Neurosci 9(1):115–124

7. Lowry OH, Rosebrough NJ, Farr AL, Randall RJ (1951) Protein measurement with the Folin phenol reagent. J Biol Chem 193(1):265–275

8. Harada A, Teng J, Takei Y, Oguchi K, Hirokawa N (2002) MAP 2 is required for dendrite elongation, PKA anchoring in dendrites, and proper PKA signal transduction. J Cell Biol 158(3):541–549. https://doi.org/10.1083/jcb.200110134

9. Pool M, Thiemann J, Bar-Or A, Fournier AE (2008) NeuriteTracer: a novel ImageJ plugin for automated quantification of neurite outgrowth. J Neurosci Methods 168(1):134–139. https://doi.org/10.1016/j.jneumeth.2007.08.029

10. Dotti CG, Sullivan CA, Banker GA (1988) The establishment of polarity by hippocampal neurons in culture. J Neurosci 8(4):1454–1468

11. Bartlett WP, Banker GA (1984) An electron microscopic study of the development of axons and dendrites by hippocampal neurons in culture. II. Synaptic relationships. J Neurosci 4(8):1954–1965

12. Fletcher TL, De Camilli P, Banker G (1994) Synaptogenesis in hippocampal cultures: evidence indicating that axons and dendrites become competent to form synapses at different stages of neuronal development. J Neurosci 14(11 Pt 1):6695–6706

13. Ziv NE, Smith SJ (1996) Evidence for a role of dendritic filopodia in synaptogenesis and spine formation. Neuron 17(1):91–102

14. Garner CC, Waites CL, Ziv NE (2006) Synapse development: still looking for the forest, still lost in the trees. Cell Tissue Res 326(2):249–262. https://doi.org/10.1007/s00441-006-0278-1

15. Banker G, Goslin K (1988) Developments in neuronal cell culture. Nature 336(6195):185–186. https://doi.org/10.1038/336185a0

16. Benson DL, Watkins FH, Steward O, Banker G (1994) Characterization of GABAergic neurons in hippocampal cell cultures. J Neurocytol 23(5):279–295

17. Biffi E, Regalia G, Menegon A, Ferrigno G, Pedrocchi A (2013) The influence of neuronal density and maturation on network activity of hippocampal cell cultures: a methodological study. PLoS One 8(12):e83899. https://doi.org/10.1371/journal.pone.0083899

18. Banker GA, Cowan WM (1977) Rat hippocampal neurons in dispersed cell culture. Brain Res 126(3):397–342

19. Yang Q, Ke Y, Luo J, Tang Y (2017) Protocol for culturing low density pure rat hippocampal neurons supported by mature mixed neuron cultures. J Neurosci Methods 277:38–45. https://doi.org/10.1016/j.jneumeth.2016.12.002

20. Banker GA (1980) Trophic interactions between astroglial cells and hippocampal neurons in culture. Science 209(4458):809–810

21. Ray J, Peterson DA, Schinstine M, Gage FH (1993) Proliferation, differentiation, and long-term culture of primary hippocampal neurons. Proc Natl Acad Sci U S A 90(8):3602–3606

22. Grabrucker A, Vaida B, Bockmann J, Boeckers TM (2009) Synaptogenesis of hippocampal neurons in primary cell culture. Cell Tissue Res 338(3):333–341. https://doi.org/10.1007/s00441-009-0881-z

23. Brewer GJ, Torricelli JR, Evege EK, Price PJ (1993) Optimized survival of hippocampal

neurons in B27-supplemented Neurobasal, a new serum-free medium combination. J Neurosci Res 35(5):567–576. https://doi.org/10.1002/jnr.490350513

24. Dotti CG, Banker GA (1987) Experimentally induced alteration in the polarity of developing neurons. Nature 330(6145):254–256. https://doi.org/10.1038/330254a0

25. Goslin K, Banker G (1989) Experimental observations on the development of polarity by hippocampal neurons in culture. J Cell Biol 108(4):1507–1516

26. Bradke F, Dotti CG (1997) Neuronal polarity: vectorial cytoplasmic flow precedes axon formation. Neuron 19(6):1175–1186

27. Basarsky TA, Parpura V, Haydon PG (1994) Hippocampal synaptogenesis in cell culture: developmental time course of synapse formation, calcium influx, and synaptic protein distribution. J Neurosci 14(11 Pt 1): 6402–6411

28. Lemmon MA, Schlessinger J (2010) Cell signaling by receptor tyrosine kinases. Cell 141(7):1117–1134. https://doi.org/10.1016/j.cell.2010.06.011

29. Ubersax JA, Ferrell JE Jr (2007) Mechanisms of specificity in protein phosphorylation. Nat Rev Mol Cell Biol 8(7):530–541. https://doi.org/10.1038/nrm2203

30. Iliuk A, Martinez JS, Hall MC, Tao WA (2011) Phosphorylation assay based on multifunctionalized soluble nanopolymer. Anal Chem 83(7):2767–2774. https://doi.org/10.1021/ac2000708

31. de Graauw M, Hensbergen P, van de Water B (2006) Phospho-proteomic analysis of cellular signaling. Electrophoresis 27(13):2676–2686. https://doi.org/10.1002/elps.200600018

32. Iliuk A, Galan J, Tao WA (2009) Playing tag with quantitative proteomics. Anal Bioanal Chem 393(2):503–513. https://doi.org/10.1007/s00216-008-2386-0

33. Kruger R, Zinn N, Lehmann WD (2009) Quantification of protein phosphorylation by microLC-ICP-MS. Methods Mol Biol 527:201–218., ix. https://doi.org/10.1007/978-1-60327-834-8_15

34. Czernik AJ, Girault JA, Nairn AC, Chen J, Snyder G, Kebabian J, Greengard P (1991) Production of phosphorylation state-specific antibodies. Methods Enzymol 201:264–283

35. Mohammadi M, Dikic I, Sorokin A, Burgess WH, Jaye M, Schlessinger J (1996) Identification of six novel autophosphorylation sites on fibroblast growth factor receptor 1 and elucidation of their importance in receptor activation and signal transduction. Mol Cell Biol 16(3):977–989

36. Belluardo N, Wu G, Mudo G, Hansson AC, Pettersson R, Fuxe K (1997) Comparative localization of fibroblast growth factor receptor-1, -2, and -3 mRNAs in the rat brain: in situ hybridization analysis. J Comp Neurol 379(2):226–246

37. Ignatoski KM (2001) Immunoprecipitation and western blotting of phosphotyrosine-containing proteins. Methods Mol Biol 124: 39–48

38. Hopkins AM, DeSimone E, Chwalek K, Kaplan DL (2015) 3D in vitro modeling of the central nervous system. Prog Neurobiol 125:1–25. https://doi.org/10.1016/j.pneurobio.2014.11.003

Chapter 6

Electrophysiological Approach to GPCR–RTK Interaction Study in Hippocampus of Adult Rats

Davide Lattanzi, David Savelli, Michael Di Palma, Stefano Sartini, Silvia Eusebi, Dasiel O. Borroto-Escuela, Riccardo Cuppini, Kjell Fuxe, and Patrizia Ambrogini

Abstract

The allosteric receptor–receptor interactions in G protein-coupled receptor (GPCR) heteroreceptor complexes provided a new dimension for the integration of signaling at plasma membrane level of neurons in brain. Neuronal plasticity processes underlying brain functions, such as learning and memory, have been proposed to be based on rearrangement of heteroreceptor complexes and protomer interactions in the postsynaptic membrane. Among the different partners for GPCRs to form heteromers, receptor tyrosine kinases (RTKs) represent an intriguing combination due to their possible involvement in brain diseases. Different methodologies are available to study heteroreceptor complexes in brain tissue, but their functional role in modulating neuron electrical activity can be properly evaluated using an electrophysiological approach. Here, we describe patch clamp technique protocol for studying GPCR–RTK interaction in hippocampus CA1 pyramidal neurons of adult rat, paying particular attention to highlight major problems that can occur using this technique and providing useful troubleshooting steps to achieve reliable results.

Key words Heteroreceptors, GPCRs, RTK, GIRK channels, Holding current, Hippocampus, Adult rat

1 Introduction: GPCR-RTK Heteromer Overview

G protein-coupled receptors (GPCRs) are ubiquitous multifaceted transmembrane proteins that respond to a wealth of extracellular stimuli, exerting a variety of physiological effects [1]. GPCRs constitute the largest family of membrane receptors, divided into A to F classes in which only A, B, and C are present in mammals, showing more than 900 GPCRs encoded by the human genome [2–4].

David Savelli and Michael Di Palma contributed equally.

Kjell Fuxe and Dasiel O. Borroto-Escuela (eds.), *Receptor-Receptor Interactions in the Central Nervous System*, Neuromethods, vol. 140, https://doi.org/10.1007/978-1-4939-8576-0_6, © Springer Science+Business Media, LLC, part of Springer Nature 2018

In the early 1980s, Fuxe and Agnati introduced the concept of intramembrane receptor–receptor interactions (RRIs) between different types of GPCRs, suggesting the existence of receptor heteromerization [5]. Two years later, the same authors proposed the existence of aggregates of multiple receptors in the plasma membrane and coined the term receptor mosaics (RM) [6] to indicate *"an assemblage of more than two receptors, which binds and decodes signals (transmitters, allosteric modulators,…) to give out an integrated input, via direct allosteric receptor-receptor interactions, to one or more than one intra-cellular cascade."*

The first evidence about GPCRs forming heteromers with other GPCRs binding different ligands dates back to 1993 from studies by Maggio and coworkers using chimeric α2-adrenergic/M3 muscarinic receptors [7] and then, further evidence followed [8, 9]. It has been suggested that GPCR heteromerization may also involve different types of ion channels/receptors among which receptor tyrosine kinases (RTKs) [9–17]. RTKs are a family of transmembrane receptors in mammals that mediate signaling from ligands, including the majority of growth factor receptors, such as neurotrophins (e.g., BDNF) and FGFs [18]. It has been proposed that GPCRs, beyond showing the capacity for transactivation of RTKs [19], may directly interact with them through an allosteric receptor–receptor interaction [12, 20].

The heteroreceptor complex formation has a very significant impact on cellular signaling due to the unique biochemical profile that receptors attain through heteromerization and it is now well established that receptor–receptor interactions may contribute to synaptic transmission regulation in the central nervous system (CNS). Indeed, in neurons, these complexes are directly involved in neuronal plasticity, which is responsible for learning, emotion, and motivation, and it may result altered in CNS diseases, inducing neural network dysfunction with deficits in brain functions [21, 22]. In this scenario, GPCR-RTK heteroreceptors have been involved in major depression, increasing its molecular basis understanding [23]. Thus, GPCR-RTK heteromer discovery has paved the way to a new field for pharmacology. In this view, it is mandatory to accurately characterize the role of the identified heteroreceptor complexes in physiology and then in pathology for developing appropriate pharmacological strategies.

In principle, to establish the physiological relevance of a heteroreceptor complex, evidence about its heteromerization in brain tissue should be provided from a morphological, biochemical, and functional point of view. Over the last decade, a large number of studies have examined various aspects of heteromerization, leading to interesting but sometimes opposing results [24, 25]. A set of standards has been proposed by researchers to allow for a thorough and critical evaluation of GPCR heteromers, that has been grouped into the following three criteria by Gomes and coworkers [26]:

1. *Heteromer components should colocalize and physically interact, either directly or via intermediate proteins acting as conduits for allosterism.*

2. *Heteromers should exhibit properties distinct from those of the protomers.*

3. *Heteromer-selective reagents should alter heteromer-specific properties.*

To address these issues a combination of different available techniques and experimental models may be used, such as coimmunoprecipitation, immunoelectron microscopy, proximity-based biophysical techniques, proximity ligation assay, transgenic animals, and heteromer-selective antibodies for recognizing and/or activating heteromers in vivo (for a review, see [26]). However, if the research goal is studying GPCR-RTK heteroreceptor activation effects on electrical function of brain neurons, electrophysiological technique appears to be the most suitable method (see for example [27, 28]), considering that several neurotransmitters act by binding to GPCRs, which may form heteromers.

In order to study receptor–receptor interaction by means of electrophysiological technique in brain neurons, a variation in currents flowing through the plasma membrane is required. These currents are mediated by ion channels and could result in cell membrane depolarization or hyperpolarization, thus regulating neuron excitability. Among all the ion channel species, K^+ channels comprise the most heterogeneous group [29] and a large number of GPCRs have been shown to affect neuron membrane potential by activating G protein-coupled inwardly rectifying K^+ (GIRK) channels [30–32]. In CA1 pyramidal neurons of hippocampus, GIRK channels allow a slow inhibitory modulation of the overall cell excitability [33] and several inhibitory neurotransmitters and neuromodulators hyperpolarize CA1 neurons, by activating GIRK channels, including serotonin (5-HT) [33]. Thus, GIRK channels are able to mediate inhibitory currents once activated by $G\beta\gamma$ or $G\alpha_{i/o}$ following the ligand binding on GPCRs and represent useful elements for RRI electrophysiological studies that involve GPCRs as a protomer. To this aim, a valid approach is represented by the patch clamp technique, which permits monitoring of voltage, current, and membrane passive property changes on single living neurons in brain slices, thus allowing evaluation of GIRK channel conductance changes. The attainment of adequate brain slices for performing patch clamp recordings represents the step limiting the good success of the experiments and, thus, the achievement of reliable results. This is particularly relevant when slices derive from adult/old animals.

Here, we describe patch clamp technique protocol for studying GPCR–RTK interaction through GIRK channels in hippocampus CA1 pyramidal neurons of adult rat, paying particular attention to highlight major problems that can occur using this technique and providing useful troubleshooting steps to achieve reliable results.

1.1 Brain Slice Preparation from Adult Rat

Living acute brain slices from adult rats were used to record GIRK currents in CA1 hippocampal neurons. In neurons, GIRK (also known as Kir3) channels are located perisynaptically outside the postsynaptic density in the dendrite branches and shafts [34]. These channels are able to hyperpolarize neurons in response to activation of numerous G protein-coupled receptors (GPCRs) by neurotransmitters, mediating a slow inhibitory postsynaptic current (sIPSC) and thus modulating neuron excitability. Four GIRK channel subunits are expressed in mammals (GIRK1, also known as Kir3.1; GIRK2, also known as Kir3.2; GIRK3, also known as Kir3.3; GIRK4, also known as Kir3.4), but in the brain GIRK1-GIRK3 subunits are the most common. In the hippocampus, the three-subunit forming GIRK channels are identified in all subfields, showing a progressively increasing expression during postnatal development, which reaches adult level between 21 and 60 postnatal days [35]. Therefore, to be sure to record mature GIRK currents, we decided to work on adult animals of about 60 days of age. However, if on one hand the choice of using adult rats allows to study GIRK channels in their adult configuration, on the other hand it results in some problems for obtaining alive slices due to tissue damage induced by brain cutting in adult rat, thus compromising the success of the experiment. In particular, the major limitation of this model is the difficulty in preparing slices with alive neurons in the superficial layers of the slice, which are easier to reach and to record. Thus, we fine-tuned a procedure, paying specific attention to minimize slice edema and damage [10, 36–41]. In this way, we increased the chances of success for targeted patch clamp recordings in hippocampal CA1 neurons of adult rats.

There are several methods described in literature able to improve neuron viability in brain slices from juvenile or young animals, using different protective cutting and recording solutions, while relatively few papers have taken into consideration the mature rat [42]. One of the major problems for obtaining healthy brain slices is neurotoxicity occurring during the slice cutting that induces a sodium influx in the cell with subsequent water entry and neuronal swelling. Therefore, neurons located in the superficial layers sustain direct trauma from the blade movement and they are the most affected by this damage. In most cases, both in juvenile and in mature rat, the researchers, for reducing neuronal damage, adopt cutting solutions having equimolar replacement of sodium chloride (NaCl) with sucrose or having sodium ion substitutes, such as choline and N-methyl-D-glucamine (NMDG). Moreover, many laboratories also reduce glutamate-induced excitotoxic damage occurring during slice preparation, by including in the cutting solution kynurenic acid, to block in a nonspecific manner all the ionotropic glutamate receptors. In addition, lowering the extracellular Ca^{2+} concentration and replacing it with Mg^{2+} ions further improve neuroprotection, minimizing calcium influx caused by glutamate-induced depolarization. An additional aspect related to brain slice preparation is represented by anoxic insult that can worsen neuron viability. Neuronal death after an acute injury, such as trauma or ischemia-reperfusion [43, 44], is closely related to oxygen-derived free radical generation. Besides, a study performed on cultured striatal neurons from mouse embryos [45] showed that pyruvate protects neurons against both exogenously and endogenously produced hydrogen peroxide (H_2O_2), acting as intracellular and extracellular H_2O_2 scavengers. Therefore, adding pyruvate to the cutting and recording solutions can defend neurons against oxidation from H_2O_2 generation. Finally, ascorbate is an antioxidant molecule localized predominantly in neuronal cytosol. After cutting procedure, neurons in brain slices rapidly leak ascorbate, take up water and swell. This water gain seems to be due to intracellular oxidative stress induced by glutamate receptor activation. Thus, ascorbate addition to cutting and recording solutions prevents neuronal swelling and death caused by slice cutting [46].

In sum, taking into account suggestions coming from others laboratories, we selected the appropriate cutting solution for preparing brain slices from adult rats and the suitable recording solution for performing the specific experiments on GIRK current modulation in CA1 hippocampal neurons, using patch clamp technique in whole-cell configuration.

2 Materials

2.1 Solutions

1. Extracellular cutting solutions: Considering the elements mentioned above, we used the following solution containing in millimolar: 110 choline Cl^-, 2.5 KCl, 1.3 NaH_2PO_4, 25 $NaHCO_3$, 0.5 $CaCl_2$, 7 $MgCl_2$, 20 dextrose, 1.3 Na^+ ascorbate, 0.6 Na^+ pyruvate, 5.5 kynurenic acid (pH = 7.4; 320 mosM). This solution was used exclusively for brain dissection and brain slice preparation procedures.

2. Extracellular recording solution: We used artificial cerebrospinal fluid (ACSF) containing in millimolar: 125 NaCl, 2.5 KCl, 1.3 NaH_2PO_4, 25 $NaHCO_3$, 2 $CaCl_2$, 1.3 $MgCl_2$, 1.3 Na^+ ascorbate, 0.6 Na^+ pyruvate, 10 dextrose (pH = 7.4; 320 mosM).

   To ensure stable pH buffering and adequate oxygenation, extracellular solutions were saturated with carbogen (95% O_2/5% CO_2) prior to use. Both cutting and recording extracellular solutions were made fresh on the day of use.

3. Intracellular solution: For patch clamp recordings in whole-cell configuration the electrodes (stimulation and recording pipettes) were filled with intracellular solution containing in millimolar: 126 potassium gluconate, 8 NaCl, 0.2 EGTA, 10 HEPES, 3 Mg_2ATP, 0.3 GTP (pH = 7.2; 290 mosM). The solution osmolarity was corrected to 290 mosM, lower

than bathing solution, to prevent neuronal swelling and burst during whole-cell recordings. The solution was aliquoted and frozen to −20 °C. Liquid junction potentials were determined to be about 10 mV for this internal solution and membrane potentials were appropriately corrected [47].

For extracellular and intracellular solutions, we used water with high purity (MilliQ water system).

2.2 Equipment

1. Dissection tools: small rodent guillotine; stainless steel clippers for cutting the skull; stainless steel tweezer to eliminate meninges; stainless steel spatula for brain removal; razor blade; and super glue (cyanoacrylate).

2. Tissue slicer: motorized advance vibroslice MA752 (Campden Instruments, UK) equipped with razor blade and tissue holder.

3. Recovering chamber: homemade air–liquid interface recovering chamber.

4. pH meter: pH Meter 3320 Jenway.

5. Osmometer: Roebling Milliosmol Osmometer.

6. Electrophysiology setup: upright microscope (Zeiss Axioskop microscope, Carl Zeiss International, Italy) equipped with infrared filter; digital camera (ORCA Flash 4.0 Hamamatsu C11440); patch clamp amplifier (Axopatch 200B, Axon instruments); A385 stimulus isolator (World Precision Instruments, USA); Shielded Rack-Mount BNC Connector Block (BNC 2090A); hydraulic micromanipulators (Narishige MD 35A); homemade vibration isolated table; Faraday's cage; gravity driven perfusion system; and peristaltic pump.

7. Data acquisition and analysis software: Strathclyde electrophysiology software for whole-cell analysis (WinWCP V 3.2.9, John Dempster, University of Strathclyde, UK).

8. Borosilicate glass capillaries: borosilicate thick septum theta capillary (OD: 1.5, ID: 1.02, World Precision Instruments, USA) for stimulating electrode and borosilicate glass capillary (OD: 1.5, ID: 1.12, World Precision Instruments, USA) for recording electrodes.

9. Pipette puller: vertical puller PP 830 (Narishige Japan).

3 Methods

3.1 Preparation and Brain Slice Maintenance

After ketamine anesthesia (65 mg/kg b.w.), animal was decapitated using a small rodent guillotine and the skull was opened to expose whole brain. Subsequently, the meninges were excised to prevent accidental brain lesions during removal phase, and brain was harvested. These operations have to be carried out as rapidly as possible to reduce tissue damage (≤1.5 min from decapitation to this stage). Brain was divided into two hemispheres, immediately immersed in oxygenated ice-cold cutting solution inside a 100 ml glass beaker and put into freezer. The cooling phase lasted 9 min for the first hemisphere and 13–15 min for the second hemisphere. Despite the different duration of cooling phase, no difference in brain slice viability and in electrophysiological responses were observed. In our model, we tested that a shorter cooling time (<9 min) compromised brain slice viability.

It is worth mentioning that a recommended practice for slices preparation in adult rats is the use of transcardial perfusion, after anesthesia induction, with modified artificial cerebrospinal fluid (ACSF) containing Na^+ ion substitutes as NMDG [42]. However, in our experimental model, transcardial perfusion with modified ACSF has not been demonstrated to be a good practice leading to massive and rapid cell death in brain slices.

After cooling, the hemisphere was trimmed, creating a block of tissue that is larger than the considered brain structure (Fig. 1); in this way, surrounding tissue confers mechanical stability to the tissue block during slicing. The reduction of brain hemisphere was made specially to obtain parasagittal brain slices, in which hippocampal resulted transversally cut. The trimmed tissue was glued to the holder, as quickly as possible, using cyanoacrylate glue. To get a proper tissue gluing, it is important that the brain surface and the microtome holder are dry. The microtome stage with glued tissue was then transferred to the slicing chamber previously partially

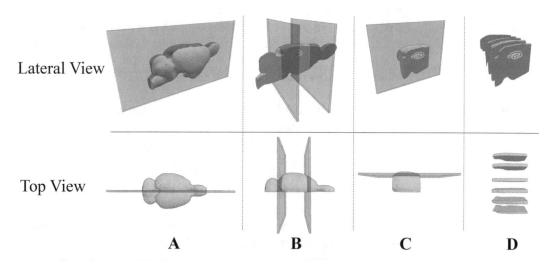

Fig. 1 Cutting method. Procedure used to obtain slices for electrophysiological recordings on the hippocampus. (**a**) Brain is sagittally cut for dividing it into two hemispheres. (**b**) After 9 min at −20 °C each hemisphere is cut again in order to refine the piece around the area of interest. (**c**) Part of the cortex is taken away so as to place the piece correctly for the subsequent slice preparation. (**d**) The brain piece is cut by means of vibratome in 400 μm-thick slices. Figure realized thanks to the rat brain 3D model shared by Bernd M. Pohl, Fernando Gasca, and Ulrich G. Hofmann (M. Pohl, Bernd; Fernando Gasca; Hofmann, Ulrich G. (2013): IGES and .stl file of a rat brain. figshare. https://doi.org/10.6084/m9.figshare.823546.v1)

filled with oxygenated ice-cold cutting solution. Subsequently, the chamber was completely filled to cover the preparation. The cutting solution was continuously oxygenated throughout the entire cutting phase, and its temperature was maintained at 1.5–2.5 °C, by adding cutting solution ice cubes. Therefore, brain slices (400 μm thickness) were obtained from each hemisphere by using vibrating microtome (Campden Instruments, UK). To prevent tissue pushing and cell damage during slices cutting, the blade vibration frequency was set near the maximum (90%) and the blade advance speed was kept the slowest possible (10–20%). Immediately after slicing, each slice was transferred to air–liquid interface recovering chamber using a cut and firepolished Pasteur pipet. The homemade recovering apparatus consisted of a square hermetic plastic box containing two small plastic containers with circular section (Fig. 2a). The bottom of the box and the containers were filled with artificial cerebrospinal fluid (ACSF). The ACSF level in square box was kept slightly lower than the height of circular containers. On the contrary, circular containers were filled completely with ACSF. A piece of filter paper was gently placed above each circular container; in this way, filter paper is able to absorb ACSF, creating a humid surface suitable to receive brain slices (Fig. 2a). The ACSF solution contained in the square plastic box was continuously bubbled with 95% O_2 and 5% CO_2 to create an oxygenated and humidified environment around the brain slices. The recovering chamber was maintained at room temperature (23–25 °C) for the duration of the experiment (about 8 h). In our model, a higher or a lower temperature during the recovering phase induced a dramatic decrease of neuronal viability in brain slices, by increasing neuronal swelling. Likewise, tissue edema and neuronal swelling considerably increase using a submerged incubation chamber where the slices were completely immersed in ACSF solution.

3.2 Recording Setup

The recording chamber was made up by a round hole (20 mm in diameter) in the center of a Plexiglas rectangular piece (4 mm thickness) glued to a microscope glass slide. A ground electrode (Ag-AgCl wire) was placed in a lateral rectangular groove connected to the recording chamber. A paper strip, positioned in the lateral rectangular groove, absorbed ACSF solution and created electrical contact between ground electrode and bath solution (Fig. 2b). This measure prevented offset shift following flux variations of perfusion bath. The recording chamber volume was 3 ml and a continuous ACSF flux (approximately 3 ml/min) was ensured by a gravity driven perfusion system coupled to a peristaltic pump aspiration. The perfusion and aspiration tubes were connected directly to the recording chamber and the capacity of aspiration was set to be higher than perfusion for obtaining a stable ACSF level. A U-shaped piece of platinum wire with a parallel array formed by two fine nylon threads was used to hold in position the brain slice. The preparation was observed using upright Zeiss

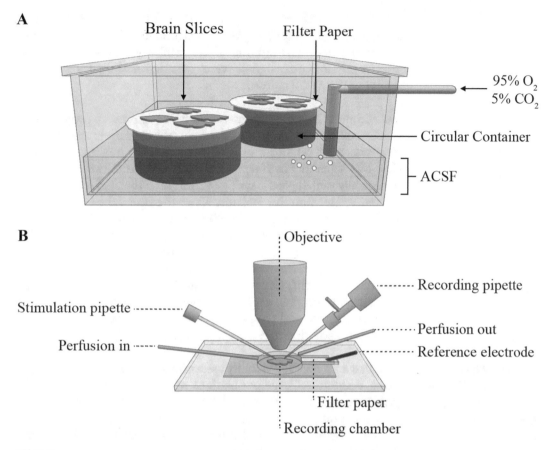

A

Brain Slices   Filter Paper

95% O_2
5% CO_2

Circular Container

ACSF

B

Objective

Recording pipette

Stimulation pipette

Perfusion out

Perfusion in

Reference electrode

Filter paper

Recording chamber

Fig. 2 Recovery chamber and patch clamp recording setup. (**a**) Interface recovery chamber consisting in a closed box partially filled with ACSF at room temperature, in which two circular containers hold up filter papers that accommodate brain slices. The bubbler supplies 95% O_2 and 5% CO_2 to the chamber for creating an adequate environment for maintaining neurons alive up to several hours. (**b**) Representing picture illustrating recording chamber and setup for patch clamp recording on brain slices

Axioskop microscope (Carl Zeiss International, Italy) with ceramic water immersion objectives. To identify individual neuron for patch clamp recording, we used an objective with 40× magnification and a working distance of about 2 mm. Furthermore, we improved the resolution using infrared illumination and infrared-sensitive camera. In fact, the infrared light has a long wavelength and it is less scattered than visible light during brain slice crossing. In this way, we can resolve cell body and principal apical dendrites of CA1 neurons.

Patch clamp recording pipette was mounted on suitable holder, which was moved in the three dimensions of space (X-Y-Z) using a hydraulic micromanipulator (Narishige MHW-3) fixed to microscope arm. A second mechanical micromanipulator, fixed to the microscope stage, was used to move extracellular stimulation electrode.

Patch clamp recording pipette holder was connected to Axopatch 200B amplifier (Axon Instruments, USA), which in turn communicated throughout BNC 2090 connector with PC. Electrophysiological signals were recorded and analyzed off-line utilizing WinWCP software.

3.3 Recording and Stimulating Electrodes

To stimulate Schaffer collateral pathway, the axon projection to CA1 neurons coming from CA3 field, we used a bipolar electrode that was built pulling a borosilicate thick septum theta capillary. The two separate chambers were filled with ACSF solution, using a manually pulled plastic syringe (1 ml), and the electrode was connected to the stimulus isolator. Then the tip of the electrode was broken using a little paint brush to obtain a size of about 30 μm. In this way, we had an appropriate tip electrical resistance that allowed us to correctly stimulate axon pathway.

Recording electrode was pulled from borosilicate glass capillary to obtain a tip resistance of about 2 MΩ when filled with intracellular solution. We preferred large electrode tip to have lower access resistance and to speed up the exchange of solutes between the pipette solution and the cell cytoplasm.

3.4 Evaluation of Brain Slice Viability and Healthy Patchable Neuron Selection

The first step for slice viability assessment was the measure of functional response of large populations of CA1 neurons. For this purpose, we carried out extracellular field recordings of excitatory post-synaptic potentials (fEPSPs) in CA1 *stratum radiatum* after Schaffer collateral electrical stimulation. A bipolar stimulating electrode (theta capillary) and recording extracellular electrode were filled with ACSF solution and placed in *stratum radiatum* at about 500 μm of distance between them (Fig. 3a). fEPSP was elicited by applying square pulses of current (300 μs in duration) to Schaffer collateral pathway at a frequency of 0.1 Hz. Usually, brain slices with a large number of living neurons showed a fEPSP with 1 mV amplitude or more and a prominent population spike (Fig. 3b). In some slices, it is possible to record a fEPSP with acceptable amplitude without population spike. In this case, the absence of population spike seems to be due to depolarized resting membrane potential of CA1 neurons and consequent Na+ channels inactivation. Thus, slices that showed fEPSP smaller than 1 mV or without population spike were rejected.

An optimal extracellular field response is not sufficient to identify a slice suitable for patch clamp recordings because the superficial neurons could be still dead. For this reason, it is important to visualize slices with high resolution to distinguish between dead, unhealthy, and healthy patchable neurons:

1. Dead neurons on brain slice surface appear swollen with a large nucleus and a prominent nucleolus; these cells burst by applying positive pressure on cell membrane through recording electrode.

2. Unhealthy neurons show shrunken appearance with visible principal dendrites. Unlike swollen dead neurons, high-resistance GΩ seals can be formed on shrunken cells, but subsequent whole-cell recording configuration is difficult to obtain even using a very big negative pressure.

3. Healthy patchable neurons are easy to distinguish since they appear three-dimensional, without a visible nucleus, and showing a cell membrane smooth and bright.

The selection of cells to be patched based on these features leads to healthy recordings from about 90% of all the neurons chosen for whole-cell recordings.

3.5 Whole-Cell Configuration

In order to perform patch clamp experiments in whole-cell configuration on CA1 neurons in brain slices, it is necessary to create a tight seal between patch recording pipette and cell membrane (cell-attached configuration) by applying gentle negative pressure in the recording electrode. After tight seal formation, it is possible to gain access to the inside of the cell (whole-cell configuration) by means of an additional gentle suction (Fig. 3c).

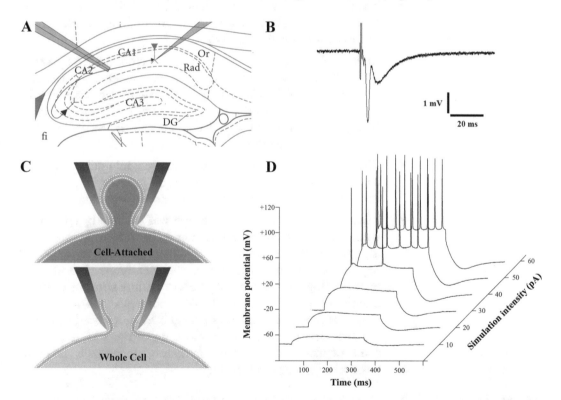

Fig. 3 Electrophysiological recordings. (**a**) Representative electrode placement in *stratum radiatum* of CA1 field to perform extracellular field recordings. (**b**) Demonstrative fEPSP recorded on a slice with a high degree of neuronal viability. (**c**) Schematic representation of cell-attached and whole-cell configurations. (**d**) Excitability profile of a typical hippocampal CA1 pyramidal neuron in response to depolarizing steps. Figure prepared with unpublished data and using slides purchased from Motifolio (http://motifolio.com)

All the steps for obtaining whole-cell configuration and for performing cell recording are described as follows, beginning from the ending of brain slicing process:

- After at least 1 h of recovering, brain slice was placed into the recording chamber; the CA1 field of hippocampus was identified using a 2.5× objective and extracellular stimulating electrode was placed in *stratum radiatum* to stimulate Schaffer collateral pathway.

- Neuron selection was performed using a 40× ceramic water objective.

- The recording pipette was mounted to the headstage and positive pressure was applied throughout a tube directly connected to the pipette holder (to do this, we used a 5 ml syringe linked to the tube).

- The recording electrode was lowered into the bath and pipette offset was manually zeroed. A 5 mV hyperpolarizing voltage step was applied in voltage clamp mode to measure pipette resistance.

- The positive pressure was augmented when the recording electrode tip approached cell membrane to eliminate cellular debris and to improve cell visualization.

- After small dimple formation on cell membrane, the positive pressure was quickly removed detaching syringe from the tube, and to obtain $G\Omega$ seal, a very gentle suction was applied by mouth; in some cells the formation of the seal was facilitated by applying a constant negative voltage to the recording electrode (about −70 mV).

- The fast and slow capacitive transients were minimized by using a very low level of intracellular solution in recording electrode and they were removed manually after seal formation.

- Therefore, whole-cell configuration was reached by applying a constant or intermittent gentle suction by mouth. Before break in, the membrane potential was kept at −70 mV to avoid cell depolarization upon rupture of the membrane. Normally, a little holding current was required in healthy neurons to bring membrane potential at −70 mV. The series resistance (Rs), usually less than 20 MΩ, was periodically monitored throughout the experiment to check the state of the seal, given that some neurons reseal the patch membrane partially or completely increasing series resistance.

3.6 CA1 Neuron Characterization

CA1 neurons may be identified by their body shape and cell body position in the hippocampus subfields, by means of microscope visualization, but also by analyzing functional features. We measured some electrophysiological properties in every recorded cell to exclude other cellular types as glial cell or interneurons:

- First, after whole-cell configuration establishment, we measured in voltage clamp mode, input resistance (IR) and membrane capacitance (C), by recording holding current in response to 300-ms, 5-mV hyperpolarizing step.

- Subsequently, we read the resting membrane potential (RMP), switching the Axopatch 200B amplifier in Vm/I0 mode (current clamp with no holding current).

- Then, cell excitability was evaluated measuring, in current clamp mode, action potential firing in response to increasing depolarizing current pulses (10 pA step, 300 ms) (Fig. 3d); to avoid differences in cell excitability due to different resting membrane potential, before the delivery of current pulses, the RMP was brought to approximately −70 mV by releasing an appropriate holding current.

- If the visually identified neuron showed typical electrophysiological features of CA1 neurons, we started to record GIRK currents as described below.

3.7 GIRK Current Recordings in Whole-Cell Configuration

3.7.1 Slice Perfusion

Generally, as previously mentioned, during patch clamp recordings the slices are persistently perfused with an oxygenated extracellular-like solution, which normally, for neurons, is the ACSF.

Many laboratories, to improve GIRK current detection, use a modified ACSF solution enriched in K^+ content in order to increase the driving force for inward K^+ current [31, 48]. Another shrewdness that helps GIRK current recordings consists in supplementing the ACSF solution with a cocktail of different antagonists of GABA/glycine and glutamate receptors to block synaptic transmission, in order to exclude any unwanted activation of the neuron under investigation and to reduce recording noise [31].

However, to characterize a receptor–receptor interaction not fully studied yet, in our opinion it is initially needed to work in an experimental condition that is close, as much as possible, to the physiological one, using an unmodified ACSF (see Sect. 2.1).

3.7.2 Gain the Access to the Cells: Whole Cell or Perforated Patch Clamp?

To study R–R interaction through patch clamp technique, the recording electrode must reach the neuron of interest and create a connection in order to isolate and record electrical signals pervading its whole membrane while keeping the cell alive and functioning. In keeping with this assumption, two different approaches are suitable for this kind of electrophysiological studies: perforated and whole-cell configuration. Both configurations allow the recording of the currents flowing through plasma membrane, thus permitting evaluation of GIRK channel conductance changes. Regarding the perforated approach, the electrical access to the cell is obtained filling patch pipette with an internal solution containing a pore-forming antibiotic, which allows the sealed patch of membrane to be perforated without affecting cytoplasm content. On the other hand, in the whole-cell approach after the tight seal a direct access to the cell is obtained following the break in, thus allowing the internal solution of patch pipette to freely diffuse into the cytoplasm. Therefore, each of these electrophysiological configurations has its advantages and drawbacks. Indeed, the former allows studying membrane currents without altering cytoplasm content; the latter allows interacting with the intracellular domains of the R–R complex, providing means for introducing drugs, such as uncoupling peptides that could disrupt these complexes [13].

3.7.3 Step-by-Step Procedure to Study R–R Interaction by GIRK Current Monitoring

An important assumption that it can be done approaching a heteroreceptor complex between an RTK and a GPCR is that the activation of the protomers establishing this interaction could alter the capability of GPCRs to interact and affect GIRK channel-mediated current. Therefore, using GIRK current as electrophysiological readout it is possible to study R–R interactions through its modulation in presence of different receptor agonists (Fig. 4a).

First step: evaluation of GPCR-induced GIRK current

1. Following the functional and morphological characterization of CA1 neurons (described above), the cell must be recorded in voltage clamp mode maintaining the membrane potential at −70 mV and monitoring the holding current to obtain a stable baseline of at least 10 min.

2. Afterward, to elicit GIRK activation, the agonist of the GPCR under investigation must be applied in the bath at a suitable concentration for at least 10 min.

3. The opening of GIRK channels could be monitored at this point using the following approaches:

    (a) IR (input resistance) Change. During the experiment, the input resistance must be constantly monitored by applying every 30 s hyperpolarizing voltage steps (5 mV, 300 ms) to verify GPCR-mediated opening of membrane channels and a following decreasing of IR (Fig. 4c).

    (b) Ih (holding current) Shift. The GPCR-mediated opening of membrane channels induces a positive shift of the holding current necessary to clamp membrane potential at −70 mV (Fig. 4b).

    (c) GIRK-mediated inward current recordings. Another method to visualize GIRK currents consists in applying hyperpolarizing voltage steps or voltage ramps (e.g., from −40 up to −140 mV) (Fig. 4d). Indeed, at very hyperpolarized potentials (about −110/120 mV) it is possible to ensure an inward K^+ current where the contribution of

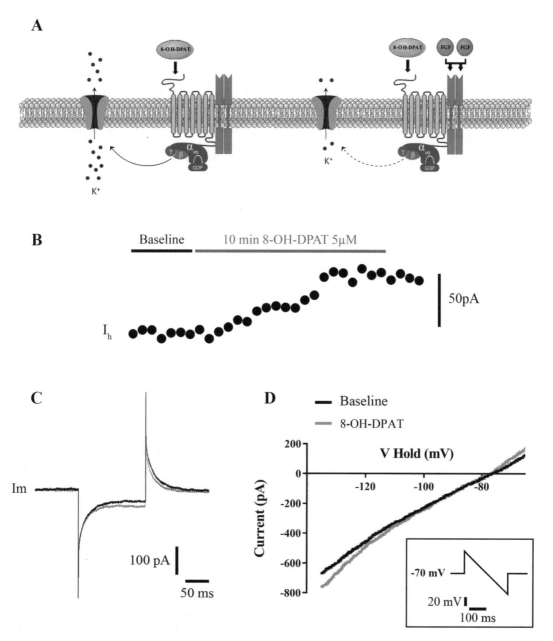

Fig. 4 Example of GPCR–RTK interaction evaluation by electrophysiological approach. (**a**) Illustration of the Fibroblast Growth Factor Receptor 1 (FGFR1)—5 Hydroxytryptamine Receptor 1A (5-HT1A) heteroreceptor complex and its modulation of GIRK current in CA1 pyramidal neurons. (**b**) Time-course of a representative experiment, showing the effect of 8-OH-DPAT (5-HT1A agonist) bath application on holding current. (**c**) 8-OH-DPAT application induces an IR decrease, suggesting a membrane channels opening. (**d**) Current voltage plot obtained by applying hyperpolarizing voltage ramps. Figure prepared with unpublished data and using slides purchased from Motifolio (http://motifolio.com)

outward rectifying K$^+$ current and voltage-gated ion channels is negligible. It is worth noting that a higher concentration of K$^+$ in extracellular solution helps to increase the amplitude of GIRK-mediated inward currents.

(d) Application of GIRK channel blockers. The application in the bath of GIRK channel blockers is advisable to fully understand the components of the elicited current. To this aim, several suitable commercial inwardly rectifying K$^+$ channel blockers, such as Ba^{2+} and Tertiapin-Q, are available [31].

Second step: RTK activation and membrane currents

1. Using the same procedure described above (see First step, point 1), other CA1 cells have to be recorded.

2. Afterward, it is firstly important to ensure that RTK activation does not induce any membrane currents. Thus, the agonist for the RTK involved in R–R interaction must be applied in the recording bath at a suitable concentration for at least 10 min.

3. The monitoring of possible RTK-induced membrane current activation must be performed using the abovementioned procedures (see First step, points 3a and 3b).

4. If the RTK activation does not elicit any membrane currents, it is possible to proceed with the third step.

Third step: Co-agonism of GPCR-RTK heterocomplexes to study R–R interaction

Once GPCR-induced GIRK current has been characterized and quantified (see First step) and no RTK-induced current has been detected, it is possible to move through R–R interaction study following the subsequent points:

1. Using the same procedure described above (see First step, point 1) other CA1 cells have to be recorded.

2. Afterward, through combined bath application (for at least 10 min) of the agonists for both receptors involved in the interaction (GPCR–RTK) it is possible to follow the procedures described above (see First step, points 3a and 3c) to study if the co-agonism affects GPCR-induced GIRK current.

3. Finally, if the co-agonism results in a detectable modification of the GPCR-induced GIRK current, to verify that this effect is specifically due to a physical R–R interaction it is possible to repeat the procedures mentioned in this step (Third step, point 1 and 2) using an intracellular solution containing a specific uncoupling peptide. In keeping with this, the uncoupling peptide should bring GIRK current to the levels observed with the GPCR agonist alone.

4   Conclusion

Patch clamp technique represents a useful means to allow studying of GPCR-RTK heteromers from an electrophysiological point of view, if the condition of a coupling with ion channel inducing a variation in currents flowing through the plasma membrane is satisfied. In this regard, GIRK channels, being G protein-gated ion channels, are able to modify their conductance upon GPCR neurotransmitter binding in CNS neurons. These channels are members of a large family of inwardly rectifying K^+ channels, which in response to activation of several GPCRs hyperpolarize neurons, thus controlling their excitability and neuronal network behavior [49]. Indeed, neurons release neurotransmitters, which, by GPCR activation, can affect GIRK channels on own dendrites causing self-inhibition (autaptic transmission) or at postsynaptic level of neighboring neurons (synaptic transmission) or on large-scale neuronal network (volume transmission) [49]. On the other hand, GIRK channels have been implicated in the pathophysiology of several diseases, such as epilepsy [50], Down's syndrome [51], and Parkinson's disease [52]. In addition, GIRK channels have been also involved in depression-related behaviors and physiology of serotonergic neurotransmission [53]. In this context, it is documented by Fuxe's works (e.g., [13, 54, 55]) that 5-HT_{1A} receptors form a large number of heterocomplexes with other receptors in the plasma membrane, such as FGFR1 RTK, and it has been found that GIRK channels are activated by 5-HT_{1A} autoreceptors in rat dorsal raphe neurons [31] and by postsynaptic HT_{1A} receptors in hippocampus ([33], our unpublished data).

Therefore, differently from the morphological, biochemical, and biomolecular techniques used to investigate receptor–receptor interactions, electrophysiology approach, analyzing the electrical properties of neurons, allows studying functional changes due to heteromer signaling that could affect nervous system activity. Indeed, examining these properties has many implications, including investigations of neurological disorders and the potential uses or effects of pharmaceutical compounds. In this scenario, the electrophysiological study of GPCR-RTK-GIRK pathway modulation on brain neurons could be of physiological, but also therapeutic interest for investigating pathologies related to an altered neurotransmission and for developing novel treatments.

References

1. Lefkowitz RJ (2013) A brief history of G-protein coupled receptors (Nobel Lecture). Angew Chem 52(25):6366–6378. https://doi.org/10.1002/anie.201301924

2. Heng BC, Aubel D, Fussenegger M (2013) An overview of the diverse roles of G-protein coupled receptors (GPCRs) in the pathophysiology of various human diseases. Biotechnol Adv 31(8):1676–1694. https://doi.org/10.1016/j.biotechadv.2013.08.017

3. Katritch V, Cherezov V, Stevens RC (2013) Structure-function of the G protein-coupled

receptor superfamily. Annu Rev Pharmacol Toxicol 53:531–556. https://doi.org/10.1146/annurev-pharmtox-032112-135923

4. Insel PA, Tang CM, Hahntow I, Michel MC (2007) Impact of GPCRs in clinical medicine: monogenic diseases, genetic variants and drug targets. Biochim Biophys Acta 1768(4):994–1005. https://doi.org/10.1016/j.bbamem.2006.09.029

5. Agnati LF, Fuxe K, Zini I, Lenzi P, Hokfelt T (1980) Aspects on receptor regulation and iso-receptor identification. Med Biol 58(4):182–187

6. Agnati LF, Fuxe K, Zoli M, Rondanini C, Ogren SO (1982) New vistas on synaptic plasticity: the receptor mosaic hypothesis of the engram. Med Biol 60(4):183–190

7. Maggio R, Vogel Z, Wess J (1993) Coexpression studies with mutant muscarinic/adrenergic receptors provide evidence for intermolecular "cross-talk" between G-protein-linked receptors. Proc Natl Acad Sci U S A 90(7):3103–3107

8. Guo W, Urizar E, Kralikova M, Mobarec JC, Shi L, Filizola M, Javitch JA (2008) Dopamine D2 receptors form higher order oligomers at physiological expression levels. EMBO J 27(17):2293–2304. https://doi.org/10.1038/emboj.2008.153

9. Fuxe K, Borroto-Escuela DO (2016) Heteroreceptor complexes and their allosteric receptor-receptor interactions as a novel biological principle for integration of communication in the CNS: targets for drug development. Neuropsychopharmacology 41(1):380–382. https://doi.org/10.1038/npp.2015.244

10. Cuppini C, Ambrogini P, Lattanzi D, Ciuffoli S, Cuppini R (2009) FGF2 modulates the voltage-dependent K+ current and changes excitability of rat dentate gyrus granule cells. Neurosci Lett 462(3):203–206. https://doi.org/10.1016/j.neulet.2009.07.029

11. Flajolet M, Wang Z, Futter M, Shen W, Nuangchamnong N, Bendor J, Wallach I, Nairn AC, Surmeier DJ, Greengard P (2008) FGF acts as a co-transmitter through adenosine A2A receptor to regulate synaptic plasticity. Nat Neurosci 11(12):1402–1409. https://doi.org/10.1038/nn.2216

12. Fuxe K, Dahlstrom A, Hoistad M, Marcellino D, Jansson A, Rivera A, Diaz-Cabiale Z, Jacobsen K, Tinner-Staines B, Hagman B, Leo G, Staines W, Guidolin D, Kehr J, Genedani S, Belluardo N, Agnati LF (2007) From the Golgi-Cajal mapping to the transmitter-based characterization of the neuronal networks leading to two modes of brain communication: wiring and volume transmission. Brain Res Rev 55(1):17–54. https://doi.org/10.1016/j.brainresrev.2007.02.009

13. Borroto-Escuela DO, Romero-Fernandez W, Mudo G, Perez-Alea M, Ciruela F, Tarakanov AO, Narvaez M, Di Liberto V, Agnati LF, Belluardo N, Fuxe K (2012) Fibroblast growth factor receptor 1- 5-hydroxytryptamine 1A heteroreceptor complexes and their enhancement of hippocampal plasticity. Biol Psychiatry 71(1):84–91. https://doi.org/10.1016/j.biopsych.2011.09.012

14. Borroto-Escuela DO, Brito I, Di Palma M, Jiménez-Beristain A, Narvaez M, Corrales F, Pita-Rodríguez M, Sartini S, Ambrogini P, Lattanzi D, Cuppini R, Agnati LF, Fuxe K (2015) On the role of the balance of GPCR homo/heteroreceptor complexes in the brain. J Adv Neurosci Res 2:36–44

15. Borroto-Escuela DO, Narvaez M, Perez-Alea M, Tarakanov AO, Jimenez-Beristain A, Mudo G, Agnati LF, Ciruela F, Belluardo N, Fuxe K (2015) Evidence for the existence of FGFR1-5-HT1A heteroreceptor complexes in the midbrain raphe 5-HT system. Biochem Biophys Res Commun 456(1):489–493. https://doi.org/10.1016/j.bbrc.2014.11.112

16. Borroto-Escuela DO, Tarakanov AO, Fuxe K (2016) FGFR1-5-HT1A heteroreceptor complexes: implications for understanding and treating major depression. Trends Neurosci 39(1):5–15. https://doi.org/10.1016/j.tins.2015.11.003

17. Di Liberto V, Borroto-Escuela DO, Frinchi M, Verdi V, Fuxe K, Belluardo N, Mudò G (2017) Existence of muscarinic acetylcholine receptor (mAChR) and fibroblast growth factor receptor (FGFR) heteroreceptor complexes and their enhancement of neurite outgrowth in neural hippocampal cultures. Biochim Biophys Acta Gen Subj 1861(2):235–245. https://doi.org/10.1016/j.bbagen.2016.10.026

18. Hubbard SR (2004) Juxtamembrane autoinhibition in receptor tyrosine kinases. Nat Rev Mol Cell Biol 5(6):464–471. https://doi.org/10.1038/nrm1399

19. Luttrell LM, Daaka Y, Lefkowitz RJ (1999) Regulation of tyrosine kinase cascades by G-protein-coupled receptors. Curr Opin Cell Biol 11(2):177–183

20. Borroto-Escuela DO, Carlsson J, Ambrogini P, Narváez M, Wydra K, Tarakanov AO, Li X, Millón C, Ferraro L, Cuppini R, Tanganelli S, Liu F, Filip M, Diaz-Cabiale Z, Fuxe K (2017) Understanding the role of GPCR heteroreceptor complexes in modulating the brain networks in health and disease. Front Cell Neurosci 11:37. https://doi.org/10.3389/fncel.2017.00037

21. Fuxe K, Agnati LF, Borroto-Escuela DO (2014) The impact of receptor-receptor interactions in heteroreceptor complexes on brain plasticity. Expert Rev Neurother 14(7):719–721. https://doi.org/10.1586/14737175.2014.922878

22. Fuxe K, Borroto-Escuela DO, Ciruela F, Guidolin D, Agnati LF (2014) Receptor-receptor interactions in heteroreceptor complexes: a new principle in biology. Focus on their role in learning and memory. Neurosci Discov 2:6

23. Borroto-Escuela DO, Pérez-Alea M, Narvaez M, Tarakanov AO, Mudó G, Jiménez-Beristain A, Agnati LF, Ciruela F, Belluardo N, Fuxe K (2015) Enhancement of the FGFR1 signaling in the FGFR1-5-HT1A heteroreceptor complex in midbrain raphe 5-HT neuron systems. Relevance for neuroplasticity and depression. Biochem Biophys Res Commun 463(3):180–186. https://doi.org/10.1016/j.bbrc.2015.04.133

24. Hansen JL, Hansen JT, Speerschneider T, Lyngso C, Erikstrup N, Burstein ES, Weiner DM, Walther T, Makita N, Iiri T, Merten N, Kostenis E, Sheikh SP (2009) Lack of evidence for AT1R/B2R heterodimerization in COS-7, HEK293, and NIH3T3 cells: how common is the AT1R/B2R heterodimer? J Biol Chem 284(3):1831–1839. https://doi.org/10.1074/jbc.M804607200

25. Frederick AL, Yano H, Trifilieff P, Vishwasrao HD, Biezonski D, Meszaros J, Urizar E, Sibley DR, Kellendonk C, Sonntag KC, Graham DL, Colbran RJ, Stanwood GD, Javitch JA (2015) Evidence against dopamine D1/D2 receptor heteromers. Mol Psychiatry 20(11):1373–1385. https://doi.org/10.1038/mp.2014.166

26. Gomes I, Ayoub MA, Fujita W, Jaeger WC, Pfleger KD, Devi LA (2016) G protein-coupled receptor heteromers. Annu Rev Pharmacol Toxicol 56:403–425. https://doi.org/10.1146/annurev-pharmtox-011613-135952

27. Liu XY, Chu XP, Mao LM, Wang M, Lan HX, Li MH, Zhang GC, Parelkar NK, Fibuch EE, Haines M, Neve KA, Liu F, Xiong ZG, Wang JQ (2006) Modulation of D2R-NR2B interactions in response to cocaine. Neuron 52(5):897–909. https://doi.org/10.1016/j.neuron.2006.10.011

28. Weiss M, Blier P, de Montigny C (2007) Effect of long-term administration of the antidepressant drug milnacipran on serotonergic and noradrenergic neurotransmission in the rat hippocampus. Life Sci 81(2):166–176. https://doi.org/10.1016/j.lfs.2007.04.039

29. Coetzee WA, Amarillo Y, Chiu J, Chow A, Lau D, McCormack T, Moreno H, Nadal MS, Ozaita A, Pountney D, Saganich M, Vega-Saenz de Miera E, Rudy B (1999) Molecular diversity of K+ channels. Ann N Y Acad Sci 868:233–285

30. Nicoll RA (1988) The coupling of neurotransmitter receptors to ion channels in the brain. Science 241(4865):545–551

31. Montalbano A, Corradetti R, Mlinar B (2015) Pharmacological characterization of 5-HT1A autoreceptor-coupled GIRK channels in rat dorsal raphe 5-HT neurons. PLoS One 10(10):e0140369. https://doi.org/10.1371/journal.pone.0140369

32. Borin M, Fogli Iseppe A, Pignatelli A, Belluzzi O (2014) Inward rectifier potassium (Kir) current in dopaminergic periglomerular neurons of the mouse olfactory bulb. Front Cell Neurosci 8:223. https://doi.org/10.3389/fncel.2014.00223

33. Luscher C, Jan LY, Stoffel M, Malenka RC, Nicoll RA (1997) G protein-coupled inwardly rectifying K+ channels (GIRKs) mediate postsynaptic but not presynaptic transmitter actions in hippocampal neurons. Neuron 19(3):687–695

34. Lujan R, Maylie J, Adelman JP (2009) New sites of action for GIRK and SK channels. Nat Rev Neurosci 10(7):475–480. https://doi.org/10.1038/nrn2668

35. Fernandez-Alacid L, Watanabe M, Molnar E, Wickman K, Lujan R (2011) Developmental regulation of G protein-gated inwardly-rectifying K+ (GIRK/Kir3) channel subunits in the brain. Eur J Neurosci 34(11):1724–1736. https://doi.org/10.1111/j.1460-9568.2011.07886.x

36. Ambrogini P, Lattanzi D, Ciuffoli S, Agostini D, Bertini L, Stocchi V, Santi S, Cuppini R (2004) Morpho-functional characterization of neuronal cells at different stages of maturation in granule cell layer of adult rat dentate gyrus. Brain Res 1017(1-2):21–31. https://doi.org/10.1016/j.brainres.2004.05.039

37. Ambrogini P, Minelli A, Lattanzi D, Ciuffoli S, Fanelli M, Cuppini R (2006) Synaptically-silent immature neurons show gaba and glutamate receptor-mediated currents in adult rat dentate gyrus. Arch Ital Biol 144(2):115–126

38. Ambrogini P, Cuppini R, Lattanzi D, Ciuffoli S, Frontini A, Fanelli M (2010) Synaptogenesis in adult-generated hippocampal granule cells is affected by behavioral experiences. Hippocampus 20(7):799–810. https://doi.org/10.1002/hipo.20679

39. Ambrogini P, Lattanzi D, Ciuffoli S, Betti M, Fanelli M, Cuppini R (2013) Physical exercise and environment exploration affect synaptogenesis in adult-generated neurons in the rat dentate gyrus: possible role of BDNF. Brain Res 1534:1–12. https://doi.org/10.1016/j.brainres.2013.08.023

40. Betti M, Ambrogini P, Minelli A, Floridi A, Lattanzi D, Ciuffoli S, Bucherelli C, Prospero E, Frontini A, Santarelli L, Baldi E, Benetti F, Galli F, Cuppini R (2011) Maternal dietary loads of alpha-tocopherol depress protein kinase C signaling and synaptic plasticity in rat postnatal developing hippocampus and promote permanent deficits in adult offspring. J Nutr Biochem 22(1):60–70. https://doi.org/10.1016/j.jnutbio.2009.11.014

41. Sartini S, Lattanzi D, Ambrogini P, Di Palma M, Galati C, Savelli D, Polidori E, Calcabrini C, Rocchi MB, Sestili P, Cuppini R (2016) Maternal creatine supplementation affects the morpho-functional development of hippocampal neurons in rat offspring. Neuroscience 312:120–129. https://doi.org/10.1016/j.neuroscience.2015.11.017

42. Ting JT, Daigle TL, Chen Q, Feng G (2014) Acute brain slice methods for adult and aging animals: application of targeted patch clamp analysis and optogenetics. Methods Mol Biol 1183:221–242. https://doi.org/10.1007/978-1-4939-1096-0_14

43. Traystman RJ, Kirsch JR, Koehler RC (1991) Oxygen radical mechanisms of brain injury following ischemia and reperfusion. J Appl Physiol 71(4):1185–1195

44. Siesjo BK, Agardh CD, Bengtsson F (1989) Free radicals and brain damage. Cerebrovasc Brain Metab Rev 1(3):165–211

45. Desagher S, Cordier J, Glowinski J, Tencé M (1997) Endothelin stimulates phospholipase D in striatal astrocytes. J Neurochem 68(1):78–87. https://doi.org/10.1046/j.1471-4159.1997.68010078.x

46. Brahma B, Forman RE, Stewart EE, Nicholson C, Rice ME (2000) Ascorbate inhibits edema in brain slices. J Neurochem 74(3):1263–1270

47. Neher E (1992) Correction for liquid junction potentials in patch clamp experiments. Methods Enzymol 207:123–131

48. Jeong HJ, Han SH, Min BI, Cho YW (2001) 5-HT1A receptor-mediated activation of G-protein-gated inwardly rectifying K+ current in rat periaqueductal gray neurons. Neuropharmacology 41(2):175–185

49. Luscher C, Slesinger PA (2010) Emerging roles for G protein-gated inwardly rectifying potassium (GIRK) channels in health and disease. Nat Rev Neurosci 11(5):301–315. https://doi.org/10.1038/nrn2834

50. Signorini S, Liao YJ, Duncan SA, Jan LY, Stoffel M (1997) Normal cerebellar development but susceptibility to seizures in mice lacking G protein-coupled, inwardly rectifying K+ channel GIRK2. Proc Natl Acad Sci U S A 94(3):923–927

51. Sago H, Carlson EJ, Smith DJ, Kilbridge J, Rubin EM, Mobley WC, Epstein CJ, Huang TT (1998) Ts1Cje, a partial trisomy 16 mouse model for Down syndrome, exhibits learning and behavioral abnormalities. Proc Natl Acad Sci U S A 95(11):6256–6261

52. Harkins AB, Fox AP (2002) Cell death in weaver mouse cerebellum. Cerebellum 1(3):201–206. https://doi.org/10.1080/14734220260418420

53. Llamosas N, Bruzos-Cidon C, Rodriguez JJ, Ugedo L, Torrecilla M (2015) Deletion of GIRK2 subunit of GIRK channels alters the 5-HT1A receptor-mediated signaling and results in a depression-resistant behavior. Int J Neuropsychopharmacol 18(11):pyv051. https://doi.org/10.1093/ijnp/pyv051

54. Borroto-Escuela DO, Narvaez M, Marcellino D, Parrado C, Narvaez JA, Tarakanov AO, Agnati LF, Diaz-Cabiale Z, Fuxe K (2010) Galanin receptor-1 modulates 5-hydroxytryptamine-1A signaling via heterodimerization. Biochem Biophys Res Commun 393(4):767–772. https://doi.org/10.1016/j.bbrc.2010.02.078

55. Tena-Campos M, Ramon E, Borroto-Escuela DO, Fuxe K, Garriga P (2015) The zinc binding receptor GPR39 interacts with 5-HT1A and GalR1 to form dynamic heteroreceptor complexes with signaling diversity. Biochim Biophys Acta 1852(12):2585–2592. https://doi.org/10.1016/j.bbadis.2015.09.003

Chapter 7

In Vivo Microdialysis Technique Applications to Understand the Contribution of Receptor–Receptor Interactions to the Central Nervous System Signaling

Sergio Tanganelli, Tiziana Antonelli, Sarah Beggiato, Maria Cristina Tomasini, Kjell Fuxe, Dasiel O. Borroto-Escuela, and Luca Ferraro

Abstract

During the last 50 years, microdialysis technique has been continuously improved to become a well-established method to monitor local concentrations of neurotransmitters. In respect to other currently used techniques, such as voltammetry, microdialysis has the advantage to be possibly applied to all measurable neurotransmitters and to allow local treatments. These properties render the technique a suitable approach to investigate, in vivo, the neurochemical consequences of receptor–receptor interactions, thus providing functional correlates to binding and other biochemical data.

This chapter is mainly focused on a general description of microdialysis technique in freely moving animals and on the application of one- or dual-probe(s) microdialysis to investigate the functional relevance of receptor–receptor interactions in rodent brain.

Key words Receptor heteromers, Agonist/antagonist local perfusion, Neurotransmission, NTSR1/D2R interactions

1 Introduction

Since the first indication, by the middle of the twentieth century, that the great majority of the information transfer between neurons involved chemical signals even in the brain, research on neurotransmitter release began to be relevant to monitor these processes. During years, the limits of in vitro neurochemical techniques, the necessity to continuously monitor changes in neurotransmitter release and the need to correlate animal behavior with changes in neuronal signaling, prompted the development of several in vivo methodologies to dynamically measure extracellular neurotransmitter levels. Many of these techniques (i.e., ventricular perfusion, cortical cups, and push-pull cannula techniques)

Kjell Fuxe and Dasiel O. Borroto-Escuela (eds.), *Receptor-Receptor Interactions in the Central Nervous System*, Neuromethods, vol. 140, https://doi.org/10.1007/978-1-4939-8576-0_7, © Springer Science+Business Media, LLC, part of Springer Nature 2018

presented remarkable limitations that, mainly due to the alteration of homeostatic transmitter dynamics and tissue damages, rendered difficult the proper interpretation of results. In comparison to these methodologies, in vivo microdialysis technique, which enables sampling and collecting of small-molecular-weight substances from the interstitial space by implanting a probe provided with a dialysing membrane into selected brain regions, offers several advantages [1]. Microdialysis technique has been largely validated as a valuable experimental tool for preclinical pharmacokinetic, pharmacodynamic, and toxicological studies [2, 3], and its possible clinical applications are also well documented [4].

From an historical point of view, the concept of using a dialysis bag to collect samples was originally proposed by and his collaborators [5] by surgically implanting for several days a small dialysis sac filled with a dextran-saline solution into brain tissue of dogs. Thereafter, the sac was removed to analyze the content of amino acids and electrolytes. However, it represented a static approach, allowing the collection of a single dialysis sample per animal. In 1972, Delgado and co-workers [6], based on the dialysis principle, developed the concept of long-term intracerebral perfusion of dialysis bags in order to subsequently collect several samples. It was realized by means of a "dialytrode," basically a push-pull cannula equipped with a dialysis membrane glued on its tip, which represented the first dialysis probe; its implantation in the brain of awake monkeys allowed the collection of several samples and the measurement of extracellular endogenous compound levels. In 1974, Ungerstedt and Pycock [7] ideated the microdialysis probe and provided a significant improvement of the technique.

During the last 50 years, microdialysis technique has been continuously improved to become a well-established method to monitor local concentrations of small-molecular-weight compounds in the extracellular space. Approximately, every interstitial tissue fluid can be investigated in freely moving animals. The main principles and methodological approach of in vivo microdialysis technique, as well as its advantages and limitations have been largely described elsewhere [1, 8–12] and we refer to those reviews for detailed descriptions. This chapter is mainly focused on a general description of the technique and on the application of one- or dual-probe(s) microdialysis to investigate the functional relevance of receptor–receptor interactions in rodent brain.

2 Microdialysis Overview

The concept of microdialysis is to mimic the passive function of a capillary blood vessel, by implanting a probe into selected brain regions. Probes of different designs have been employed but at present, those of a concentric shape are generally used in experimental brain research. The microdialysis probe ends with a semipermeable dialysis membrane (Fig. 1) and is continuously perfused with a physiological fluid (i.e., artificial cerebrospinal fluid). As a result of passive diffusion, under continuous probe perfusion conditions, molecules migrate across the membrane along their concentration gradient. Molecules found in high concentrations within the tissue compartment migrate across the membrane into the dialysis tubing where they can be collected for subsequent quantification. Thus, brain dialysate typically contains various neurotransmitters and their metabolites, energy metabolites, metabolic precursors, and waste products. On the other hand, molecules found in high concentrations within the membrane diffuse outward into the surrounding tissue compartment. As a consequence, microdialysis probe can be used to infuse drugs or known compounds in selected brain regions (retrodialysis or inverse microdialysis).

In vivo microdialysis experiments were initially carried out in anesthetized animals. The evidence that anesthetic drugs can have strong effects on experimental outcome due to their influence on physiological parameters, and the necessity to directly correlate neurochemical changes with animal behavior prompted the development of methods allowing microdialysis in awake, freely moving, animals. During an in vivo microdialysis experiment in freely moving animal, the probe is connected to an infusion pump by means of tubings and a balance arm provided with a liquid swivel system mounted onto the testing cage (Fig. 1). The tubing line connected with the probe outlet allows the collection of perfusate samples.

Microdialysis is invasive and damage to brain tissue has been repeatedly described. It is well known that in the early phase after probe implantation, blood–brain barrier integrity is impaired, blood flow in the area is decreased, and neurotransmitters release is altered [13]. However, 18–24 h after the implantation these processes seem to have recovered. Thus, neurochemical experiments carried out 1 or 2 days after probe implantation provide reproducible results. It is worth noting that after 3 days, glial cells can surround the probe altering neurotransmitter extracellular levels. Therefore, development of gliosis limits the duration of the experiment [14, 15]. In case microdialysis must be performed over longer time spans (i.e., several days or weeks in rodents), a guide cannula which allows the probe insertion several times over a period of weeks is often used. It is generally implanted 4–7 days before the release experiment. However, guide cannulae are larger, thus providing more damages after their implantation, especially in small areas of the brain.

The possibility of continuous perfusion reduces tissue damage, and the presence of microdialysis membrane, which constitutes a physical barrier between the perfusate and tissue, protects brain structures from the turbulent perfusate flow and excludes the

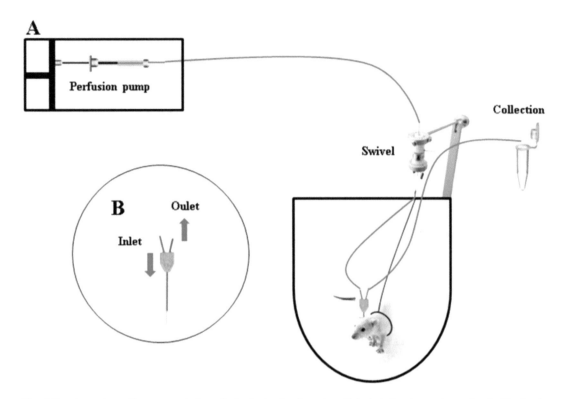

Fig. 1 Panel **a**: schematic representation of a typical setup for microdialysis in freely moving animals. The basic elements are the perfusion pump, a balance arm provided with swivel avoiding tubing twisting during the experiment and the microdialysis animal bowl. An automatic sample collector could also be used. Panel **b**: an example of concentric microdialysis probe

collection of high-molecular-weight substances (typically compounds with molecular masses >20,000 Da), such as proteins and bacteria. Therefore, collected samples do not require purification before analyte measurement. Furthermore, the possibility to use a low perfusate flow rate reduces analyte depletion, thus permitting a quite long collection time. It is worth noting that the probe allows the collection of only a fraction of the actual extracellular concentration [1] and, therefore, the concept of probe recovery has been introduced. The probe recovery represents the percent of analytes moved and retained in the perfusion solution relative to the respective external concentration. It generally depends on the perfusion flow rate, with higher flow rates giving smaller relative recoveries. For neurotransmitter and pharmacologically active substance optimal flow rates are generally in the range of 0.3–2.0 μl/min (relative in vivo recovery = 5–20%, depending on the analytes). The used flow rate is generally a compromise between recovery considerations and the necessity to collect a sufficient dialysate volume for analyte detection. Recovery also depends on diffusion resistance of the tissue matrix, physical features of the analytes, and dimensions of the membrane surface (i.e., probe length and diameter). To a lesser extent, recovery is influenced by the physicochemical characteristics of the probe membrane and the composition of the perfusion medium. The relative and absolute probe recovery can be evaluated in vitro to possibly estimate the recovery of a particular substance under in vivo conditions. However, in vivo recovery only approximates, but never reaches, the in vitro recovery [1, 16].

The choice of the probe mainly depends on the dimensions of the targeted area and the neurotransmitter under study. Membranes with a high cut-off should be preferred if large or lipophilic molecules are evaluated. Concerning the perfusion medium, artificial cerebrospinal fluid, artificial extracellular fluid, or Ringer's solution is generally used. The medium should be freshly prepared and passed through 0.2 μm bacterial filters. The perfusion medium can be modified due to specific experimental necessity. For example, ascorbic acid is commonly used to prevent monoamine degradation in the collected samples, while bovine serum albumin and cyclodextrane can be added to facilitate retention of lipophilic analytes and/or prevent adsorption of the neuroactive substance on plastic surfaces. In case of very low basal neurotransmitter concentration (below the analytical method sensitivity), the perfusion medium can be enriched with metabolizing enzymes and uptake inhibitors. However, it should be considered that this procedure might disturb the homeostatic balance of the extracellular environment, thus affecting the data meaning. Finally, the perfusion medium can be loaded with the sodium channel blocker tetrodotoxin (TTX; 1 μM) or impoverished of Ca^{2+} [17, 18]. The resulting decrease in neurotransmitter content proves its neuronal origin [19]. This is particularly relevant when GABA, and especially glutamate, are measured, as a large percentage (about 70–80%) of their extracellular levels is of metabolic origin. TTX and low-calcium medium also allow to assess whether a specific drug-induced effect on neurotransmitter levels could be due to neuronal depolarization.

The time resolution of microdialysis is relatively low: sampling times below 1 min are rarely reported and most of the studies report sampling times between 5 and 20 min.

3   Using the In Vivo Microdialysis Technique to Understand the Neurochemical Consequences of Receptor–Receptor Interaction

With the discovery of intramembrane receptor–receptor interactions with the likely formation of receptor aggregates of multiple receptors (heteroreceptor complexes), the entire recognition-transduction decoding process became a branched process already at the receptor level in the surface membrane. By acting on receptor heteromers, a neurotransmitter or a modulator, by binding to its receptor, modifies the characteristics of the receptor for another transmitter or modulator [20]. In neurons, these complexes are directly involved in the neuronal transmission, which is responsible for learning, memory, behavior, and many other CNS functions. Thus, once the existence of a specific receptor heteromer has been demonstrated or postulated, its functional relevance into the brain can be assessed by evaluating how ligands of one receptor protomer modify the changes in neurotransmitter levels induced by an agonist/antagonist of another receptor protomer. Due to the complexity of neuronal networks and to possibly exclude indirect effects, local application of the selected treatments is generally necessary. The microdialysis technique appears a suitable approach to investigate, in vivo, the neurochemical consequences of receptor–receptor interactions, as it allows the infusion by means of the probe of known compounds in selected brain regions (retrodialysis). By in vivo microdialysis, receptor–receptor interactions are generally investigated by locally perfusing selected agonists/antagonists of the specific heteromer receptor protomers, either alone or in combination. Thus, microdialysis allows determining the functional relevance of receptor heteromers into the brain, but also whether the heteromer function is altered under pathological conditions or following specific drug treatments and genetic manipulations. It is worth noting that by measuring different neurotransmitters it is also possible, in some cases, to evaluate whether a specific receptor heteromer is mainly localized and/or mainly acts at pre- or postsynaptic levels. Furthermore, the possibility to simultaneously implant probes in different brain regions allows investigating the functional consequences of receptor–receptor interactions on a specific neuronal pathway. Finally, in vivo microdialysis also permits to directly correlate neurochemical results with animal behavior [12, 21].

It is expected that before carrying out in vivo microdialysis in freely moving animal, the researcher has already made several decisions regarding microdialysis parameters (e.g., flow rate and perfusate composition, experiment duration, collection time/sample). These aspects are not considered in this chapter as they are fully detailed elsewhere [1, 16]. However, when microdialysis is used to investigate the neurochemical consequences of receptor–receptor interaction, the use of commercial or home-made probes should also be considered. An undoubled advantage of commercially available probes is their high reliability, which ensures a good data reproducibility. In fact, as stated above, the microdialysis probe allows only the collection of a fraction of the actual extracellular concentration mainly because at most flow rates (i.e., >0.1 μl/min) the rate of neurotransmitter removal from the inside of the probe is higher than the rate of neurotransmitter replacement to the probe membrane surface. Thus, mechanically assembled probes guarantee very low recovery variability. This is relevant when microdialysis is used to investigate on receptor–receptor interactions as the experimental protocol generally requires local perfusion of drugs through the probe. Furthermore, the experimental setup for commercial probes and their characteristics are well established and validated for many well-known neurotransmitters and metabolites. However, the commercially available microdialysis probes and equipment are quite expensive and for this reason several techniques to prepare home-made probes have been published [10, 16, 22]. In this case, a number of probes are necessary at one time because ~10% will be discarded due to leakage or breakage, and the researchers typically prepare probes no more than 4 days prior to intended use. Although the choice to use home-made probe should be acceptable, it is worth noting that manual construction of probes cannot guarantee an optimal reproducibility in terms of membrane length and conditions, thus leading to quite high data variability. Furthermore, mainly due to construction inaccuracies, insertion trauma might be higher than that induced when a commercial probe is implanted. Thus, when microdialysis is used to evaluate receptor–receptor interactions we strongly recommend the use of commercial probes, to reduce results variability and animal group sizes.

In the following part of this chapter, practical details related to probe implantation and experiment setup are described. Furthermore, part of results we obtained by using one- or dual-probe microdialysis to evaluate neurotensin (NT) 1 receptor subtype (NTSR1) and dopamine (DA) D2 receptor (D2R) interaction in rat basal ganglia is described as an example of use of microdialysis to understand the neurochemical consequences of receptor–receptor interaction.

3.1 Probe Implantation

Prior beginning surgery, the microdialysis membrane is generally washed by soaking in 70% ethanol for about 5 min. Then, the probe is carefully checked for leaks or air bubbles inside the membrane by perfusing it for several minutes with distilled water or perfusion fluid (flow rate 2–3 μl/min). If leaks are noted, the probe must be discarded. Air bubbles may be dislodged by gently prodding the probe while continuing to perfuse. The probe should be perfused during surgery in order to prevent the membrane from drying out.

The microdialysis probe is stereotaxically implanted into the target area (directly, in the case of acute experiments; when used sub-chronically, via a guide cannula that should be implanted 4–7 days before). In case of dual-probe microdialysis, it is strongly recommended to ascertain before surgery that, based on brain coordinates, there is sufficient distance from the target regions to allow the implantation of two probes. Stereotaxic surgery is a common and widely used method, which renders detailed recommendations unnecessary ([16], for specific details). Briefly, following induction of anesthesia (generally 1.5% mixture of isoflurane/air, but other anesthetics can be used) the animal is mounted in a stereotaxic frame with the upper incisor bar set at the appropriate level. After exposing the skull, fascia and connective tissue are scraped away and the entire area is dried with a cotton sponge or Q-tips. Dilute hydrogen peroxide can be applied to facilitate drying of the skull and aid bregma (the intersection of the coronal suture with midline) visualization (this step is also critical for the proper adherence of dental acrylic to the skull at the end of the surgical procedure). Then, microdialysis probe/guide cannula is mounted on the stereotaxic holder and gently placed over the bregma (instrument zero). The respective coordinates obtained from the atlas (for example, [23]) are then added or subtracted from the instrument readings to determine the coordinates to be used for implantation. In case of dual-probe implantation, it is recommended to proceed in parallel at this step with the two probes. After marking the right point, a hole is drilled in the skull to allow probe implantation. Bleeding can be reduced by several methods, including application of pressure with gauze or a Q-tip, cauterization, or application of a gelatin sponge. Three additional holes around the cannula/probe hole are generally made to allow the insertion of 3/16-in. bone screws, which serve to anchor the cannula/probe assembly when dental acrylic is applied (see below). The screws should be rigid but should not extend to the dura. Thereafter, after centring the probe over the drill hole, by using the stereotaxic holder the probe is slowly (about 1 mm/min) lowered into the target region. To reduce trauma this procedure should be slow and continuous. It is recommended to verify that the probe is regularly perfusing (i.e., not damaged) during and at the end of this step. Then, the outlet and inlet tubings are clamped and cut in the proximity of the probe. To permanently fix the probe, dental acrylic cement is applied to the base of the probe, the screws, and the exposed skull. Two steps are here recommended: the first one is necessary to fix the probe and screws before removing the holder; a second one is required to place the cement in a more extensive area and on the probe tube to avoid that the animal damages it. It is relevant to smooth any jagged edges by applying additional dental acrylic. Sutures or staples might be or not necessary. Then, the animal is placed in its cage until consciousness is regained. The animals should be housed individually to avoid damage to the implant. Furthermore, it is recommended to house the animals directly in the microdialysis bowel, to allow them to familiarize with the environment, thus reducing stress during the experiment. Twenty-four to thirty-six hours after implantation, necessary to recover from surgical trauma, the in vivo microdialysis experiment in freely moving animal can be carried out. Here below is reported, as example, the surgical protocol we followed to evaluate NTS1R/D2R interactions in rat basal ganglia by dual-probe microdialysis [24, 25].

Following induction of anesthesia under isoflurane (1.5% mixture of halothane and air) animals were mounted in a David Kopf stereotaxic frame with the upper incisor bar set at −2.5 mm below the interaural line. After exposing the skull and drilling two holes, two microdialysis probes of concentric design (molecular weight cut-off, 20 kDa; CMA 12; Carnegie Medicine, Stockholm, Sweden) were implanted. One probe was implanted into the right striatum (outer diameter, 0.5 mm; length of dialysing membrane, 2 mm), and the other (outer diameter, 0.5 mm; length of dialysing membrane, 1 mm) was implanted into the ipsilateral lateral globus pallidus. The coordinates relative to the bregma were as follows: striatum, anterior (A) +0.3 mm; lateral (L) +3.1 mm; ventral (V) −7.5 mm; globus pallidus, A −1.3 mm; L +3.3 mm; V −7.0 mm [23]. After the implantation, the probes were permanently secured to the skull with methacrylic cement and 36 h later microdialysis was performed in the awake, freely moving rats.

3.2 Animal Connection to the Microdialysis System

Prior to run the experiment, it is important to be sure that the inlet and outlet lines are long enough to reach the pump and allow an unobtrusive removal of the collected perfusate. The inlet and outlet lines must be protected against strain and torque, as well as chewing, by means of a tether and liquid swivel system mounted onto the testing cage. Therefore, the length of the lines is a compromise between unrestrained movement of the animals during the experiment and minimization of inlet and outlet line volumes. Before connecting the animal, it is useful to directly connect, by an adapter, the inlet and outlet lines in order to wash them and successively ascertain the regular flow of the perfusion medium. When reverse dialysis is required, the lag time between introduction of a substance to the perfusate and its entrance into the dialysis probe should be measured before, and considered during, sample collection.

On the day of microdialysis, a syringe is filled with the perfusion medium, placed into the infusion pump, and connected to the inlet line. To hold the animal in one hand while connecting the probe to the system, it may be wrapped in a towel intruded by animal odors. This also reduces animal stress. To avoid the entrapment of air bubble into the line during probe connection we suggest a "counterflow strategy": (1) the probe outlet is directly connected to running infusion pump by means of tubing; (2) as soon as the perfusion fluid drops out from the probe the inlet line is connected and the probe is maintained with the opposite flows for few seconds; (3) the outlet is then disconnected rapidly. This procedure can be repeated twice, if necessary. Thereafter, the outlet line is connected and the animal is gently placed in its bowl. In a conscious animal, the standard equilibration time is typically

5–7 h, especially for amino acid. To shorten equilibration, the flow rate of perfusate can be reduced until 1–2 h prior to sample collection.

Here below is reported, as example, the experimental protocol we followed to evaluate NTS1R/D2R interactions in rat basal ganglia by dual-probe microdialysis [24, 25].

On the day of microdialysis the probes were perfused with Ringer's solution (in millimolar: Na^+ 147, K^+ 4, Ca^{2+} 1.4, Cl^- 156, and glucose 2.7) at a constant flow rate (2 μl/min) by using a microinfusion pump. The striatal probe was employed for both local perfusion with the ligands under investigation and for recovery of endogenous neurotransmitters, while the pallidal probe was employed exclusively for endogenous extracellular neurotransmitter measurement. The collection of dialysate samples commenced 300 min after the onset of perfusion to achieve stable dialysis levels and dialysates were collected every 20 min thereafter. Following the collection of three stable basal values, treatments were included in the Ringer's solution perfusing the striatum (15 min). This medium was then replaced with the original perfusate until the end of the collection period. The total sample collection time was 180 min.

3.3 Experimental Procedure: NTSR1/D2R Interactions in Rat Basal Ganglia as Investigated by In Vivo Microdialysis

In order to highlight the general aspects to consider when in vivo microdialysis is used to investigate on receptor–receptor interactions, the procedure we followed to evaluate NTSR1/D2R interactions in rat basal ganglia by one- and dual-probe microdialysis is described.

3.3.1 Rationale

It is well known that the striatum regulates the activity of the GABAergic projection neurons of the substantia nigra pars reticulata (SNr) via two main functional opposing pathways, the direct and indirect pathways. The former is represented by the striatonigral GABAergic projection and monosynaptically inhibits the nigrothalamic GABA pathway, leading to a thalamic disinhibition of the glutamatergic drive to the cortex and therefore leading to movement initiation, whereas the latter encompasses a trisynaptic link including (1) a GABAergic projection from the striatum to the globus pallidus (GP), (2) a GABAergic projection from the GP to the subthalamic nucleus (STN), and (3) a glutamatergic projection from the STN to the SNr, which also innervates the GP. Activation of the striopallidal "link" in the indirect pathway is associated with an increase in STN glutamatergic input to the SNr and GP, and this results in changes in both thalamic activity and locomotor behavior opposite to those elicited by activation of the direct pathway [26, 27].

In the rat striatum, D2Rs are colocalized with NTSR1 either on DA nerve terminals (i.e., presynaptic level) or on striatopallidal GABA neurons (i.e., postsynaptic level). NT, a 13-amino acid peptide, is widely distributed in the striatum [28, 29]. The existence of a reciprocal modulation between neurotensinergic and dopaminergic systems in several brain areas has been demonstrated by many functional studies. In particular, several data indicated that NT reduces the affinity of D2R agonist binding sites and possibly their transduction through a receptor–receptor interaction at both presynaptic and postsynaptic levels in the striatum [29–32]. Based on this evidence we performed microdialysis study aimed to demonstrate the possible neurochemical and functional relevance of this receptor–receptor interaction.

3.3.2 Presynaptic Level

As reported above, previous studies had demonstrated the colocalization of D2R and NTSR1 on DA nerve terminals in the rat striatum, and that NT induced a reduction of D2R agonist binding site affinity through a NTSR1/D2R interaction. Thus, the hypothesis was that NT, by acting on NTSR1/D2R heteromer, could modulate the D2R-mediated changes of striatal DA release. To verify this possibility, we used intrastriatal monoprobe microdialysis to evaluate the effects of intrastriatal perfusion with D2R and NTSR1 ligands, either alone or in combination (Fig. 2), on DA release from local terminals [29]. We observed that intrastriatal perfusion with the preferential DA D2R agonist pergolide (500 nM) decreased local DA outflow (Fig. 2), an effect which reflects a stimulation of terminal D2 autoreceptors causing the inhibition of striatal DA release. When perfused alone at 1 μM or 100 nM concentrations, NT significantly increased striatal DA outflow; this effect was absent in the presence of TTX (1 μM) and thus possibly due to depolarizing mechanism(s). At a lower nanomolar concentration (10 nM), NT was by itself ineffective on striatal DA levels. This threshold concentration was therefore selected and tested for its possible effects on D2R-mediated reduction of striatal DA levels. Interestingly, we observed that when NT (10 nM) was co-perfused in combination with pergolide, the inhibitory effect of the preferential dopamine D2R agonist on DA release was counteracted (Fig. 2). The involvement of the NTSR1 receptor in modulating DA receptor function was also obtained, since in the presence of the selective NTSR1 antagonist SR48692 [33], the antagonistic action of NT on D2R-mediated inhibition of striatal DA release was fully blocked. All together these results strongly supported the view that the prejunctional NTSR1 located on the striatal DA terminals antagonize the inhibitory action exerted by pergolide on DA release mediated by activation of D2 autoreceptors. This provided a functional in vivo correlation to the binding results indicating the existence of antagonistic NTSR1/D2R interactions previously shown in neostriatal membranes and sections. These findings also gave the first

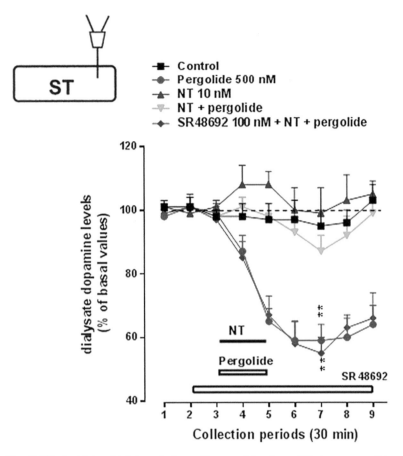

Fig. 2 Effect of intrastriatal perfusion with pergolide alone, NT alone, pergolide plus NT, and pergolide plus NT plus the NT receptor antagonist SR48692 on striatal dopamine extracellular levels using a monoprobe implantation [29, 38]. The solid bar indicates the periods of perfusion with pergolide and/or NT (60 min). SR48692 was added to the striatal perfusion medium 20 min before the peptide and maintained until the end of the experiment (open bar). Control rats were perfused with normal Ringer perfusion medium throughout the experiment. The results are expressed as percent of the mean of the three basal values before treatment. Mean ± SEM are shown. The significance for the peak effects (maximal responses) is represented. The statistical analysis was carried out according to a single one-way ANOVA followed by Newman–Keuls test for multiple comparisons performed considering all groups. Pergolide group and pergolide plus NT plus SR48692 group are significantly different (**$P < 0.01$) from control, NT alone and pergolide plus NT

indications of the existence of prejunctional NT receptors interacting with D2 autoreceptors at the level of the DA nerve terminal, representing a novel type of integrative mechanism. Finally, by using a similar experimental approach and measuring striatal glutamate levels, it was possible to also demonstrate that possibly through a

NTSR1/D2R interaction, NT also modulated D2R-mediated regulation of glutamate terminals of cortical and/or thalamic afferents [25].

3.3.3 Postjunctional Level

As stated above, the striatopallidal GABA pathway represents the first link in the chain of the so-called "*indirect*" efferent pathway to the substantia nigra. Postsynaptic D2Rs in the neostriatum exist predominantly on the medium sized striatal GABAergic neurons which project to the GP and exert an inhibitory influence on striopallidal GABA transmission [34]. As NTSR1 had been demonstrated to be colocalized with D2R on striopallidal GABA neurons, the possibility existed that NT, by reducing the transduction of the postsynaptic D2R and thus diminishing striatal D2 receptor mediated inhibition, could modulate striatopallidal GABA release. Thus, by employing the dual-probe microdialysis technique whereby one probe was implanted into the striatum and the other into the ipsilateral globus pallidus we investigated the functional role of the striatal NT receptor in changing the aminoacidergic transmission of the *indirect* pathway by monitoring simultaneously alterations in striatal and pallidal GABA release induced by striatal NT receptor activation and blockade. Firstly, it was possible to provide evidence that intrastriatal perfusion with pergolide (100, 500, and 1500 nM) induced a concentration-dependent inhibition of pallidal GABA release [24]. On the contrary, at 100 nM concentration either NT or its biologically active NT(8–13) fragment [35] increased endogenous striatal GABA release, probably derived from the collaterals of medium spiny GABA neurons in the striatum and endogenous pallidal GABA release from the terminals present in the ipsilateral globus pallidus [36, 37]. At a lower concentration (10 nM), both peptides were ineffective. The involvement of NTSR1 was demonstrated by the antagonistic action of SR48692, which at a concentration that by itself did not affect striatal or pallidal GABA release fully counteracted the facilitatory effects of NT (100 nM) on striatal and pallidal GABA release. The 10 nM NT concentration was, therefore, selected and tested for its effects on D2R-mediated reduction of striatopallidal GABA levels (Fig. 3). Thus, it was possible to demonstrate that the reductions of striatopallidal GABA levels associated with intrastriatal pergolide 500 nM perfusion were fully antagonized when NT at a low threshold concentration (10 nM) was co-perfused with the D2 agonist (Fig. 3). On the contrary, NT (1 nM) was ineffective. Such an antagonistic action of NT was mediated by NTSR1 activation since in the presence of the NTSR1 antagonist SR48692, NT failed to antagonize the inhibition of pallidal GABA release induced by pergolide. Thus, these findings provided a functional correlate to biochemical evidence suggesting the existence of antagonistic NTSR1/D2R interaction in the neostriatum at a postsynaptic level [29]. Furthermore, it was also possible to postulate that the cataleptic (and neuroleptic) activity of NT could be in part related to the reduction in affinity of postsynaptic D2R probably being part of an uncoupling of the D2R through receptor–receptor interaction with striatal NTSR1, thus also proposing a behavioral consequence of this interactions. In fact, the finding that NT increased both striatal and pallidal GABA release suggested that activation of striatopallidal GABA transmission is a primary target

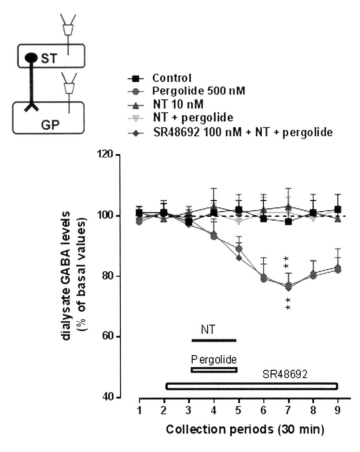

Fig. 3 Effect of intrastriatal perfusion with pergolide alone, NT alone, pergolide plus NT, and pergolide plus NT plus the NT receptor antagonist SR48692 on pallidal GABA extracellular levels using a dual-probe implantation in the awake rat. The solid bar indicates the periods of perfusion with pergolide and/or NT (60 min) [24, 25]. SR48692 was added to the striatal perfusion medium 20 min before the peptide and maintained until the end of the experiment (open bar). Control rats were perfused with normal Ringer perfusion medium throughout the experiment. The results are expressed as percent of the mean of the three basal values before treatment. Mean ± SEM are shown. The significance for the peak effects (maximal responses) is represented. The statistical analysis was carried out according to a single one-way ANOVA followed by Newman–Keuls test for multiple comparisons performed considering all groups. Pergolide group and pergolide plus NT plus SR48692 group are significantly different (**$P < 0.01$) from control, NT alone and pergolide plus NT

for NT, and that the peptide may exert its cataleptic action by activating striatopallidal GABA transmission, leading to a decrease in the excitatory drive of the thalamocortical (motor cortex) projections via the STN and the zona reticulata of the SNr. This hypothesis has been successively validated by measuring pallidal glutamate levels and by using a dual-probe microdialysis approach with one probe implanted in the SNr and the other in the ventral thalamus [24, 25, 38, 39].

4 Conclusions

During years, microdialysis has become a versatile "in vivo" technique that offers the possibility to sample and monitor neurotransmitters. In respect to other currently used techniques, such as voltammetry, microdialysis has the advantage to be possibly applied to all measurable neurotransmitters and to allow local treatments. These properties render microdialysis technique a suitable approach to investigate, in vivo, the neurochemical consequences of receptor–receptor interactions, thus providing functional correlates to binding and other biochemical data. Furthermore, this technique also allows the possibility to establish whether heteromer functions are modified during the development of a specific pathology, thus suggesting these complexes as emerging therapeutic targets [40]. The possibility to perform experiments in freely moving animals and thus to directly correlate neurochemical with behavioral data represents a further added value of this technique. Finally, recent technological developments which improve the temporal and spatial resolution of the technique [41] will enable the potential of microdialysis in the next future.

References

1. Chefer VI, Thompson AC, Zapata A et al (2009) Overview of brain microdialysis. Curr Protoc Neurosci Chapter 7:Unit7.1
2. Pan YF, Feng J, Cheng QY et al (2007) Intracerebral microdialysis technique and its application on brain pharmacokinetic-pharmacodynamic study. Arch Pharm Res 30: 1635–1645
3. Höcht C, Opezzo JA, Taira CA (2007) Applicability of reverse microdialysis in pharmacological and toxicological studies. J Pharmacol Toxicol Methods 5:3–15
4. Hutchinson PJ, Jalloh I, Helmy A et al (2015) Consensus statement from the 2014 International Microdialysis Forum. Intensive Care Med 41:1517–1528
5. Bito L, Davson H, Levin E et al (1966) The concentration of free amino acids and other electrolytes in cerebrospinal fluid: In vivo dialysis of brain and blood plasma of the dog. J Neurochem 13:1057–1067
6. Delgado JMR, DeFeudis FV, Roth RH et al (1974) Dialytrode for long term intracerebral perfusion in awake monkeys. Arch Int Pharmacodyn 198:9–21
7. Ungerstedt U, Pycock C (1974) Functional correlates of dopamine neurotransmission. Bull Schweiz Akad Med Wiss 30:44–55
8. Westerink BHC, Justice JB Jr (1991) Microdialysis compared with other in vivo release models. In: Robinson TE, Justice JB Jr (eds) Microdialysis in the Neurosciences. Elsevier Science Publishing, New York, pp 23–43
9. Ungerstedt U (1991) Microdialysis--principles and applications for studies in animals and man. J Intern Med 230:365–373

10. Thompson AC, Shippenberg TS (2001) Microdialysis in rodents. Curr Protoc Neurosci Chapter 7:Unit7.2

11. Anderzhanova E, Wotjak CT (2013) Brain microdialysis and its applications in experimental neurochemistry. Cell Tissue Res 354:27–39

12. König M, Thinnes A, Klein J (2017) Microdialysis and its use in behavioural studies: focus on acetylcholine. J Neurosci Methods S0165-0270(17):30294–30297

13. Benveniste H, Drejer J, Schousboe A et al (1987) Regional cerebral glucose phosphorylation and blood flow after insertion of a microdialysis fiber through the dorsal hippocampus in the rat. J Neurochem 49:729–734

14. Benveniste H, Diemer NH (1987) Cellular reactions to implantation of a microdialysis tube in the rat hippocampus. Acta Neuropathol 74:234–238

15. Georgieva J, Luthman J, Mohringe B et al (1993) Tissue and microdialysate changes after repeated and permanent probe implantation in the striatum of freely moving rats. Brain Res Bull 31:463–470

16. Zapata A, Chefer VI, Shippenberg TS (2009) Microdialysis in rodents. Curr Protoc Neurosci Chapter 7:Unit7.2

17. Herrera-Marschitz M, Meana JJ, O'Connor WT et al (1992) Neuronal dependence of extracellular dopamine, acetylcholine, glutamate, aspartate and gamma-aminobutyric acid (GABA) measured simultaneously from rat neostriatum using in vivo microdialysis: reciprocal interactions. Amino Acids 2:157–179

18. Morari M, O'Connor WT, Darvelid M et al (1996) Functional neuroanatomy of the nigrostriatal and striatonigral pathways as studied with dual probe microdialysis in the awake rat–I. Effects of perfusion with tetrodotoxin and low-calcium medium. Neuroscience 72:79–87

19. LaLumiere RT, Kalivas PW (2008) Glutamate release in the nucleus accumbens core is necessary for heroin seeking. J Neurosci 28:3170–3177

20. Fuxe K, Canals M, Torvinen M et al (2007) Intramembrane receptor-receptor interactions: a novel principle in molecular medicine. J Neural Transm (Vienna) 114:49–75

21. Hernández L, Paredes D, Rada P (2011) Feeding behavior as seen through the prism of brain microdialysis. Physiol Behav 104:47–56

22. Lietsche J, Gorka J, Hardt S et al (2015) Custom-made Microdialysis Probe Design. J Vis Exp 101:e53048

23. Paxinos G, Watson C (1986) The rat brain in stereotaxic coordinates. Academic, New York

24. Ferraro L, O'Connor WT, Antonelli T et al (1997) Differential effects of intrastriatal neurotensin(1-13) and neurotensin(8-13) on striatal dopamine and pallidal GABA release. A dual-probe microdialysis study in the awake rat. Eur J Neurosci 9:1838–1846

25. Ferraro L, O'Connor WT, Beggiato S et al (2012) Striatal NTS1, dopamine D2 and NMDA receptor regulation of pallidal GABA and glutamate release–a dual-probe microdialysis study in the intranigral 6-hydroxydopamine unilaterally lesioned rat. Eur J Neurosci 35:207–220

26. Alexander GE, Crutcher MD (1990) Functional architecture of basal ganglia circuits: neural substrates of parallel processing. Trends Neurosci 13:266–271

27. Chevalier G, Deniau JM (1990) Disinhibition as a basic process in the expression of striatal functions. Trends Neurosci 13:277–280

28. Quirion R, Chiueh CC, Everist HD et al (1985) Comparative localization of neurotensin receptors on nigrostriatal and mesolimbic dopaminergic terminals. Brain Res 327:385–389

29. Fuxe K, O'Connor WT, Antonelli T et al (1992) Evidence for a substrate of neuronal plasticity based on pre- and postsynaptic neurotensin-dopamine receptor interactions in the neostriatum. Proc Natl Acad Sci U S A 89:5591–5595

30. Agnati LF, Fuxe K, Benfenati F et al (1983) Neurotensin in vitro markedly reduces the affinity in subcortical limbic [3H] N-propylnorapomorphine binding sites. Acta Physiol Scand 117:299–301

31. von Euler G, Fuxe K, Benfenati F et al (1987) Neurotensin modulates the binding characteristics of dopamine D2 receptors in rat striatal membranes following treatment with toluene. Acta Physiol Scand 135:442–448

32. Fuxe K, Agnati LF, von Euler G (1992) Neuropeptides, excitatory amino acid and adenosine A2 receptors regulate D2 receptors via intramembrane receptor-receptor interactions. Relevance for Parkinson's disease and schizophrenia. Neurochem Int 20:215S–224S

33. Gully D, Canton M, Boigegrain R et al (1993) Biochemical and pharmacological profile of a potent and selective nonpeptide antagonist of the neurotensin receptor. Proc Natl Acad Sci U S A 90:65–69

34. Reid MS, O'Connor WT, Herrera-Marschitz M et al (1990) The effects of intranigral GABA and dynorphin A injections on striatal dopamine and

GABA release: evidence that dopamine provides inhibitory regulation of striatal GABA neurons via D2 receptors. Brain Res 519:255–260

35. Granier C, van Rietschoten J, Kitabgi P et al (1982) Synthesis and characterization of neurotensin analogues for structure/activity relationship studies. Acetyl-neurotensin-(8-13) is the shortest analogue with full binding and pharmacological activities. Eur J Biochem 124:117–24

36. Sirinathsinghji DJ, Heavens RP (1989) Stimulation of GABA release from the rat neostriatum and globus pallidus in vivo by corticotropin-releasing factor. Neurosci Lett 100:203–209

37. Parent A, Hazrati LN (1995) Functional anatomy of the basal ganglia. I. The cortico-basal ganglia-thalamo-cortical loop. Brain Res Brain Res Rev 20:91–127

38. Ferraro L, Tomasini MC, Fernandez M et al (2001) Nigral neurotensin receptor regulation of nigral glutamate and nigroventral thalamic GABA transmission: a dual-probe microdialysis study in intact conscious rat brain. Neuroscience 102:113–120

39. Antonelli T, Tomasini MC, Fuxe K et al (2007) Focus on NTR/D2 interactions in the basal ganglia. J Neural Transm (Vienna) 114:105–113

40. Guidolin D, Agnati LF, Marcoli M et al (2015) G-protein-coupled receptor type A heteromers as an emerging therapeutic target. Expert Opin Ther Targets 19:265–283

41. Kennedy RT (2013) Emerging trends in in vivo neurochemical monitoring by microdialysis. Curr Opin Chem Biol 17:860–867

Chapter 8

Behavioral Methods to Study the Impact of Receptor–Receptor Interactions in Fear and Anxiety

Miguel Pérez de la Mora, José del Carmen Rejón-Orantes, Minerva Crespo-Ramírez, Dasiel O. Borroto-Escuela, and Kjell Fuxe

Abstract

An important step forward in understanding synaptic transmission has been the discovery that within the realm of G-protein coupled receptors (GPCRs) agonist activation of a receptor type favors physical interactions with different types of neighboring GPCRs. Such receptor interactions lead to receptor oligomerization and formation of receptor heteromers, in which through allosteric mechanisms original protomers change kinetic properties, G-protein recognition and trafficking. Since such changes have important influence on behavior, the aim of this chapter is to describe general strategies to study in the rat the behavioral consequences linked to the formation of molecular heteromers as well as to the application of some commonly used behavioral methods to evaluate the receptor oligomerization on anxious behavior.

Key words Elevated plus maze, Shock-probe burying test, Marble burying test, Open-field test, Anxiety, Fear, Heteromers, Receptor oligomerization

1 Introduction

One of the largest accomplishments during the last years has been the realization that synaptic transmission both at synaptic and at extrasynaptic sites does not occur as an isolated event just contributing to change by algebraic summation the membrane potential but rather involves a series of events at membrane level that have allosteric effects on neighboring receptors. See for reviews [1, 2]. Although this seems to occur mostly in the realm of G-protein coupled receptors (GPCRs), it can also involve interactions between GPCRs and ion channel receptors [3–5]. Adapter proteins also participate in this process to achieve an optimal allosteric interaction between the participating receptors. Thus, heteroreceptor complexes are formed [6].

Agonist activation of a particular GPCR protomer leads to alterations in receptor oligomerization [7]. Within heteromers, allosteric receptor–receptor interactions take place giving rise to

Kjell Fuxe and Dasiel O. Borroto-Escuela (eds.), *Receptor-Receptor Interactions in the Central Nervous System*, Neuromethods, vol. 140, https://doi.org/10.1007/978-1-4939-8576-0_8, © Springer Science+Business Media, LLC, part of Springer Nature 2018

important changes in their dynamic properties [8, 9] like their G-protein recognition and trafficking [7].

Although important observations were made from 1980 till 1993 [1, 2, 10, 11], suggesting the formation of homo- and heteromers, it was the discovery of the GABAB heterodimer [12] that unequivocally established the existence of receptor oligomerization between isoreceptors using the same neurotransmitter. It was shown that $GABA_B$ receptors in order to be active required the assembly of a dimer formed by GABAB1 and GABAB2 protomers, which otherwise were inactive by themselves. Now considerable work has been done by several groups and as a result a large number of receptor heteromers have been described involving among others the physical interactions between A2A receptors and dopamine D2 receptors and between A1 receptors and dopamine D1 receptors [13], A2A receptors and mGlu5 receptors [14, 15], dopamine D2 receptors and sigma 1 receptors [16], dopamine D2 receptors and oxytocin receptors [9, 17], delta opioid receptor and kappa and μ-opioid receptors [18], μ opioid and δ opioid receptors [19], dopamine D1 and dopamine D2 receptors [20, 21], dopamine D1 and dopamine D3 receptors, suggesting that formation of receptor heterodimers is a general feature in at least the world of GPCRs [22].

On the other hand, since receptor–receptor interactions in heteroreceptor complexes are now recognized as a very important feature in integration of synaptic transmission its impact on behavior is increasingly studied. Thus, effects of dopamine D2 receptor-oxytocin receptor heteromers [17], dopamine-D1-D2 receptor heteromers [20, 21], and 5-hydroxytryptamine 1A-galanine receptor 1 heteromers [23] were reported to modulate anxiety and depression. In addition, dopamine D1-D2 receptor heteromers increased grooming behavior and its sensitivity was reported to be upregulated both in the rat striatum and in the globus pallidus from schizophrenic patients following chronic amphetamine treatment [24]. Activity of delta-kappa and delta μ opioid heteromers were shown to have an important role in nociception and alcohol consumption, respectively [18].

The aim of this chapter is to describe general strategies to study in the rat the behavioral consequences linked to receptor oligomerization and formation of heteroreceptor complexes. This includes the application of some commonly used behavioral methods used to evaluate fear and anxiety such as open-field test, elevated plus maze and shock-probe burying test.

2   Materials

2.1   Facilities

2.1.1   Surgical Room

Since most behavioral experiments involve intracerebral drug administration, a surgical room having a table where to have stereotaxic frames and all other surgical equipment, i.e., temperature controller, is needed. The room has to be only used for surgery and needs to be thoroughly cleaned before surgery to prevent postsurgical infections.

2.1.2 Behavioral Room

It will be absolutely needed to carry out all behavioral experiments in a sound-attenuated two-compartment room having good air exhaust. In such a facility, rats will be allowed to be undisturbed in one room (experimental compartment) during their behavioral performance, whereas experimenter within the other room (recording compartment) will watch the rats by using a monitor and video-record their behavior for its off-line analysis. Video cameras are usually placed on top of the behavioral devices but can be installed in different locations. For video-recording, a Panasonic video camera (WV-BP1444) is connected via an AC adaptor to a PC DELL (Optilex 780). Video images are captured using Pinnacle Studio for Dazzle software. Within the experimental compartment only a little table to hold animal cages together with the behavioral dispositive used is present. Since illumination is crucial for most behavioral models, particularly to those used to evaluate anxiety-like behavior the behavioral room should be equipped with a rheostat to regulate lighting.

2.1.3 Stereotaxical Cannula Implantation and Microinjection

Rats. Male Wistar rats (250–280 g) bred in the Instituto de Fisiología Celular, Universidad Nacional Autónoma de México, Mexico City, Mexico are used. Rats are housed in a controlled environment (temperature 22 °C, lights on 6:00–18:00 h) with water and food at libitum.

Rat stereotaxical frame. It can be obtained either from Kopf Instruments, Tujunga, CA, USA or from Stoelting Co, Wood Dale, IL, USA. There exist several models, including the automatized ones. Cannula holders (MH300) can be obtained from Plastics One, Roanoke, VA, USA.

Stereotaxical Atlas. The Paxinos and Watson atlas has been so far the most used stereotaxical atlas (George Paxinos and Charles Watson "The Rat Brain in Stereotaxical Coordinates" Elsevier).

Rechargeable Micro Drill. 58610 Cordless Micro Drill equipped with 58609 Carbide Burr Drill Bits. Stoelting Co, Wood Dale, IL, USA. We have found that for this procedure 0.8 mm burr is the most suitable.

Temperature controller CMA 150. It can be obtained from CMA Microdialysis, Stockholm, Sweden.

Since bilateral micro-infusions will be done, a couple of micro-infusion pumps will be needed. We have found that both CMA 100 and CMA 400 models from CMA Microdialysis, Stockholm, Sweden are suitable for this purpose. Similar micro-infusion pumps can also be purchased from Stoelting Co, Wood Dale, IL, USA.

Ketamine hydrochloride/xylazine hydrochloride solution (80 mg/ml/12 mg/ml) Catalog number K113-10ML Sigma/

Aldrich Chemical Co (St. Louis, MO, USA). As a rule of thumb, good anesthetic effects lasting for 20–60 min are attained with 80 mg kg ketamine/12 mg kg xylazine. For practical purposes, 1 ml i.p. injections of a fourfold dilution of the concentrated ketamine/xylazine solution in 0.9% NaCl is administered to rats weighing 250 ± 10 g rats. Anesthetics should be stored at 4 °C in a place with exclusive access for only authorized personnel. It will be however advisable to check up in advance the effectiveness of the anesthetic dose given above.

Electrical razor. Any domestic razor is suitable for shaving the head of the rats.

Surgical type latex gloves.

Cover-mouths.

Surgical instruments (scalpel, 11 cm straight operating scissors, 11 cm mosquito curved forceps, curved 10 cm forceps, freer). Instruments should be preferentially sterilized and/or kept in an antiseptic solution during surgery.

Bone anchor screws. These screws can be purchased from CMA Microdialysis, Stockholm, Sweden (Catalog number 7743 1021) or from Stoelting Co, Wood Dale, IL, USA (Catalog number 51457). Jeweler screws also are suitable for this type of task.

Mini screwdriver set.

Dental acrylic cement.

Antiseptic solution.

Long-acting antibiotic like streptomycin works usually well.

Guide cannulae 0.46 mm outer diameter (C315G), Plastics One, Roanoke, VA, USA.

Dummy cannulae (C315DC), Plastics One, Roanoke, VA, USA.

Injection cannulae 0.20 mm outer diameter (C3151), Plastics One, Roanoke, VA, USA.

Hamilton microsyringes 50 μl or syringes fitting CMA micro-infusion pumps.

Connector cannula 0-SPR 40 cm. Plastics One (C313C), Plastics One, Roanoke, VA, USA. These connectors allow connecting syringes from micro-infusion pump to injection cannulae. Buy one connector for each syringe. Connectors can be reused provided that they are thoroughly flushed with water. Buy from the same source C313CT polyethylene PE50 to fit cannula connectors.

Saline solution (0.9 g/100 ml).

2.1.4 Histology

Any microscope equipped with either a 4 or 10× objective to look at cannulae placements within the brain. It will be more convenient to use a stereotaxical microscope, which can feed its information into a PC such as the EZ4HD model from Leica Instruments, Nussloch, Germany.

Cryostat. Any cryostat with a good temperature controller and cutting precision can be used. We have successfully used CM-1510-3, Leica Instruments, Nussloch, Germany.

Sodium pentobarbital. Catalog number: P3761-5G. Sigma/Aldrich Chemical Co, St Louis, MO, USA. As with ketamine pentobarbital has to be kept in a safe place with only access for authorized personnel.

Pontamine blue. Catalog number B-0126. Sigma/Aldrich Chemical Co, St Louis, MO, USA.

Nitric acid. Catalog number 438073-2.2LP. Sigma/Aldrich Chemical Co, St Louis, MO, USA.

Chromium(III) potassium sulfate dodecahydrate. Catalog number 101036. Merck Millipore, Massachusetts 01821, USA.

Gelatine. Catalog number G6550. Sigma/Aldrich Chemical Co, St Louis, MO, USA.

Finally, slices mounted in slide racks are transferred, using a Tissue-Tek® II set placed in the hood. Catalog number SA62540-01. Science Services Gmbh, Unterhachinger Straße 75, Munchen, Germany.

Gelatinized microscope slides. To gelatinize microscope slides, 2.5 g gelatin + 0.25 g Chromium (III) potassium sulfate dodecahydrate are dissolved in 500 ml water kept at 50 °C under stirring. The solution is filtered and maintained at the same temperature using a Maria bath. Meanwhile, a 70% (v/v) nitric acid solution is prepared in the hood. Finally, slices mounted in slide racks are transferred using the Tissue –Ten Manual Slide Staining Set placed in the hood, from 70% nitric acid solution (3 min), to running water (10 min), to distilled water (5 min), and finally to gelatin/Chromium (III) potassium sulfate solution (3 min). Finally, slides are dried in an oven set at 57 °C for 24–41 h.

Coverslips.

Permount. FL-10-0505. Fisher Scientific Co, Fair Lawn, 074109, NJ, USA. To avoid bubble formation when mounting brain slices we have had used to dilute eight parts resin with two parts xylene.

2.1.5 Cresyl Violet Staining

Distilled water.

Absolut ethanol. 9000-03 JT Baker S.A. de C.V. 55320, Xalostoc, Mexico. Working ethanol solutions are prepared from absolute ethanol to contain a concentration of 70%, 95%, or 100% (v/v). Remember that 70% and 95% solutions contain 70 or 95 ml ethanol in 100 ml solution, respectively. Keep the alcohol solutions tightly closed in a fresh place and never use them for more than 15 times.

Xylene. Catalog number: 9490-03, JT Baker S.A. de C.V. 55320, Xalostoc, Mexico.

1.0% (w/v) Cresyl violet. Sigma Chemical Co Catalog number: C5040. 0.1 g cresyl violet is dissolved in 100 ml distilled water at 40 °C. Add either 250 µl (approx. two drops) concentrated acetic acid or 16 drops 10% acetic acid. Keep overnight the solution under stirring. For storage, maintain the solution in the dark and filter it before using. Colorant solution is degraded as acetic acid is being volatilized. As a consequence, use cresyl violet during the staining procedure for 1 min if solution has not been used more than 15 times. If necessary, increase the staining step for up to 4 min to have good staining results. Adding some drops of 10% acetic acid can solve slight precipitation.

2.1.6   Open-Field Test

Apparatus, illumination and recording conditions. For rats, a $50 \times 50 \times 30$ cm acrylic box is used (Fig. 1) (OMNIALVA, Mexico City, Mexico). In order to record the movement of the animals in the arena, each wall of the box contains 10 photoelectric cells separated by 5 cm from each other. Cells are located 4.0 cm above the bottom of the box. The box is interphased with a PC that allows estimating quantitatively the number of beam interruptions through the photocells during the locomotion of the animal in the arena and transforming them into arbitrary locomotion events with a sample frequency of 10 Hz. A similar apparatus can be obtained from Harvard Apparatus, Holliston, 01746, Massachusetts, USA. Video camera is positioned on the device at a distance of 87 cm from the bottom of the arena. Light intensity at the center of the device is adjusted to 138 lux.

Animals: Cannulated and treated Wistar rats (250 ± 30 g). See Sect. 3.2.1.

Light meter 840020 SPER Scientific Ltd., Scottsdale, 085260, AZ, USA.

Camera. Panasonic WV-BP1444.

PC DELL (Optilex 780) equipped with Pinnacle Studio for Dazzle software.

Paper cards.

Mild detergent.

Odor-free paper.

Surgical type latex gloves.

A statistical package. Both GraphPad Prism 4.0 (GraphPad software, USA) and IBM SPSS can be used.

2.1.7   Elevated Plus Maze

Apparatus illumination and recording conditions. For rats, the maze (Fig. 2) is usually done as described by Pellow et al. [25]. It is built of wood and consisted of two open (50×10 cm) and two enclosed arms ($50 \times 10 \times 40$ cm) with an open roof. The arms are intersected at the central square (10×10 cm). The maze is elevated 50 cm from the floor by a pedestal joined to the central square (*see* **Note 8**). Video camera is positioned on the device at a distance of 1.2 m from the floor of central square. Illumination level at the central square is 5.1 lux during testing.

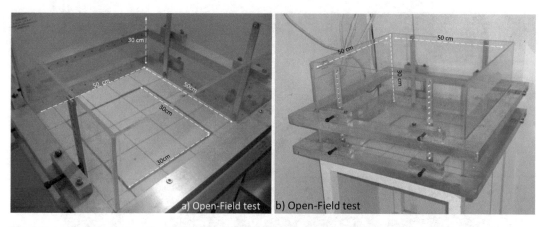

Fig. 1 Open-field apparatus. An acrylic box with the dimensions depicted in the figure is used for evaluation of the impact of heteromer formation on both locomotion and anxiety responses (panels **a** and **b**). Squares drawn at its bottom allow the experimenter to distinguish between a central and a peripheral part of the open field, which allows to evaluate anxiety and locomotion. Open-field exploration is ascertained by interruption of light beans between photoelectrical cells located in its walls

Animals: Cannulated and treated Wistar rats (250 ± 28 g). See Sect. 3.2.1.

Lux meter. Light meter 840020 SPER Scientific Ltd., 085260, AZ, USA.

Camera. Panasonic WV-BP1444.

PC DELL (Optilex 780) equipped with Pinnacle Studio for Dazzle software.

Paper cards.

Mild detergent.

Odor-free paper.

Surgical type latex gloves.

A statistical package. Both GraphPad Prism 4.0 (GraphPad software, USA) and IBM SPSS can be used.

2.1.8 Shock-Probe Burying Test

Apparatus illumination and recording conditions. A behavioral dispositive (27 × 16 × 23 cm) having an open roof made of acrylic is used. The floor of the dispositive was uniformly covered by a 5 cm layer of fine sawdust. An acrylic probe (7 cm long and 0.5 cm diameter) having two 18 gauge single stranded insulated copper wires wrapped along its full length and leaving gaps of 3 mm approximately is built according to Treit et al. [26] and used. The probe protruded into the behavioral dispositive through a hole practiced in one of its small walls placed 2.0 cm above the sawdust bed (Fig. 3). Insulated wires coming from the probe are fed into a shocker, which deliver a 0.6 mA shock any time that either rat's spout or forepaws touch the probe. Video camera is positioned on top of the dispositive at a distance of

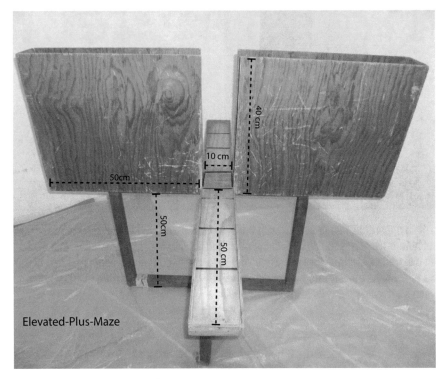

Fig. 2 Elevated plus maze. A maze with two open and two enclosed arms which intercepts at a central square and is elevated 50 cm from the floor is used to measure the impact of heteromer formation on anxiety in the rat. In the maze rats experience a conflict between either remain in a save place (closed arms) or to explore open spaces devoid of thigmotactic cues (open arms). Increased open arms exploration signal for decreased anxiety

75 cm from the top of sawdust bed. Illumination intensity was 4.7 lux at the center of sawdust bed using a red light bulb (40 W).

Animals: Cannulated and treated Wistar rats (250 ± 28 g). See Sect. 3.2.1.

Shocker. LE 100-26 Harvard Apparatus, Holliston, 01746, Massachusetts, USA.

Light meter 840020 SPER Scientific Ltd., 085260, AZ, USA.

Camera. Panasonic WV-BP1444.

PC DELL (Optilex 780) equipped with Pinnacle Studio for Dazzle software.

Paper cards.

Sawdust.

Odor-free paper.

Surgical type latex gloves.

A statistical package. Both GraphPad Prism 4.0 (GraphPad software, USA) and IBM SPSS can be used.

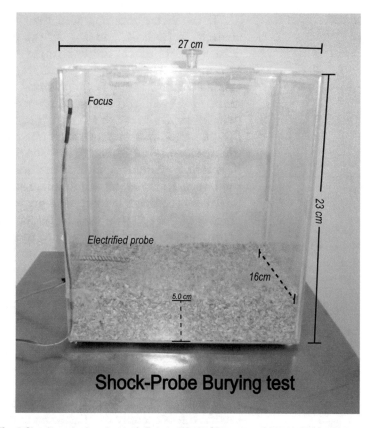

Fig. 3 Shock-probe burying test. Rats in this test bury any object in their surroundings which may harm them such as the electrified probe which is shown protruding from one of its walls. In this test, rats spray sawdust from the bed of the cage to the probe after they are shocked by the probe when exploring the cage. Any anxiogenic effect resulting from heteromer formation will be detected in this test by an enhancement of burying behavior

3 Methods

3.1 General Strategies

Physical receptor–receptor interactions within the membrane domain are by definition occurring between neighboring receptors, in which the binding of an agonist by one receptor triggers its interaction with another receptor molecule giving rise to heteromer formation. Under these conditions, it is clear that when trying to promote specific heteromer formation a comprehensive knowledge of the brain distribution of participating protomers should be attained and agonists should be delivered *locally*, near the interacting receptors. Furthermore, to make easier the interpretation of the results local injections should be given using experimental conditions (low volume and low rate of administration), which will favor a low diffusion of the compound microinjected from the sites of its administration.

It has been demonstrated that upon heteromer formation changes in both K_d and B_{max} of the participating protomers do exist [8, 9]. Since these changes will affect the behavioral response to their agonists, one way to demonstrate the heteromer formation and the behavioral consequences of this oligomerization would be the demonstration of changes in the dose–response curve of a receptor agonist, recognized by one of the putative protomers, carried out either in the presence or in the absence of an ineffective dose of an agonist for the second interacting protein. A shift of the dose–response curve either to the left or to the right will indicate an interaction between both putative protomers leading to a greater or to a lesser behavioral efficacy of the receptor agonist under study, respectively. A more comprehensive study will require the demonstration of the lack of effects of the second agonist under the presence of a pure competitive antagonist *see* **Note 1**).

A second approach to demonstrate a receptor–receptor interaction with relevance for a given behavior, i.e., anxiety, will involve the demonstration of behavioral effects associated with the activity of one receptor and their suppression by an antagonist of the putative interacting second receptor. To control for the behavioral response under study, the effects of the active receptor should be suppressed by one of its antagonists.

Finally, since in principle receptor–receptor interactions are carried out bilaterally this has to be confirmed experimentally performing the same experiments but now involving the second receptor.

3.2 Stereotaxical Cannula Implantation and Microinjection

Stereotaxical surgery was developed as earlier as 1908 at University College London Hospital by Victor Horsley and Robert H. Clark to reach specific brain targets in the brain taking advantage of a three-dimensional coordinate system. This type of surgery is minimally invasive and allows substance infusion as well as other brain interventions such as region stimulation and lesion. Cannula implantation site must be carefully verified.

3.2.1 Detailed Procedure

(Day 1). Bilateral cannulae placement

1. Rats are taken from the animal house in polycarbonate cages and kept in racks near the surgical room. Before cannulation, rats are anesthetized with ketamine (80 mg/kg) and xylazine (12 mg/kg) (see Sect. 2.1.3. Ketamine/xylazine solution) and their heads are shaved.

2. The rat to be operated is placed in the stereotaxical frame introducing gently the ear bars of the frame into its ears and setting the incisive bar at 3.3 mm in order to keep its skull plane (*see* **Note 2**). Its body temperature is kept at 37 °C with the aid of a temperature controller.

3. After application of an antiseptic solution on the skin of its head, an incision is done with a scalpel along the longitudinal axis of the head and the skin is pulled away laterally from the middle line to expose bregma.

4. The selected anteroposterior and lateral coordinates of the target, according to the stereotaxical atlas, are used to position the anteroposterior and lateral arms of the stereotaxic frame on top of one side of the head of the rat. A label scratch having these coordinates is done on the skull in order to position afterwards a drill and to make a hole (main hole) on the skull of the rat avoiding damaging the underlying brain tissue particularly at the middle line where blood vessels and sinuses are located. The same procedure is repeated for the other side of the head. Additional drills around the main one are done to screw anchor screws, which will help to affix guide cannulae in its position inside the brain.

5. The dorso-ventral arm of the stereotaxical frame having attached a guide cannula protruding from its ventral end is positioned on top of one of the main bilateral holes and is descended with the aid of the micrometric button controlling the movement of the arm. The same procedure is repeated for the main hole located on the other side of the head. In order to avoid damage of the chosen target, guide cannulae are descended until their tips reach a position of 1 mm dorsal to the target (*see* **Note 2**).

6. Cannulae are affixed into the skull of the rats by means of four small stainless-steel screws introduced into the supplementary holes made around the main holes and an overlay of dental acrylic cement.

7. Dummy cannulae are introduced into guide cannulae in order to avoid infection and keep them patent.

8. A long-acting antibiotic is given to prevent infection.

9. (Days 1–8). Recovery and handling. After cannulae implantation, animals are taking into a vivarium where they are kept under standard conditions (temperature 22 °C; lights on 6:00–18:00 h) with water and food at libitum. Animals are singly housed in polycarbonate $42 \times 25 \times 20$ cm cages to avoid biting of implanted cannulae. Dummy cannulae are everyday taken out and reintroduced again into the guide cannulae to keep them permeable.

10. (Days 9–13). Habituation. See sections "Detailed Experimental Procedure" and "Detailed Procedure" under Sects. 3.3.2 and 3.3.3.

11. (Day 14). Drug infusion. Drug solutions are prepared in advance in a proper vehicle according to their chemical composition, concentration, and solubility. Look at the storage conditions recommended by suppliers.

12. Syringes are attached to injection cannulae through the use of connectors (see Sect. 3.2.1) and loaded with proper drug solutions. For this purpose, please aspirate first a sweep fluid (saline or an artificial CSF) and then the drug to be injected leaving a bubble of air between them to monitor the movement of the solutions during injection. Syringes are placed into the infusion pumps. *See* **Note 4**.

13. Within the monitoring compartment the rat is gently held by the experimenter wearing surgical type latex gloves and their dummy cannulae are substituted by injection cannulae.

14. The animal is placed in its own cage and both pumps are simultaneously started. During infusion, the rat is left undisturbed.

15. Infusion is carried out during 3–5 min avoiding infusing a total volume larger than 0.5 µl on each side.

16. Cannulae after infusion are kept in their place for 1 additional minute to allow for diffusion and to prevent back-flowing.

17. Following infusion animals are taken into the behavioral compartment to initiate the behavioral testing.

3.2.2   Histology

1. After drug infusion, rats are deeply anesthetized with sodium pentobarbital (65 mg/rat) and afterwards 0.2 µl pontamine blue is bilaterally injected through their injection cannulae, as an aid to locate cannulae tips in their brains.

2. Brains are removed from the rats and kept in 10% formaldehyde for 1–2 weeks.

3. Before sectioning, brains are successively submerged in a 10%, 20%, and 30% sucrose solution where they remain for 24 h in each solution until they go to the bottom of the beaker containing the sucrose solution.

4. Pieces of the brain having the selected regions are frozen in dry ice and sectioned in a cryostat at 40–50 µm intervals.

5. As sections are being obtained, they are layered on gelatinized slides.

6. Sections are counterstained with cresyl violet (see below).

7. Verification of cannula location is done using a low power microscope with the aid of illustrations depicted in the stereotaxical atlas used.

3.2.3   Cresyl Violet Staining

1. For staining using this method, consecutively expose brain slices on their supporting slides using the Tissue –Ten Manual Slide Staining Set for a period of 2 min each to (1) water, (2) 70% ethanol, (3) 70% ethanol, (4) 95% ethanol, (5) 95% ethanol, (6) 100% ethanol, (7) 100% ethanol, (8) 95% ethanol, (9) 95% ethanol, (10) 70% ethanol, (11) 70% ethanol, (12) water, (13) Cresyl violet, (14) water, (15) repeat steps 1–7, (16)

xylene, (17) xylene, (18) cover the sections with diluted Permount avoiding formation of bubbles, (19) allow the slides to air dry.

3.3 Behavioral Testing

In order to avoid circadian variations, experiments must be done during a specified day cycle (i.e., normal or inverted) and during a defined time period, usually between 9:00 and 18:00 h. Rats should be selected at random for each experimental group and tested also at random. Experimenter must wear surgeon type latex gloves any time he touches the rats. Finally, it is worth saying that since anxiety is elicited when testing rats in anxiety paradigms ceiling effects may be developed in those experiments in which anxiety is developed. In view of that, it will be preferable to choose experimental conditions in which the expected outcome will be an anxiolytic state.

3.3.1 Open-Field Test

This test was initially described by Hall [27] to monitor the degree of emotionality (anxiety) of rodents when they were staying in a novel environment. Since he considered that this test was modulated by the autonomic nervous system, defecation and urination were taken as measures of anxiety. The test since then has been nevertheless enriched with contributions of many authors. See for a review [28]. As indicated above, open-field test exploits the conflict that a rodent experiences between staying at the periphery of an open space limited by walls or to explore its central part devoid of thigmotactic cues [29, 30]. This behavioral paradigm can be used both in rats and mice [31] and has been mainly used to evaluate anxiety and locomotion. The test has become very popular due to its low cost and easiness of implementation. At the beginning of the test, the rodent is placed either in the center of the box or in one of its corners and it is allowed to explore the arena for 5 min. Although a large number of measures (rearing, defecation, urination, stretching, etc.) can be made during the rodent exploration of the open field, both the number of beam interruptions between photoelectric cells placed in opposite walls of the cage, as a measure of locomotion, and the time spent in the central part of the arena, as an index of anxiety, are the main measures taken. Since no motor impairment is required for the correct interpretation of results in most unconditioned models of anxiety, the open-field test has been widely used to evaluate locomotion either immediately before or following the exposure of animals to the test (*See* **Note 5**).

Detailed Experimental Procedure

1. (Days 1–3). Habituation. Cannulated rats (see Sect. 3.2), kept in a vivarium in single cages to avoid biting of their implanted cannulae, are moved into the behavioral room (see Sect. 3.3) where they stay in racks for 4 days in the recording compartment. During this time, rats are everyday gently handled by

the experimenter in the experimental compartment for 3–5 min simulating the drug-infusion step. During this procedure, dummy cannulae are taken out and taken in from the guide cannulae to keep them clear (*see* **Note 3**).

2. (Day 5). Following drug infusion (see Sect. 3.2.1, steps 11–16), rats are taken out from their cages by the experimenter and introduced into the experimental compartment of the behavioral room. Rats are gently placed either in the center of the box or in one of its corners allowing them to explore the box undisturbed for 5 min.

3. After the rodent is placed in the open field, the experimenter starts a chronometer and moves into the recording compartment to watch the rat behavior. Behavior is video-recorded for 5 min for its posterior off-line analysis. Identification of the rats is done using a paper card placed under the camera view containing date, experiment name, and rat code (*see* **Note 6**).

4. Following the removal of the animals from the open-field box, the device is carefully cleaned using a soft detergent and then dried out with the aid of odor-free paper.

5. For evaluation of locomotor activity, researcher measures locomotion by estimating the number of beam interruptions between photoelectric cells located in opposite walls of the open-field device.

6. For evaluation of anxiety, researcher registers the time spent by the animal in the central part of the box.

7. Behavioral evaluation is carried out off-line using those video-recordings obtained during the experiment (*see* **Note 7**).

8. For statistical evaluation of the results, statistical normality distribution of data is first ascertained for each experimental measurement and according to the results obtained either parametric or non-parametric statistics is used to further analyze the data. Use a suitable statistical package (i.e., GraphPad Prism) to perform all these computations.

9. Results are presented as either means ± SEM or as medians with their respective interquartile range depending upon if they follow or not a normal statistical distribution, respectively.

3.3.2 Elevated Plus Maze   The Elevated Plus Maze is the most used unconditioned methods in the field of anxiety [32–36]. This test is based on the observation of Montgomery [37] showing that rats exploring different types of mazes persistently show an avoidance for their open alleys. This observation led to the conclusion that such avoidance was most probably due to the higher fear that open alleys induce in rodents in comparison with the enclosed ones. Although the first trials using this maze were done in rats, posterior experiments showed that it also can be used with mice [38]. Full validation involving behavioral, physiological, pharmacological, and hormonal aspects was done by Pellow et al. [25] who showed that the maze can be used to detect both anxiolytic and anxiogenic conditions, including effects of drugs which are effective in humans (predictive validity). In view of that, it is not surprising that the Elevated Plus Maze had been widely used to study the role of different neurotransmitter systems in the modulation of fear/anxiety such as the GABAergic [35, 39], cholecistokinergic [40], and dopaminergic [41, 42] among others.

Behavioral characterization of the maze has showed that what produces avoidance to enter its open arms is actually their lack of thigmotactic cues [43], which otherwise are present within the preferred closed arms [35, 36, 44]. Surprisingly, the height of the maze does not seem to have any influence on this avoidance [44, 45]. On the other hand, unlike other anxiety tests, estimation of entries into either the closed arms or closed arms + open arms of the maze represents an in-built index of locomotion [33, 35]. The main parameters scored in this test include: total arm entries, number of entries into both open and enclosed arms, total time spent in each arm.

The popularity of this test is due to its simplicity and low cost, as well as to the fact that it does not require neither a tedious previous training nor food/water deprivation [25]. Its popularity has however led to a wide methodological variation both in its construction (see **Note 8**) and the behavioral scoring [35, 46], which have led to numerous inconsistencies in the results published from several laboratories. In spite of its advantages, a proper interpretation of the results obtained is complicated by the fact that numerous factors affect the performance of rodents in the maze making difficult to attain a suitable baseline that allows the experimenter to detect both anxiogenic and anxiolytic effects. Thus, factors such as daytime testing, illumination level during testing, and repeated exposure to the maze were found to affect the behavioral baseline in this model [47]. It has also been found that under certain experimental conditions such as those, which induce intense thirst (i.e., diabetes), rats may exhibit bizarre behaviors. Thus, anxious and thirsty diabetic rats display an increased motivation and an enhanced drive to explore the anxiogenic open arms of the maze in their search of water [48]. Furthermore, a previous 5 min non-drugged exposure to the maze is able to elicit a particular state called one-trial tolerance, in which rodents do not show any longer the anxiolytic effects of benzodiazepines [49, 50] suggesting the induction of important changes involving GABA-A receptor kinetics. It is clear that such effects preclude the use of this test prior the application of another anxiety test.

1. (Days 1–3). Habituation. Cannulated rats (see Sect. 3.2), kept in a vivarium in single cages to avoid biting of their implanted cannulae, are moved into the behavioral room (see Sect. 3.3) where they stay in racks for 4 days in the recording compartment. During this time, rats are everyday gently handled by the experimenter in the monitoring compartment for 3–5 min simulating the drug-infusion step. During this procedure, dummy cannulae are taken out and taken in from the guide cannulae to keep them clear (*see* **Note 3**).

2. (Day 5). On the day of the experiment following drug infusion (see Sect. 3.2.1, steps 11–16), rats are taken out from their cages by the experimenter and introduced into the experimental compartment of the behavioral room. Rats are gently placed on the central square of the maze looking at one of its open arms and avoiding that the rat will fall off the maze.

3. After the rodent is placed on the maze, the experimenter starts a chronometer and moves into the recording compartment to watch the rat behavior. Behavior is video-recorded during 5 min for its posterior off-line analysis. Identification of the rats is done using a paper card placed under the camera view containing date, experiment name, and rat code (*see* **Note 6**).

4. Avoid the temptation of continue testing those rats that have fall off from the maze rats that have fall off an arm to the floor, since most probably their anxiety levels have been considerably modified.

5. Following the removal of the animal from the maze, the device is carefully cleaned using a soft detergent and then dried out with the aid of odor-free paper.

6. During off-line analysis, the main parameters scored will include (*see* **Note 9**):

   Number of entries and time spent in the open arms of the maze.

   Number of entries and time spent in the closed arms of the maze.

   Total arm entries.

7. Behavioral evaluation, statistical analysis and results depiction is done as described in steps 8 and 9 of the open-field detailed procedure (see section "Detailed Experimental Procedure") but using video-recordings obtained using the Elevated Plus Maze (*see* **Note 7**).

3.3.3 Shock-Probe Burying Test

This method, also called Defensive Burying Test [51], was based on the pioneering observation of Hudson [52] showing that rats receiving a low intensity electrical shock in their home cages pushed wood shavings from their cage bedding to bury the perceived source of shock. Further work indicated that rodents had actually the innate tendency to bury any object that may harm them including not only electrified probes [53] but many other objects including also their own burrows when they are in danger [51]. In the shock-probe burying test as developed by Pinel and Treit [26, 53, 54], rats while exploring the cage used in this model are shocked by an electrified probe protruding from one of its walls. Shocked rats retreat to the opposite side of the cage and following a variable period of time (latency to bury) they approach the probe and start pushing and spraying bedding material with its snout and forepaws in its direction in an attempt to bury it (burying behavior). During the whole time of the test, rats usually receive additional electrical shocks and engage themselves in further burying episodes. Although shock-probe burying test can also be adapted to mice, their response to the test has been shown to be less robust and more variable [55].

Burying as shown by Pinel and Treit [26, 53, 54] is an important component of the behavioral repertoire of shocked rats as long as proper bedding material does exist and has proved to be rather specific. Thus, rats bury the probe when they have only been shocked by it and not by another source, i.e., a wire grid at the floor of the cage [53]. In addition, in the presence of two probes located on opposite sides of the cage but being only one of them electrified rats only bury the probe where they were shocked indicating that burying response is highly adaptive since it is directed to an aversive source [53]. Furthermore, as noted by Pinel et al. [56] burying behavior seems to be an instinctive defensive response since it is developed during ontogeny in rats deprived of particulate material.

In summary, shock-probe burying test is a low cost and well validated [51] unconditioned fear/anxiety paradigm that does not require long training procedures and in which rats exhibit clear and enduring responses that make easy and straightforward its quantitative evaluation.

Detailed Procedure

1. (Days 1–3). Habituation. Cannulated rats (see Sect. 3.2), kept in a vivarium in single cages to avoid biting of their implanted cannulae, are moved into the behavioral room (see Sect. 3.3) where they stay in racks for 4 days in the recording compartment. During this time, rats are everyday gently handled by the experimenter for 3–5 min simulating the drug-infusion step. During this procedure dummy cannulae are taken out and taken in from the guide cannulae to keep them clear (*see* **Note 3**).

2. (Days 5–9). Rats are gently moved by the experimenter wearing surgeon type latex gloves from the monitoring to the behavioral compartment where they are exposed to the dispositive used in this test but lacking the probe from which animals will be shocked on the day of the experiment. Rats are

allowed to stay in the dispositive for 30 min and then they are returned to their home cages (*see* **Note 10**).

3. (Day 10). Following drug infusion (see Sect. 3.2.1, steps 11–16), rats are taken out from their home cages by the experimenter and introduced into the experimental compartment of the behavioral room, which contains the behavioral dispositive with its probe already electrified to deliver a 0.6 mA shock any time that either rat's spout or forepaw touch it (*see* **Note 11**). Rats are gently placed in the behavioral dispositive facing away from the probe. Identification of the rats is done using a paper card placed under the camera view containing date, experiment name, and rat code (*see* **Note 6**).

4. (Day 10). After the rodent is placed on the experimental dispositive, the experimenter leaves the experimental compartment to watch the rat from the recording compartment. The experimenter activates video camera, starts chronometer, and electrifies the probe. Behavior is video-recorded for 10 min.

5. (Day 10). After the removal of the animal from the behavioral dispositive, the device is carefully cleaned using a soft detergent and then dried out with the aid of odor-free paper. Probe is also cleaned from any material that were deposited on its surface during the test. After feces are removed, bedding material is smoothed before a new rat is tested.

6. Off-line analysis (*see* **Notes 12** and **13**) will include:

Latency to bury (period of time from the first clear shock to the first burying episode).

Cumulative burying time (burying behavior).

Number of electrode approaches (probe avoidance).

Freezing/immobility episodes.

Total time spent in freezing/immobility. Freezing is defined as the absence of body movements with the exception of those needed for breathing.

Number of shocks delivered.

Reaction to the shocks.

7. Behavioral evaluation, statistical analysis and presentation of results is done as described in steps 6–8 of the open-field detailed procedure using those video-recordings obtained during the experiment. For statistical evaluation and presentation of results, follow steps 7 and 8 of the open-field detailed procedure (see section "Detailed Experimental Procedure").

4   Notes

1. In those experiments in which receptor antagonists are used in order to facilitate that agonists and antagonists will reach their targets at nearly the same concentration ratio in which they were injected, and to try that their binding will mostly depend of their receptor affinity we prefer to infuse both compounds simultaneously. Moreover, under these conditions a double microinjection is avoided reducing the risk of tissue damage and preventing that the second compound will follow a different diffusion path due to the increased extracellular volume and pressure elicited by the first injection, particularly if the inter-injection period is short.

2. Since flat stereotaxical coordinates are given in stereotaxic atlas, it will be important to set the incisor bar at 3.1 mm. Make sure that the ear bars are properly placed and that the skull is actually flat before localizing the sites of injection.

3. Habituation has to be carried out by the people who will actually inject the animals. Since rats are very smell sensitive, care has to be taken to avoid different odors, i.e., perfumes, during both the whole habituation period and the experimental procedure. Keep the same context during habituation and injection times.

4. Since both volume and rate of infusion are usually very low it will be fundamental to monitor the displacement of the air bubble interposed between the experimental solution and the flushing fluid to be sure that the drug used is actually been injected into the brain.

5. Open-field test has also been employed in mice using an acrylic box ($48 \times 48 \times 30$ cm) having at its bottom black lines forming $16 \times 12 \times 12$ cm^2. Behavior is evaluated by counting the number of lines crossed during the open-field exploration.

6. Since behavioral evaluation must be done by a person blind to the experimental group to which rats belong, rat code should be written in such a way that it won't be guessed by the experimenter.

7. Video tracking software such as AnyMaze and EthoVision® XT which allows an automatized evaluation of behavior can be purchased from Stoelting Co, Wood Dale IL, USA and Noldus Information Technology BV, Netherlands, respectively.

8. Numerous maze designs do exist which vary essentially in the materials used for its construction. In some mazes, to avoid rats falling from the maze wooden sledges are attached along

the edges of the open arms, see however Fernandez and File [57]. Miniaturized mouse elevated plus maze having a central square of 5×5 cm, 15×5 cm arms and 15 cm wall height with an elevation of 38.5 cm can also be constructed. Fat rats can be successfully tested with a larger maze having dimensions $27 \times 16 \times 23$ cm [58]. Video tracking software such as AnyMaze and EthoVision® XT which allows an automatized evaluation of behavior can also be purchased from Stoelting Co. Wood Dale IL, USA and Noldus Information Technology BV, Netherlands, respectively.

9. For statistical evaluation, both number of entries and time spent in each arm are transformed in the percentage of entries or time spent into a particular arm relative to the total number of entries or total time spent into open + closed arms. i.e., % time spent in open arms = [time spent in open arms/time spent in open + closed arms] $\times$ 100. Number of entries into the open arms of the maze and total time spent in these arms is taken as an anxiety index (the higher the index, the lower the anxiety). Either the total number of arm entries (open + closed) or the number of entries into the closed arms of the maze is taken as a measure of locomotion. An entry into any arm is only counted when the four paws of the rat are placed in that arm.

10. Optimal basal levels of burying are required in this test to have both anxiogenic and anxiolytic responses. Since low levels of fear are associated with a decrease of burying behavior, it is clear that very low burying levels preclude or at least render difficult the study of anxiolytic states. Habituation increases burying behavior and reduces burying variability making easier the study of anxiolytic situations. However since in this test, as in many other fear/anxiety paradigms, i.e., the Elevated Plus Maze, fear is elicited by the test itself, the experimental conditions used in any experiment should be tailored to induce experimentally a reduction in fear.

11. Shock intensities to be used in this model need to be established to get a proper burying behavior. Although shock intensities in the order of 1–2 mA or more have been recommended [26, 53, 54], shock intensities higher than 0.6 mA reduce, in our hands, considerably burying behavior.

12. Although burying behavior is the most sensitive measure of fear in this test, extra measures such as a decrease in latency to bury, avoidance to the electrified probe, or an increase in freezing activity also indicate the presence of a fear state. In addition, in spite that it is not fully proved both a high number of shocks received during the test and an elevated reaction score to them based in a 4-point scale developed by Treit et al. [26] have been considered indicatives of nociception during shock

delivery. On the other hand, since in this test active coping mechanisms are mostly activated evidence for an intact locomotion is required for the correct interpretation of results.

13. Interestingly, in spite that shock-probe burying test as developed by Treit et al. [26, 53, 54] and depicted here is very dependent on active coping mechanisms a somewhat similar test having an electrified probe but devoid of any bedding material has been developed to give information on passive coping mechanisms related to fear and anxiety [51].

Acknowledgments

This work was supported by Vetenskaskapsrådet in year 2015–2017 (No 348-2014-4396) and by grants IN205217 from Dirección General de Asuntos del Personal Académico (DGAPA) de la Universidad Nacional Autónoma de México and 2013-01-220173 from Consejo Nacional de Ciencia y Tecnología (CONACYT), Mexico.

References

1. Fuxe K, Agnati LF (1985) Receptor-receptor interactions in the central nervous system. A new integrative mechanism in synapses. Med Res Rev 5:441–482

2. Zoli M, Agnati LF, Hedlund PB, Li XM, Ferre S, Fuxe K (1993) Receptor-receptor interactions as an integrative mechanism in nerve cells. Mol Neurobiol 7:293–334

3. Perez de la Mora M, Ferre S, Fuxe K (1997) GABA-dopamine receptor-receptor interactions in neostriatal membranes of the rat. Neurochem Res 8:1051–1054

4. Liu F, Wan Q, Pristupa ZB et al (2000) Direct protein-protein coupling enables cross-talk between dopamine D5 an gamma-aminobutyric acid A receptors. Nature 403:274–280

5. Liu F, Chu X, Mao L et al (2006) Modulation of D2R-NR2B interactions in response to cocaine. Neuron 52:897–909

6. Fuxe K, Borroto-Escuela DO (2016) Heteroreceptor complexes and their allosteric receptor-receptor interactions as a novel biological principle for integration of communication in the CNS: targets for drug development. Neuropsychopharmacology 41:380–382

7. Fuxe K, Borroto-Escuela DO, Romero-Fernandez W et al (2014) Moonlighting proteins and protein-protein interactions as neurotherapeutic targets in the G protein-coupled receptor field. Neuropsychopharmacology 39:131–155

8. Fiorentini C, Busi C, Spano P, Missaele C (2010) Dimerization of dopamine D1 and D3 receptors in the regulation of striatal function. Curr Opin Pharmacol 10:87–92

9. Romero-Fernández W, Borroto-Escuela DO, Agnati LF et al (2013) Evidence for the existence of dopamine D2-oxytocin receptor heteromers in the ventral and dorsal striatum with facilitatory receptor-receptor interactions. Mol Psychiatry 18:849–850

10. Fuxe K, Agnati LF, Benfenati F et al (1981) Modulation by cholecystokinins of 3H-spiroperidol binding in rat striatum: evidence for increased affinity and reduction in the number of binding sites. Acta Physiol Scand 113:567–569

11. Fuxe K, Agnati LF, Benfenati F et al (1983) Evidence for the existence of receptor–receptor interactions in the central nervous system. Studies on the regulation of monoamine receptors by neuropeptides. J Neural Transm Suppl 18:165–179

12. Marshal FH, White J, Main M et al (1999) GABA (B) receptors function as heterodimers. Biochem Soc Trans 27:530–535

13. Fuxe K, Ferre S, Zoli M et al (1998) Integrated events in central dopamine transmission as analyzed at multiple levels. Evidence for intramembrane adenosine A2A/dopamine D2 and adenosine A1/dopamine

D1 receptor interactions in the basal ganglia. Brain Res Brain Res Rev 26:258–273

14. Ferré S, Karcz-Kubicha M, Hope BT et al (2002) Synergistic interaction between adenosine A2A and glutamate mGlu5 receptors: implications for striatal neuronal function. Proc Natl Acad Sci U S A 99:11940–11945

15. Nishi A, Liu F, Matsuyama S et al (2003) Metabotropic mGlu5 receptors regulate adenosine A2A receptor signaling. Proc Natl Acad Sci U S A 100:1322–1327

16. Navarro G, Moreno E, Bonaventura J et al (2013) Cocaine inhibits dopamine D2 receptor signaling via sigma-1-D2 receptor heteromers. PLoS One 8(4):e61245. https://doi.org/10.1371/journal.phone.0061245.print2013

17. Perez de la Mora M, Pérez Carrera D, Crespo-Ramírez M et al (2016) Signaling in dopamine D2 receptor-oxytocin receptor heterocomplexes and its relevance for the anxiolytic effects of dopamine and oxytocin interactions in the amygdala of the rat. Biochim Biophys Acta 1862:2075–2085

18. Van Rijn RM, Whistler JL, Waldhoer M (2010) Opioid-receptor-heteromer-specific trafficking and pharmacology. Curr Opin Pharmacol 10:73–79

19. Kabli N, Nguyen T, Balboni G et al (2014) Antidepressant-like and anxiolytic-like effects following activation of the µ-δ opioid receptor heteromer in the nucleus accumbens. Mol Psychiatry 19:986–994

20. Shen MY, Perreault MN, Bambico FR et al (2015a) Rapid anti-depressant and anxiolytic actions following dopamine D1-D2 receptor heteromer inactivation. Eur Neuropsychopharmacol 25:2437–2448

21. Shen MY, Perreault ML, Fan T et al (2015) The dopamine D1-D2 receptor heteromer exerts a tonic inhibitory effect on the expression of amphetamine-induced locomotor sensitization. Pharmacol Biochem Behav 128:33–40

22. Borroto-Escuela DO, Brito I, Romero-Fernandez W et al (2014) The G protein-coupled receptor heterodimer network (GPCR-HetNet) and its hub components. Int J Mol Sci 15:8570–8590

23. Tena-Campos N, Ramon E, Lupala CS et al (2016) Zinc is involved in depression by modulating-protein coupled receptor heteromerization. Mol Neurobiol 53:2003–2015

24. Perreault ML, Hasbi A, Alijaniaram M et al (2010) The dopamine D1-D2 heteromer localizes in dynorphin/enkephalin neurons: increased high affinity state following amphet-

amine and schizophrenia. J Biol Chem 285:36625–36634

25. Pellow S, Chopin P, File SE et al (1985) Validation of open:closed arm entries in an elevated plus-maze as a measure of anxiety in the rat. J Neurosci Methods 14:149–167

26. Treit D, Menard J, Pesold C (1994) The shock-probe burying test. Neurosci Protoc Module 3:9–17

27. Hall CS (1934) Emotional behavior in the rat. 1. Defecation and urination as measures of individual differences in emotionality. J Comp Psychol 18:382–403

28. Prut L, Belzung C (2003) The open field as a paradigm to measure the effects of drugs on anxiety-like behaviors: a review. Eur J Pharmacol 463:3–33

29. Barnett SH (2007) The rat: a study in behavior. Aldline Publishing Co, Chicago, pp 31–32

30. Geyer MA (1990) Approaches to the characterization of drug effects on locomotor activity in rodents. In: Adler MW, Cowan A (eds) Modern methods in pharmacology, vol 6. Testing and evaluation of drugs of abuse. Wiley-Liss, New York, pp 81–100

31. Archer J (1973) Tests for emotionality in rats and mice: a review. Anim Behav 21:205–235

32. Dawson GR, Tricklebank MD (1995) Use of the elevated plus maze in the search for novel anxiolytic agents. Trends Pharmacol Sci 16:33–36

33. Hogg S (1997) A review of the validity and variability of the elevated plus-maze as an animal model of anxiety. Pharmacol Biochem Behav 54:21–30

34. Rodgers RJ (1997) Animal models of "anxiety": where next? Behav Pharmacol 8:477–496

35. Rodgers RJ, Dalvi A (1997) Anxiety, defence and the elevated plus-maze. Neurosci Biobehav Rev 21:801–810

36. Pawlak CR, Karrenbauer BD, Schneider P et al (2012) The elevated plus-maze test: differential psychopharmacology of anxiety-related behavior. Emotion Rev 4:98–108

37. Montgomery KC (1955) The relation between fear induced by novel stimulation and exploratory behavior. J Comp Physiol Psychol 48:254–260

38. Lister RG (1987) The use of a plus-maze to measure anxiety in the mouse. Psychopharmacology 92:180–185

39. Rägo L, Kiivet RA, Harro J et al (1988) Behavioral differences in an elevated plus-maze: correlation between anxiety and decreased number of GABA and benzodiaze-

pine receptors in mouse cerebral cortex. Naunyn Schmiedeberg's Arch Pharmacol 337:675–678

40. Perez de la Mora M, Hernández-Gómez AM, Fuxe K et al (2007) Role of the amygdaloid cholecystokinin (CCK)/gastrin-2 receptors and terminal networks in the modulation of anxiety in the rat. Effects of CCK-4 and CCK-8S on anxiety-like behavior and [³H] GABA release. Eur J Neurosci 26:3614–3630

41. Perez de la Mora M, Cárdenas-Cachón L, Fuxe K et al (2005) Anxiolytic effects of intra-amygdaloid injection of the D1 antagonist SCH23390 in the rat. Neurosci Lett 377:101–105

42. Perez de la Mora M, Gallegos-Cari A, Fuxe K et al (2012) Distribution of dopamine D(2)-like receptors in the rat amygdala and their role in the modulation of unconditioned fear and anxiety. Neuroscience 201:252–266

43. Simon P, Dupuis R, Costentin J (1994) Thigmotaxis as an index of anxiety in mice. Influence of dopaminergic transmission. Behav Brain Res 61:59–64

44. Treit D, Menard J, Royan C (1993) Anxiogenic stimuli in the elevated plus-maze. Pharmacol Biochem Behav 44:463–469

45. Falter U, Gower AJ, Gobert J (1992) Resistance on baseline activity in the elevated plus-maze to exogenous influence. Behav Pharmacol 2:123–128

46. Rodgers RJ, Cole JC (1994) The elevated plus-maze: pharmacology, methodology and ethology. In: Cooper SJ, Hendrie CA (eds) Ethology and psychopharmacology. Wiley, Chichester, pp 9–14

47. Griebel G, Moreau J-L, Jenk F et al (1993) Some critical determinants of the behavior of rats in the elevated plus-maze. Behav Processes 29:37–48

48. Rebolledo-Solleiro D, Crespo-Ramírez M, Roldán-Roldán G et al (2013) Role of thirst and visual barriers in the differential behavior displayed by streptozotocin-treated rats in the elevated plus-maze and the open-field test. Physiol Behav 120:130–135

49. File SE (1990) One trial tolerance to the anxiolytic effect of chlordiazepoxide in the plus-maze. Psychopharmacology 100:281–282

50. Rodgers SJ, Shepherd JK (1993) Influence of prior maze experience on the behavior and response to diazepam in the elevated plus-maze and light/dark tests of anxiety in mice. Psychopharmacology 113:327–342

51. De Boer SF, Koolhaas JM (2003) Defensive burying in rodents: ethology, neurobiology and psychopharmacology. Eur J Pharmacol 463:145–161

52. Hudson BB (1950) One-trial learning in the domestic rat. Genet Psychol Monogr 41:99–145

53. Pinel JPJ, Treit D (1978) Burying as defense response in rats. J Comp Physiol Psychol 92:708–712

54. Treit D, Pinel JPJ, Fibiger HC (1981) Conditioned defensive burying: A new paradigm for the study of anxiolytic agents. Pharmacol Biochem Behav 15:619–626

55. Treit D, Terlecki LJ, Pinel JPJ (1980) Conditioned defensive burying: organismic variables. Bull Psychonom Soc 16:451–454

56. Pinel JPJ, Symons LA, Christensens BK et al (1989) Development of defensive burying in *Rattus norvegicus*: experience and defensive responses. J Comp Physiol Psychol 103:359–365

57. Fernandez C, File SE (1996) The influence of open arm ledges and maze experience in the elevated plus-maze. Pharmacol Biochem Behav 54:31–40

58. Rebolledo-Solleiro D, Roldán-Roldán G, Díaz D et al (2017) Increased anxiety-like behavior is associated with metabolic syndrome in non-stressed rats. PLoS One 12(5):e0176554. https://doi.org/10.1371/journal.pone.0176554

Chapter 9

Small Interference RNA Knockdown Rats in Behavioral Functions: GALR1/GALR2 Heteroreceptor in Anxiety and Depression-Like Behavior

Antonio Flores-Burgess, Carmelo Millón, Belén Gago, José Angel Narváez, Kjell Fuxe, and Zaida Díaz-Cabiale

Abstract

The gene silencing by RNA interference (RNAi) represents a new approach for inhibition of gene expression in cell culture and in vivo and it is a potential clinical tool for the treatment of illnesses, including neuropsychiatric disorders. In this chapter, we described an animal model using this technology to study the role of Galanin (1–15) [GAL(1–15)] through GALR1-GALR2 heteroreceptor complexes in anxiety and depression-like behavior. We have generated siRNA GALR1 and siRNA GALR2 rats to analyze the involvement of GALR1-GALR2 heteroreceptor complexes in the behavioral effects induced by GAL(1–15). In this chapter, we provide a protocol for using these animal models to investigate the role of heterodimers in vivo when a behavioral test is needed.

Key words siRNA, Galanin (1–15), GALR1-GAL2 heteroreceptor

1 Introduction

1.1 Small Interference RNA Technology

The gene silencing by RNA interference (RNAi) represents a new approach for inhibition of gene expression in cell culture and in vivo. The RNAi gene silencing strategy is a process by which a specific mRNA is targeted for degradation as a means of inhibiting the synthesis of the encoded protein. The RNAi response is activated by the presence of double-stranded RNA (dsRNA) in cells. The dsRNA is degraded into short double-stranded fragments known as small Interference RNA (siRNA). The siRNA generated enters the RNA-induced silencing complex (RISC), which becomes activated upon guide (antisense) strand selection. The incorporated strand acts as a guide for the activated RISC complex to selectively degrade the complementary mRNA [1, 2] (Fig. 1).

Antonio Flores-Burgess and Carmelo Millón contributed equally to the manuscript.

Kjell Fuxe and Dasiel O. Borroto-Escuela (eds.), *Receptor-Receptor Interactions in the Central Nervous System*, Neuromethods, vol. 140, https://doi.org/10.1007/978-1-4939-8576-0_9, © Springer Science+Business Media, LLC, part of Springer Nature 2018

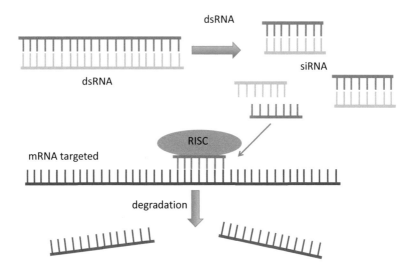

Fig. 1 Schematic image of the gene silencing strategy by RNAi. The RISC complex uses siRNA as templates to remove a specific mRNA by complementarity. These siRNAs can be obtained from the degradation of a dsRNA or from an exogenous source

The mammalian cells lack the ability to cleave dsRNA into siRNA [2]. Unfortunately, introduction of long dsRNA into mammalian cells leads to the initiation of the antiviral immune response and global protein expression shutdown [3]. Thus, the applications with siRNA must be made through the delivery of siRNA produced synthetically ex vivo, leading to highly specific and potent gene silencing in vivo [1].

Naked siRNAs do not freely cross the cell membrane, and thus delivery systems are required to access to its intracellular sites of action. Moreover, they are relatively unstable because they are degraded rapidly by endo- and exonucleases [4]. To overcome such problems, chemical modifications to promote metabolic stability and improve target cell penetration have been introduced (see reviews [1, 5]).

In this chapter, we will focus in a new type of naked siRNA chemically modified to allow for delivery without requiring transfection reagents developed by Dharmacon (Accell siRNA); this Accell siRNA results in robust silencing of the selected genes and knockdown of the associated proteins [6]. Habitually, Accell siRNA has been used exclusively in in vitro cell culture studies [7, 8]; however, in recent reports Accell siRNA was used in animals by single intracerebroventricular (icv) injections [6].

The siRNA technology has been applied in experimental investigations as a potential clinical tool for the treatment of illnesses, including neuropsychiatric disorders [1, 6, 9]. We have used this technology to study in animals models the role of Galanin (1–15) [GAL(1–15)] in anxiety and depression-like behavior. We have generated siRNA GALR1 and siRNA GALR2 in rats to analyze the involvement of GALR1-GALR2 heteroreceptor complexes in the behavioral effects induced by GAL(1–15).

1.2 Galanin and Galanin N-Terminal Fragment (1–15)

Galanin (GAL) is a 29 amino acids neuropeptide widely distributed in the central nervous system (CNS) [10]. Three GAL receptor (GALR) subtypes, GALR1–3, with high affinity for GAL have been cloned [11, 12]. GAL participates in many functions as the stimulation food intake, the central cardiovascular control and has an inhibitory role of learning and memory among others [12–14]. GAL participates in mood regulation and depression [15–18]. The activation of GALR1 and GALR3 results in a depression-like behavior while stimulation of GALR2 leads to anti-depressant-like effects [19–21]. Moreover, GAL modulates serotonin receptor 5-HT1A function at autoreceptor and postsynaptic level in the brain [22–24].

Not only GAL but also the N-terminal fragments like GAL(1–15) are active in the CNS [14, 25, 26]. Recently, we described that GAL(1–15) induces strong depression-related and anxiogenic-like effects in rats and these effects were significantly stronger than the ones induced by GAL [27]. GAL(1–15) is also able to enhance the antidepressant effects induced by the 5-HT1AR agonist 8-OH-DPAT in the forced swimming test (FST) [28], effect that was again significantly stronger than the ones induced by GAL. With the proximity ligation assay we obtained indications for the existence of GALR1-GALR2 heteroreceptor complexes [29–31] in the dorsal hippocampus and especially in the dorsal raphe nerve cells [27], areas rich in GAL fragment binding sites.

In this chapter, we will describe a selective siRNA GALR1 and siRNA GALR2 animals developed by our group, to study the behavioral function of heterodimer GALR1/GALR2. The ability of these models to perform different behavioral test has enabled us to investigate in detail the role of the heterodimers in vivo [27, 32, 33].

2 Materials

2.1 Accell siRNA System

We used a commercial reagent, the Accell Smartpool siRNA to reduce the expression of target proteins by reducing the expression of messenger RNA through the mechanism of gene silencing interference by small RNA (siRNA). The Accell Smartpool siRNA consists in four sequences of 20 nucleotides each complementary to the messenger RNA of the protein that we want to downregulate. In the methods described below, we used the siRNA system with complementary sequences for the GALR1 and GALR2 receptors.

In addition we used a control siRNA, which contains a pool of four sequences that do not have complementarity with any messenger RNA (non-targeting siRNA) to evaluate the possible non-specific interference that these reagents may have in vivo.

Following the manufacturer's instructions, Accell Smartpool siRNAs dry pellets are resuspended in 10× buffer (5× Buffer: 300 mM KCl, 30 mM HEPES pH 7.5, 1 mM $MgCl_2$) free of RNase. Then aliquoted and stored at −80 °C in small volumes until use. For the administration of the siRNAs in animals, on the same day, we diluted the Accell Smartpool aliquots in the Accell siRNA Delivery Media culture medium so that the final concentration is 1 μg/μL, being maintained at 4 °C.

2.2 Animals

Male Sprague-Dawley rats obtained from CRIFFA, Barcelona (200–250 g) were maintained under a 12 h dark/light cycle in temperature- and humidity-controlled conditions (22 ± 2 °C, 55–60%). The animals had free access to food pellets and tap water. Behavioral tests were performed during the light phase of the diurnal cycle. All experimental procedures were approved by the Institutional Animal Ethics Committee of the University of Málaga, Spain.

2.2.1 Intracerebroventricular Injections

This protocol has been used previously [34]. Briefly, the rats were anesthetized intraperitoneally with Equitesin (3.3 mL/kg body weight) and stereotaxically implanted with a unilateral chronic 22-gauge stainless-steel guide cannula into the right lateral cerebral ventricle using the following coordinates: 1.4 mm lateral and 1 mm posterior to the bregma, and 3.6 mm below the surface of the skull (Paxinos G, 1986). After surgery, animals were individually housed and allowed a recovery period of 7 days. The injections in the lateral ventricle were performed using a 26-gauge stainless-steel injection cannula connected via a PE-10 tubing to a Hamilton syringe. The total volume was 5 μL per injection and the infusion time was 1 min.

2.3 Real-Time Quantitative PCR

RNA extraction from the different samples was obtained by using the RNeasy Lipid Tissue Mini Kit (Qiagen) purification kit following the manufacturer's instructions with the samples always at 4 °C to avoid RNA degradation. The purity and quantity of the purified RNA was determined with a Nanodrop spectrophotometer (Thermo Scientific) by measuring the absorbance ratios at 260/280 nm and 260/230 nm. The RNA was aliquoted and stored at −80 ° C until use. The cDNA was obtained using the Reverse Transcriptase Core kit (Eurogentec) following the manufacturer's instructions, obtaining a final concentration of 20 ng/μL.

All polymerase chain reaction (PCR) were conducted in triplicate using Power SYBR Green PCR Master Mix (Applied Biosystems) in a 7500 Real-Time PCR system (Applied Biosystems). The amplification protocol was as follows: 2 min at 50 °C and 10 min at 95 °C. Then 40 cycles of 15 s at 95 °C and 1 min at 60 °C. At the end of the 40 cycles, a melting curve was obtained. We also performed negative controls without cDNA and controls with RNA samples to rule out genomic DNA contamination.

The primer sequences used in this study are: GAPDH-Forward: 5′-GCTCTCTGCTCCTCCCTGTTC, GAPDH-Reverse: 5′-GAGGCTGGCACTGCACAA, GALR2-Forward: 5′-AACAGGAATCCACAGACC, GALR2-Reverse: 5′-CCCTTTGGTCCTTTAACAAG, GALR1-Forward: 5′-AAAACTGGACAAAACTTAGCC, GALR1-Reverse: 5′-GGATACCTTTGTCTTTGCTC. The data were analyzed using the Comparative Ct method ($\Delta\Delta$Ct) and normalized to measures of glyceraldehyde-3-phosphate dehydrogenase (GAPDH) mRNA.

The efficiencies of the primers were determined by calibration lines under the same conditions described above: GAPDH 95%, GALR1 90%, GALR2 93%.

2.4 Behavioral Assessment

2.4.1 Forced Swimming Test

Animals were individually placed in a vertical glass cylinder (50 cm height, 20 cm diameter) containing water (25 °C) to a height of 30 cm. Two swimming sessions were conducted: a 15 min pre-test followed 24 h later by a 5 min test [27, 33]. The total duration of immobility behavior and climbing were recorded during the second, 5 min. Immobility was defined as floating passively in an upright position in water, with only small movements necessary to keep the head above the water surface. Climbing was defined as forepaw movements directed toward the walls of the cylinder.

2.4.2 Open Field Test

Rats were individually placed and allowed to freely explore, recording the behavior over a 5 min period by a ceiling-mounted video camera. Activity was analyzed using the video tracking software Ethovision XT (Noldus, SL). After each trial, all surfaces were cleaned with a paper towel adding 70% ethanol solution. For the open field ($100 \times 100 \times 50$ cm), total time spent in and entries into the inner square were recorded.

2.5 Immunohisto-chemical Analysis

The procedures have been previously used [27, 33]. The knockdown rats were perfused transcardially with 0.1 M phosphate buffered saline (PBS), pH 7.4, followed by 4% paraformaldehyde (wt/vol) in 0.1 PBS, pH 7.4. The brains were then dissected and postfixed in the same solution overnight at 4 °C, then placed into cryoprotectant (30% sucrose in 0.1 M PBS) and frozen in dry ice. Serial coronal free-floating sections (30 μm thick) were collected in cryostat (Micron) and stored in Hoffman buffer at −20 °C until GALR1 or GALR2 immunohistochemistry. The sections were processed free-floating. Endogenous peroxidase activity was removed by incubating the sections and permeabilization with 0.3% H_2O_2 for 20 min. After blocking with 1% normal goat serum (Sigma, Spain) with 0.3% Triton X-100 (10 min, room temperature), the sections were incubated with primary antibody goat anti-GALR1 (Santa Cruz, 1/250) or rabbit anti-GALR2 (Alomone Lab, 1/250). Overnight incubation at 4 °C was performed in 0.01 M PBS containing 0.3% Triton X-100. The sections were washed three times in PBS and incubated with biotinylated specific secondary antibodies (1:200; Vector Labs Inc., Burlingame, CA) for 1 h at room temperature. The immunostaining was performed according to the ABC method using the Vectastain kit (Vector, Burlingame, CA). The chromogen used was 0.03% 3-30-diaminobenzidine tetrahydrochloride (DAB) (Sigma, Spain) and 0.03% fresh H_2O_2 in 0.1 M Tris-HCI (pH 7.6). After mounting the sections on gelatin-chromalum coated slides, the sections were dehydrated and coverslipped with DPX (Panreac, Barcelona, Spain).

Analysis of GALR1 and GALR2 IR were performed under light microscopy (Nikon Optiphot-2) in the neuro-anatomical area of interest. Four sections of two animals per day were examined and captured by video camera (Olympus UC30) linked to a PC computer and GALR1 and GALR2 IR were quantified by Optical Density using the computer software ImageJ (NIH, USA).

2.6 Double Immunofluorescence

The procedures have been previously used [27]. The brain tissue was obtained and treated as explained above. To double immunofluorescence GALR1/GALR2 an initial incubation with blocking (2% Albumin Human) and permeabilization (0.3% triton X100 in PBS) were done during 60 min. Primary antibody rabbit anti-GALR2 (Alomone Lab, 1/250) was incubated for 24 h at 4 °C and detected with the red secondary antibody mouse anti-rabbit DyLight 549 (Jackson ImmunoResearch Laboratories, 1/100). Goat anti-GalR1 (Santa Cruz Biotechnology INC, EEUU, 1:250) was incubated in a similar manner as described above and detected with the secondary antibody rabbit biotinylated anti-goat (Vector Labs Inc., Burlingame, CA) and Alexa Fluor 488-conjugated Streptavidin (Jackson Laboratories ImmunoResearch, 1:1000). Sections were mounted on slides with fluorescent mounting medium (Dako) and visualized by using a spectral confocal microscope Leica SP5 (Leica). The double immunolabeling was performed in tissue of knockdown rats of GALR2 (Fig. 4).

2.7 Proximity Ligation In Situ Assay

Proximity ligation assay (PLA) and quantification were carried out as described previously using a Duolink in situ PLA detection kit [27, 28]. The brain tissue was obtained by the similar way as described in the above sections. The primary antibodies of different species directed to the GALR1 (goat polyclonal, Santa Cruz, 1/250) and GALR2 (rabbit polyclonal, Alomone Lab, 1/250) were used. Sections were mounted on slides with fluorescent mounting medium containing 4′,6-diamino-2-phenylindole (DAPI; Sigma-Aldrich), staining the nuclei with blue color. The in situ PLA-positive signals were visualized using a Leica SP5 confocal microscope and the Duolink Image Tool software. Quantification was made in 10–20 cells per photo and performed eight photos per brain zone. Thus, each brain nuclei is represented by the mean of 80–160 cells.

3 Methods

3.1 Generation of siRNA GALR1 and siRNA GALR2 Rats by siRNA Accell Smartpool Injection

For knockdown rats, animals were operated with the same procedure as described in Sect. 2. When the cannula was implanted, rats were intracerebroventricularly injected with 5 µg (0.35 nmol) of Accell Smartpool siRNA GALR2 or siRNA GALR1 (Dharmacon) or 5 µL the vehicle (Accell siRNA Delivery Media). Also, some animals were injected with siRNA-Control (Accell Non-targeting pool, Dharmacon) to verify that the Accell system has no effect on animals.

The injection was performed during 1 min period with a Hamilton syringe of 10 µL volume attached to a polypropylene tube (P10) fitted with a stainless-steel injection cannula (26G) (C313I, Plastic One) with a length of 0.3 mm greater than the cannula guide to avoid back diffusion of the substances.

3.2 Time-Course Curve in siRNA GALR1 and siRNA GALR2 Animals

We analyzed the mRNA and protein expression of GALR1 and GALR2 by RT-PCR and immunohistochemical analysis to determine the time-course curve of GALR1 and GALR2 after siRNA administration.

Animals injected with siRNA GALR2 or siRNA GALR1 were killed by decapitation 4, 6, 8, and 14 days after injection to measure the reduction in messenger RNA levels. Four animals injected with vehicle, followed the same process as controls. Hippocampus of all animals were quickly extracted and stored at −80 ° C until use. The isolation of the RNA from the dorsal hippocampus and the RT-PCR were conducted as described under the material section.

A strong reduction ($p < 0.05$) of mRNA GALR2 expression was induced by siRNA GALR2, 4 and 6 days after the injection (Fig. 2a) [27]. The injection of siRNA GALR2 lacked effect on mRNA GALR1 expression (Fig. 2a) [32].

The same pattern of response in mRNA expression was shown after the administration of siRNA GALR1. A single injection of siRNA GALR1 induced a reduction of mRNA GALR1 expression 4 days after the injection and was maintained throughout the day 8 ($p < 0.01$) [27] (Fig. 3a). The injection of siRNA GALR1 lacked effect in mRNA GALR2 expression (Fig. 3a) [32].

We have also conducted a time-course curve of GALR2 or GALR1 protein expression. Animals injected with siRNA GALR2 or siRNA GALR1 were killed by decapitation 4, 6, 8, 10, 12, and 14 days after injection to measure the protein expression levels of the receptors. The immunohistochemical analysis was conducted as described in the material section.

The single injection of siRNA GALR2 produced the strongest decrease in GALR2 expression 8 days after the injection by 35% in the CA1 of the dorsal hippocampus and by 50% in the piriform cortex as shown in Fig. 2b, c [27].

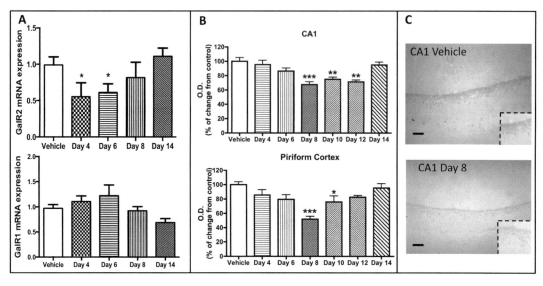

Fig. 2 (a) Expression levels of GALR1 and GALR2 mRNA at different days obtained after the icv injection of Accell Smartpool siRNA for GALR2. The qPCR results were normalized to the expression levels of GAPDH and expressed as arbitrary units *$p < 0.05$ vs Vehicle, Student's t-test versus respective control. **(b)** Time course of GALR2 protein expression following a single injection of siRNA GALR2 or vehicle into the rat brain in hippocampus (CA1) and piriform cortex. GALR2 immunoreactivity was determined after immunohistochemistry by measuring the Optical Density (O.D.). Vertical bars represent mean ± SEM of percentage of change from respective controls. *$p < 0.05$; **$p < 0.01$ and ***$p < 0.001$ versus vehicle group according to one-way ANOVA followed by Newman–Keuls Multiple Comparison Test. **(c)** Representative photographs of GALR2 immunohistochemistry. Magnifications of selected portions of these images are shown in the insets. Scale bar, 50 μm

In Fig. 3b, c it is shown that siRNA GALR1 also reduced by 30% the expression of GALR1 8 days after of its administration in CA1 and dentate gyrus ($p < 0.05$) of the dorsal hippocampus [33].

The time-course curve indicated a maximal reduction of GALR1 and GALR2 protein expression 8 days after the injection, and this was the time point selected for the behavioral tests.

3.3 siRNA GALR1 and siRNA GALR2 Rats in Behavioral Test Related to Depression and Anxiety

Once determined the time-course curve of receptor expression in GALR1 and GALR2 knockdown rats, we have used the following protocol for the behavioral tests (Fig. 4a).

Chronic cannula and icv siRNA administration was performed as we described in the material section. After the surgery, the rats have a post-surgery period for 7 days to allow the siRNA to reduce the GALR expression in the brain. The seventh day the animals performed the first behavioral test, the open field test (OFT), 15 min after the GAL(1–15) administration. Previously, the animals had two sessions of Handling by the researcher to minimize the stress of the procedures. Finally, the ninth day after siRNA injection, the rats perform the forced swimming test (FST) 15 min after GAL(1–15) injection.

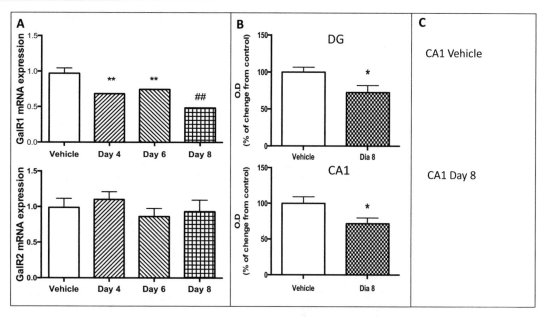

Fig. 3 (**a**) Expression levels of GALR1 and GALR2 mRNA at different days obtained after the icv injection of Accell Smartpool siRNA for GALR1. The qPCR results were normalized to the expression levels of GAPDH and expressed as arbitrary units **p < 0.01 vs Control group; ##p < 0.01 vs rest of the groups according to one-way ANOVA followed by Newman–Keuls Multiple Comparison Test. (**b**) GALR1 protein expression 8 days after a single icv injection of siRNA GALR1 or vehicle into the rat brain. Data are represented as mean ± SEM of the percentages with respect to the control value (100%) of optic density (O.D) *p < 0.05 according to Student's unpaired t-test. (**c**) Representative photographs of GALR1 immunohistochemistry

To verify the Accell system employed, we generated siRNA-Control rats (Accell Non-targeting pool) and vehicle rats (Accell siRNA Delivery Media). We did not observe any difference between these groups compared to artificial cerebrospinal fluid normal rats in any behavioral test (Table 1).

Downregulation of GALR1 or GALR2 did not affect any parameter of the behavioral tests neither in the FST nor in the OFT (Fig. 4b–e). However, the decrease in GALR1 or GALR2 was sufficient to block the effect of GAL(1–15) in the siRNA GALR1 or siRNA GALR2 rats (Fig. 4b–e; [27]). Thus, GAL(1–15) at dose of 3 nmol lacked effect on immobility, climbing, and swimming time in the FST in both knockdown rats (Fig. 4c, e). In the OFT, the same effect was observed in the number of entries and in the time spent into the central square (Fig. 4b, d; [27]).

The use of knockdown animals for GALR1 and GALR2 demonstrates that both GALR1 and GALR2 are involved in the GAL(1–15) effects in behavioral test, confirming that GAL(1–15) effects depend on the existence of GALR1/GALR2 heteroreceptor complexes to exert its strong depression- and anxiogenic-like effects.

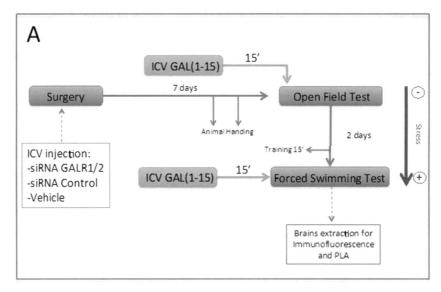

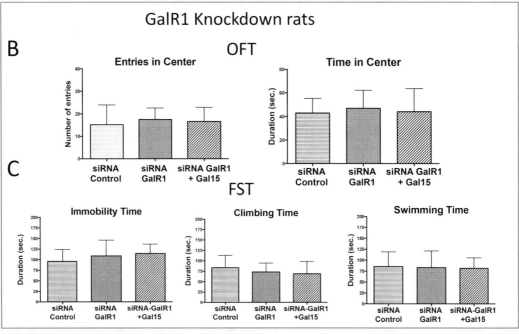

Fig. 4 Analysis of Galanin (1–15) [GAL(1–15)] in GALR1 or GALR2 knockdown rats in the open field test (OFT) and forced swimming test (FST). (**a**) Experimental design of use of the GALR1 or GALR2 knockdown animals in behavioral tests. (**b–e**) Behavioral effects of an effective dose of GAL(1–15) (3 nmol) in the GALR1 (**b, c**) or GALR2 (**d, e**) knockdown in OFT and FST. Treatments were injected intracerebroventricularly 15 min prior to the tests. siRNA control–injected rats were used as the control group. Data represents mean ± standard error of the mean of the time in the center and entries in the open field test (**b–d**) and immobility, climbing, and swimming time in the FST (**c–e**) during the 5 min test period (6–14 rats per group). No differences were found according to one-way analyses of variance

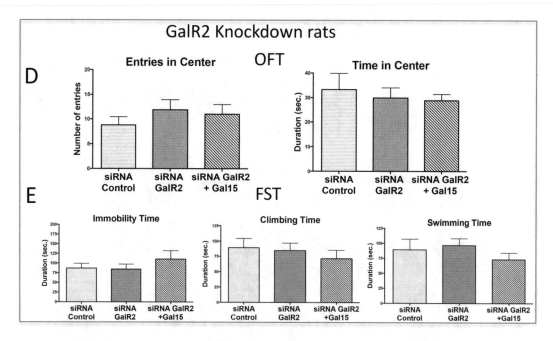

Fig. 4 (continued)

Table 1

Behavioural effects of the administration of aCSF (Control), Delivery Media (DM; Vehicle) or siRNA non-targeting (Control siRNA) in the forced swimming test

	Control (aCSF)	Vehicle (DM)	Control siRNA
Immmobility time (s)	123 ± 9	102.3 ± 49	86.8 ± 12
Climbing time (s)	110.8 ± 9	67.3 ± 22	88.9 ± 15
Swimming time (s)	50.5 ± 8	95.2 ± 40	89 ± 18

Treatments were injected 15 min before the test. Data represents mean ± SEM of immobility, climbing and swimming time in FST during the 5 min test period ($n = 4$–6 rats per group). No differences were found according to one-way ANOVA followed by Newman–Keuls Multiple Comparison Test

3.4 Neurochemical Studies Validate the Existence of GALR1-GALR2 Heteroreceptor Complexes in siRNA GALR1 and siRNA GALR2 Animals

We have previously shown that GALR1 and GALR2 co-localized in the dorsal hippocampus as well as in the neurons of the dorsal raphe of naïve rats. Moreover, the PLA-positive signaling confirmed that GALR1 and GALR2 are in close proximity indicating the formation of GALR1/GALR2 heteroreceptor complexes. The quantification of PLA demonstrated that the highest number of PLA clusters were present in the nerve cells of the dorsal raphe [27, 32].

We have analyzed in the brain tissue of the siRNA GALR2 rats with in situ PLA and GALR1-GALR2 double immunofluorescence the existence of GALR1/GALR2 heteroreceptor complexes.

In the siRNA GALR2 treated animals, PLA-positive red clusters were still observed in the dorsal hippocampus and dorsal raphe (Fig. 5; [27]). However, the number of GALR1/GALR2 complex

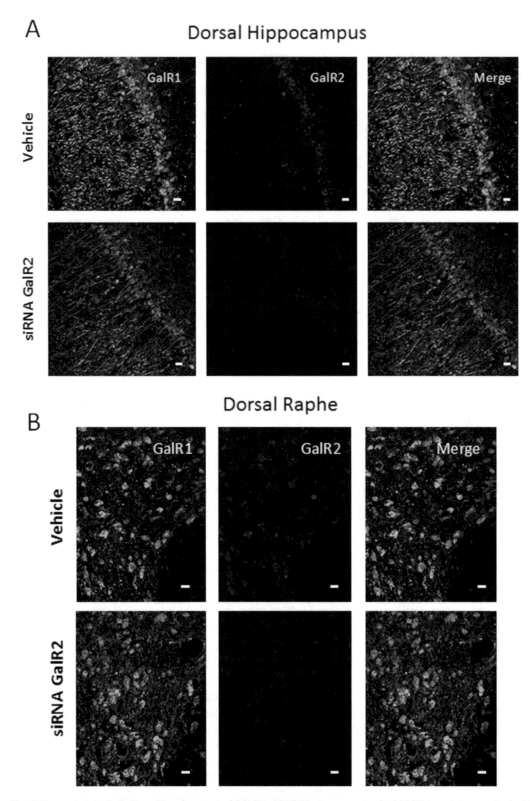

Fig. 5 Neurochemical studies of involvement of GALR1-GALR2 heteroreceptor in GALR2 knockdown rats. (**a, b**) Representative laser scanning confocal micrographs illustrating the decrease of GALR2 immunoreactivity in GALR2 knockdown rat at dorsal hippocampus and dorsal raphe. Double-immunolabeled for GALR1 (green) and

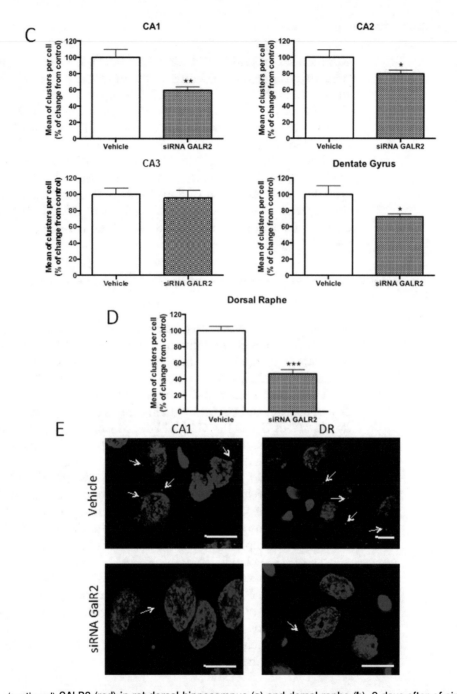

Fig. 5 (continued) GALR2 (red) in rat dorsal hippocampus (**a**) and dorsal raphe (**b**), 9 days after of single icv injection of siRNA GALR2 or vehicle. Escale bar, 20 μm. (**c–e**) Detection of close proximity between GALR1 and GALR2 in the dorsal hippocampus and dorsal raphe in GALR2 knockdown rats by in situ proximity ligation assay (PLA). Quantification of red clusters/6-diamidino-2-phenylindole-positive nuclei in the dorsal hippocampus (**c**) and dorsal raphe (**d**) 9 days after a single intracerebroventricular injection of siRNA GALR2 or vehicle. Vertical bars represent mean ± standard error of the mean of percentage of change from respective controls. *$p < 0.05$, **$p < 0.01$, and ***$p < 0.001$ versus respective control according to student's *t*-test. (**e**) Representative photographs for decrease of detection of cytoplasmatic GALR1 and GALR2 via in situ PLA (seen as red clusters indicated by arrows) in the dorsal hippocampus (CA1) and dorsal raphe in GALR2 knockdown rats. Nuclei are shown in blue (6-diamidino-2-phenylindole). Scale bar, 10 μm. *CA1* Ammon's horn 1, *CA2* Ammon's horn 2, *CA3* Ammon's horn 3, *DR* dorsal raphe nucleus

was reduced around 40% in CA1 ($p < 0.01$), CA2 ($p < 0.05$), and dentate gyrus ($p < 0.05$) in the dorsal hippocampus and by 60% in the dorsal raphe nucleus ($p < 0.001$) compared with vehicle group (Fig. 5c, d; [27]). In agreement with these results, in the siRNA GALR2 animals the colocalization of GALR1 and GALR2 was reduced in the both areas (Fig. 5a, b; [27]).

These results strongly indicate that the PLA signals obtained are specific and represent the GALR1 and GALR2 heteroreceptor complex. This reduction of the PLA signal was sufficient to block the depression- and anxiogenic-like effects of GAL(1–15), linking them to its actions at the GALR1/GALR2 heteroreceptor complex [27, 32].

4   Conclusions

In this chapter, we analyzed selective siRNA GALR1 or GALR2 animals developed in our laboratory to study GALR1-GALR2 heteroreceptor complexes in the anxiety and depression-like effects induced by GAL(1–15).

Using the Accell siRNA system we described a maximal reduction of GALR1 and GALR2 protein expression 8 days after the injection in siRNA GALR1 or GALR2 knockdown animals; this reduction allowed us to use this time point in the behavioral tests. Moreover in these animals the quantification of PLA showed a maximal reduction of around 40% in the CA1 area and 60% in the dorsal raphe compared with the vehicle group. This reduction of the PLA signal was sufficient to block the depression- and anxiogenic-like effects of GAL(1–15) linking them to its actions at the GALR1-GALR2 heteroreceptor complex. Moreover, we also blocked the behavioral effects of GAL(1–15) in siRNA GALR1 and siRNA GALR2 knockdown rats, confirming that GAL(1–15) effects depend on the existence of GALR1-GALR2 heteroreceptor complexes.

The ability of these models to perform different behavioral test with the possibility of administration or coadministration of different treatments represents an ideal system to investigate the role of heterodimers in vivo.

Acknowledgements

This work was supported by grants awarded by Spanish Ministry of Economy (SAF2016-79008-P), PSI2013-44901-P (Grant BES-2014-068426).

References

1. Akhtar S, Benter IF (2007) Nonviral delivery of synthetic siRNAs in vivo. J Clin Invest 117(12):3623–3632. https://doi.org/10.1172/JCI33494

2. Kole R, Krainer AR, Altman S (2012) RNA therapeutics: beyond RNA interference and antisense oligonucleotides. Nat Rev Drug Discov 11(2):125–140. https://doi.org/10.1038/nrd3625. nrd3625 [pii]

3. de Fougerolles A, Vornlocher HP, Maraganore J, Lieberman J (2007) Interfering with disease: a progress report on siRNA-based therapeutics. Nat Rev Drug Discov 6(6):443–453. https://doi.org/10.1038/nrd2310. nrd2310 [pii]

4. Shim MS, Kwon YJ (2010) Efficient and targeted delivery of siRNA in vivo. FEBS J 277(23):4814–4827. https://doi.org/10.1111/j.1742-4658.2010.07904.x

5. Walton SP, Wu M, Gredell JA, Chan C (2010) Designing highly active siRNAs for therapeutic applications. FEBS J 277(23):4806–4813. https://doi.org/10.1111/j.1742-4658.2010.07903.x

6. Nakajima H, Kubo T, Semi Y, Itakura M, Kuwamura M, Izawa T, Azuma YT, Takeuchi T (2012) A rapid, targeted, neuron-selective, in vivo knockdown following a single intracerebroventricular injection of a novel chemically modified siRNA in the adult rat brain. J Biotechnol 157(2):326–333. https://doi.org/10.1016/j.jbiotec.2011.10.003

7. Dolga AM, Granic I, Blank T, Knaus HG, Spiess J, Luiten PG, Eisel UL, Nijholt IM (2008) TNF-alpha-mediates neuroprotection against glutamate-induced excitotoxicity via NF-kappaB-dependent up-regulation of K2.2 channels. J Neurochem 107(4):1158–1167. https://doi.org/10.1111/j.1471-4159.2008.05701.x. JNC5701 [pii]

8. Dreses-Werringloer U, Lambert JC, Vingtdeux V, Zhao H, Vais H, Siebert A, Jain A, Koppel J, Rovelet-Lecrux A, Hannequin D, Pasquier F, Galimberti D, Scarpini E, Mann D, Lendon C, Campion D, Amouyel P, Davies P, Foskett JK, Campagne F, Marambaud P (2008) A polymorphism in CALHM1 influences Ca2+ homeostasis, Abeta levels, and Alzheimer's disease risk. Cell 133(7):1149–1161. https://doi.org/10.1016/j.cell.2008.05.048. S0092-8674(08)00751-4 [pii]

9. Davidson BL, McCray PB Jr (2011) Current prospects for RNA interference-based therapies. Nat Rev Genet 12(5):329–340. https://doi.org/10.1038/nrg2968. nrg2968 [pii]

10. Jacobowitz DM, Kresse A, Skofitsch G (2004) Galanin in the brain: chemoarchitectonics and brain cartography—a historical review. Peptides 25(3):433–464. https://doi.org/10.1016/j.peptides.2004.02.015. S0196978104000981 [pii]

11. Branchek TA, Smith KE, Gerald C, Walker MW (2000) Galanin receptor subtypes. Trends Pharmacol Sci 21(3):109–117. doi:S0165-6147(00)01446-2 [pii]

12. Mitsukawa K, Lu X, Bartfai T (2008) Galanin, galanin receptors and drug targets. Cell Mol Life Sci 65(12):1796–1805. https://doi.org/10.1007/s00018-008-8153-8

13. Mikulaskova B, Maletinska L, Zicha J, Kunes J (2016) The role of food intake regulating peptides in cardiovascular regulation. Mol Cell Endocrinol 436:78–92. https://doi.org/10.1016/j.mce.2016.07.021. S0303-7207(16)30261-1 [pii]

14. Diaz-Cabiale Z, Parrado C, Narvaez M, Millon C, Puigcerver A, Fuxe K, Narvaez JA (2010) Neurochemical modulation of central cardiovascular control: the integrative role of galanin. EXS 102:113–131

15. Bellido I, Diaz-Cabiale Z, Jimenez-Vasquez PA, Andbjer B, Mathe AA, Fuxe K (2002) Increased density of galanin binding sites in the dorsal raphe in a genetic rat model of depression. Neurosci Lett 317(2):101–105

16. Juhasz G, Hullam G, Eszlari N, Gonda X, Antal P, Anderson IM, Hokfelt TG, Deakin JF, Bagdy G (2014) Brain galanin system genes interact with life stresses in depression-related phenotypes. Proc Natl Acad Sci U S A 111(16):E1666–E1673. https://doi.org/10.1073/pnas.1403649111. 1403649111 [pii]

17. Wang P, Li H, Barde S, Zhang MD, Sun J, Wang T, Zhang P, Luo H, Wang Y, Yang Y, Wang C, Svenningsson P, Theodorsson E, Hokfelt TG, Xu ZQ (2016) Depression-like behavior in rat: Involvement of galanin receptor subtype 1 in the ventral periaqueductal gray. Proc Natl Acad Sci U S A 113(32):E4726–E4735. https://doi.org/10.1073/pnas.1609198113. 1609198113 [pii]

18. Weiss JM, Bonsall RW, Demetrikopoulos MK, Emery MS, West CH (1998) Galanin: a significant role in depression? Ann N Y Acad Sci 863:364–382

19. Bartfai T, Lu X, Badie-Mahdavi H, Barr AM, Mazarati A, Hua XY, Yaksh T, Haberhauer G, Ceide SC, Trembleau L, Somogyi L, Krock L, Rebek J Jr (2004) Galmic, a nonpeptide galanin receptor agonist, affects behaviors in seizure, pain, and forced-swim tests. Proc Natl Acad Sci U S A 101(28):10470–10475. https://doi.org/10.1073/pnas.0403802101. 0403802101 [pii]

20. Lu X, Barr AM, Kinney JW, Sanna P, Conti B, Behrens MM, Bartfai T (2005) A role for galanin in antidepressant actions with a focus on the dorsal raphe nucleus. Proc Natl Acad Sci U S A 102(3):874–879. https://doi.org/10.1073/pnas.0408891102. 0408891102 [pii]

21. Kuteeva E, Hokfelt T, Wardi T, Ogren SO (2008) Galanin, galanin receptor subtypes and depression-like behaviour. Cell Mol Life Sci 65(12):1854–1863. https://doi.org/10.1007/s00018-008-8160-9

22. Fuxe K, Ogren SO, Jansson A, Cintra A, Harfstrand A, Agnati LF (1988) Intraventricular injections of galanin reduces 5-HT metabolism in the ventral limbic cortex, the hippocampal formation and the fronto-parietal cortex of the male rat. Acta Physiol Scand 133(4):579–581. https://doi.org/10.1111/j.1748-1716.1988.tb08444.x

23. Hedlund PB, Fuxe K (1996) Galanin and 5-HT1A receptor interactions as an integrative mechanism in 5-HT neurotransmission in the brain. Ann N Y Acad Sci 780:193–212

24. Razani H, Diaz-Cabiale Z, Misane I, Wang FH, Fuxe K, Ogren SO (2001) Prolonged effects of intraventricular galanin on a 5-hydroxytryptamine(1A) receptor mediated function in the rat. Neurosci Lett 299(1–2):145–149. doi:S0304394000017882 [pii]

25. Diaz-Cabiale Z, Narvaez JA, Finnman UB, Bellido I, Ogren SO, Fuxe K (2000) Galanin-(1-16) modulates 5-HT1A receptors in the ventral limbic cortex of the rat. Neuroreport 11(3):515–519

26. Hedlund PB, Finnman UB, Yanaihara N, Fuxe K (1994) Galanin-(1-15), but not galanin-(1-29), modulates 5-HT1A receptors in the dorsal hippocampus of the rat brain: possible existence of galanin receptor subtypes. Brain Res 634(1):163–167

27. Millon C, Flores-Burgess A, Narvaez M, Borroto-Escuela DO, Santin L, Parrado C, Narvaez JA, Fuxe K, Diaz-Cabiale Z (2015) A role for galanin N-terminal fragment (1–15) in anxiety- and depression-related behaviors in rats. Int J Neuropsychopharmacol 18(3):pii: pyu064. https://doi.org/10.1093/ijnp/pyu064

28. Millon C, Flores-Burgess A, Narvaez M, Borroto-Escuela DO, Santin L, Gago B, Narvaez JA, Fuxe K, Diaz-Cabiale Z (2016) Galanin (1-15) enhances the antidepressant effects of the 5-HT1A receptor agonist 8-OH-DPAT: involvement of the raphe-hippocampal 5-HT neuron system. Brain Struct Funct 221:4491–4504. https://doi.org/10.1007/s00429-015-1180-y. 10.1007/s00429-015-1180-y [pii]

29. Borroto-Escuela DO, Narvaez M, Di Palma M, Calvo F, Rodriguez D, Millon C, Carlsson J, Agnati LF, Garriga P, Diaz-Cabiale Z, Fuxe K (2014) Preferential activation by galanin 1-15 fragment of the GalR1 protomer of a GalR1-GalR2 heteroreceptor complex. Biochem Biophys Res Commun 452(3):347–353. https://doi.org/10.1016/j.bbrc.2014.08.061

30. Fuxe K, Borroto-Escuela DO, Romero-Fernandez W, Tarakanov AO, Calvo F, Garriga P, Tena M, Narvaez M, Millon C, Parrado C, Ciruela F, Agnati LF, Narvaez JA, Diaz-Cabiale Z (2012) On the existence and function of galanin receptor heteromers in the central nervous system. Front Endocrinol 3:127. https://doi.org/10.3389/fendo.2012.00127

31. Fuxe K, Marcellino D, Rivera A, Diaz-Cabiale Z, Filip M, Gago B, Roberts DC, Langel U, Genedani S, Ferraro L, de la Calle A, Narvaez J, Tanganelli S, Woods A, Agnati LF (2008) Receptor-receptor interactions within receptor mosaics. Impact on neuropsychopharmacology. Brain Res Rev 58(2):415–452. https://doi.org/10.1016/j.brainresrev.2007.11.007

32. Millon C, Flores-Burgess A, Narvaez M, Borroto-Escuela DO, Gago B, Santin L, Castilla-Ortega E, Narvaez JA, Fuxe K, Diaz-Cabiale Z (2017) The neuropeptides Galanin and Galanin(1-15) in depression-like behaviours. Neuropeptides 64:39–45. https://doi.org/10.1016/j.npep.2017.01.004. S0143-4179(16)30108-1 [pii]

33. Flores-Burgess A, Millon C, Gago B, Narvaez M, Borroto-Escuela DO, Mengod G, Narvaez JA, Fuxe K, Santin L, Diaz-Cabiale Z (2017) Galanin (1-15) enhancement of the behavioral effects of Fluoxetine in the forced swimming test gives a new therapeutic strategy against depression. Neuropharmacology 118:233–241. https://doi.org/10.1016/j.neuropharm.2017.03.010. S0028-3908(17)30093-X [pii]

34. Diaz-Cabiale Z, Parrado C, Narvaez M, Puigcerver A, Millon C, Santin L, Fuxe K, Narvaez JA (2011) Galanin receptor/Neuropeptide Y receptor interactions in the dorsal raphe nucleus of the rat. Neuropharmacology 61(1–2):80–86. https://doi.org/10.1016/j.neuropharm.2011.03.002. S0028-3908(11)00111-0 [pii]

Chapter 10

Double Fluorescent Knock-In Mice to Investigate Endogenous Mu-Delta Opioid Heteromer Subcellular Distribution

Lyes Derouiche, Stéphane Ory, and Dominique Massotte

Abstract

The heteromerization of mu (MOP) and delta (DOP) opioid receptors has been extensively studied in heterologous systems. These studies demonstrated significant functional interaction of MOP and DOP evidenced by new pharmacological properties and intracellular signaling in transfected cells co-expressing the receptors. Due to the lack of appropriate tools for receptor visualization, such as specific antibodies, the pharmacological and functional properties of MOP-DOP heteromers in cells naturally expressing these receptors remain poorly understood. To address endogenous MOP-DOP heteromer trafficking and signaling in vivo and in primary neuronal cultures, we generated a double knock-in mouse line expressing functional fluorescent versions of DOP and MOP receptors. This mouse model has successfully been used to map the neuroanatomic distribution of the receptors and to identify brain regions in which the MOP-DOP heteromers are expressed. Here, we describe a method to quantitatively and automatically analyze changes in the subcellular distribution of MOP-DOP heteromers in primary hippocampal culture from this mouse model. This approach provides a unique tool to address specificities of endogenous MOP-DOP heteromer trafficking.

Key words Heteromers, Opioid receptors, mu-delta, Trafficking, Primary neuronal culture, GPCR

1 Introduction

Opioid receptors belong to the subfamily of Class A G protein-coupled receptors (GPCRs). Four subtypes of opioid receptors mu (MOP), delta (DOP), kappa (KOP), and nociceptin (NOP) receptor, respectively, encoded by the *OPRM1*, *OPRD1*, *OPRK1*, and *OPRL1* genes have been identified several decades ago (for review, see [1–3]). These seven transmembrane domain receptors associate among themselves to generate a larger assembly or with different subtypes of opioids or non-opioid receptors. In the latter case, the new entity is called heteromer and may exhibit specific functional properties.

Kjell Fuxe and Dasiel O. Borroto-Escuela (eds.), *Receptor-Receptor Interactions in the Central Nervous System*, Neuromethods, vol. 140, https://doi.org/10.1007/978-1-4939-8576-0_10, © Springer Science+Business Media, LLC, part of Springer Nature 2018

In the case of the opioid system, MOP and DOP functional interactions are well documented among others, they are essential for the development of opiate tolerance [4, 5]. Numerous studies indicate that co-expression of the two receptors in heterologous systems promotes the formation of MOP-DOP heteromers, which affects binding and signaling properties [6, 7]. However, in spite of a growing body of evidence in favor of the presence of MOP-DOP heteromers in vivo, the molecular mechanisms underlying functional interactions between these two receptors remain poorly characterized [8]. This is mainly due to the lack of appropriate tools, especially specific antibodies.

To deal with this issue and study MOP-DOP heteromers in vivo, we generated a double fluorescent knock-in mouse line co-expressing DOP and MOP receptors, respectively, fused to their C-terminus to the enhanced green fluorescent protein (DOP-eGFP) or mcherry (MOP-mCherry). The DOP-eGFP and MOP-mCherry functional fusions allow highly specific and simultaneous visualization of endogenously expressed receptors with subcellular resolution and proved to be unique tools for neuroanatomical studies [9]. Mapping of MOP and DOP receptors in the central and peripheral nervous systems indeed revealed MOP-DOP neuronal co-expression in discrete neuronal networks essential for survival such as the nociceptive pathway (see also mouse brain atlas at http://mordor.ics-mci.fr/). Specific targeting using the fluorescent tags also revealed MOP-DOP physical proximity in the hippocampus providing strong rationale for the existence of endogenous MOP-DOP heteromers [9]. In addition, the double fluorescent knock-in mice represent unique tools to explore the dynamics of this complex under physiological or pathological conditions and to characterize the functional impact of MOP-DOP heteromers in the central and peripheral nervous system.

In this chapter, we describe optimized conditions for visualization of endogenous MOP-DOP heteromers in primary hippocampal neurons obtained from the double fluorescent knock-in mice. We also provide a protocol for automatic quantitative analysis of confocal images with an open source software to determine changes in receptor subcellular localization. This method allowed MOP-DOP heteromers monitoring and specific determination of their intracellular fate upon pharmacological activation.

2   Material and Methods

2.1   Animals

Double knock-in mice co-expressing fluorescent DOP and MOP receptors were obtained by crossing previously generated DOP-eGFP and MOP-mcherry knock-in mice. Briefly, DOP-eGFP knock-in mice expressing the delta opioid receptor fused to its

C-terminus to the eGFP were generated by homologous recombination by inserting the eGFP cDNA into exon 3 of the delta opioid receptor gene, in frame and 5′ from the stop codon [10]. MOP-mCherry knock-in mice expressing the mu opioid receptor fused to its C-terminus to the red fluorescent protein mCherry were generated by homologous recombination following a procedure similar to the one used for DOR-eGFP knock-in mice [9]. The construct transfected into ES cells comprised a Gly-Ser-Ile-Ala-Thr linker followed by the cDNA sequence encoding the fluorescent protein (eGFP or mCherry). For subsequent clone selection, a resistance gene was included that corresponded to neomycin flanked by loxP sites for DOP-eGFP or to hygromycin flanked by FRT sites for MOP-mCherry (Fig. 1). The resistance gene was removed by microinjection of a plasmid expressing the recombinase. Blastocysts were implanted in pseudo gestant BalbC females. Chimeric mice were crossed with C57Bl6/J mice to obtain F1 heterozygous generation. Heterozygous animals were crossed to generate mice homozygous for *Oprd1*-eGFP or *Oprm1*-mCherry that are fertile and develop normally. Double knock-in animals were obtained by crossing the single knock-in mouse lines. The genetic background of all mice was C57/BL6/J:129svPas (50:50%).

Mice were housed in animal facility under controlled temperature (21 ± 2 °C) and humidity (45 ± 5%) on a 12-h dark–light cycle with food and water ad libitum. All experiments were performed in accordance to the European legislation (directive 2010/63/EU acting on protection of laboratory animals) and the local ethical committee.

2.2 Primary Hippocampal Culture

2.2.1 Material and Reagents

1. Double knock-in new born mice pups (P0-P3).

2. 70% ethanol solution.

3. MilliQ water (autoclaved, or sterile filtered 0.22 μm).

4. Borate buffer (*see* Sect. 2.3).

5. Poly-L-lysine hydrobromide (Sigma cat. No. P2623).

6. 13 mm coverslips (Sigma, cat. no. P6407) coated with poly-L-lysine (*see* Sect. 2.3).

7. 24-well sterile culture plates (Falcon cat. no. 353047).

8. Pasteur pipets (flamed at the extremity, cotton plugged and autoclaved).

9. Hibernate minus phenol red (BrainBits SKU: HAPR).

10. Papain (Worthington, cat. no. LS003126).

11. Dulbecco's Modified Eagle's Medium (DMEM) with 4.5 g glucose (GIBCO, cat. no. 71966-029).

12. Neurobasal A (GIBCO, cat. no. A13710-01).

152   Lyes Derouiche et al.

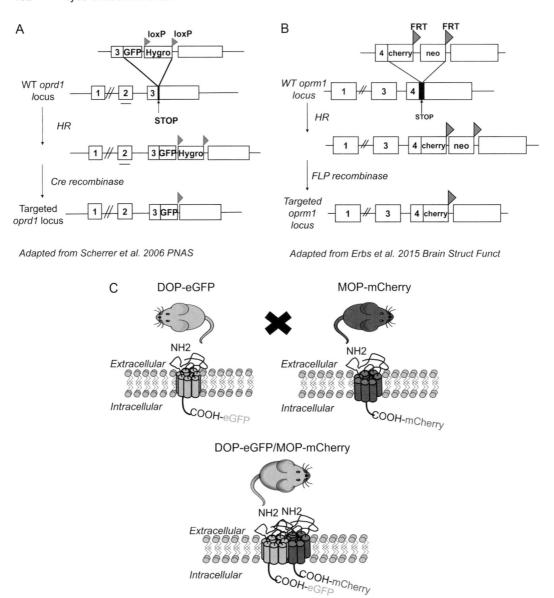

Fig. 1 Double Knock-in DOP-eGFP/MOP-mCherry mouse engineering. (**A**) Construction of the DOP-eGFP mouse. A cDNA sequence corresponding to the eGFP cDNA, and the FRT flanked neomycin (neo) cassette was inserted by homologous recombination (HR) into the *Oprd1* locus. HR was followed Cre recombinase treatment in ES cells. (**B**) Construction of the MOP-mCherry mouse. A cDNA sequence corresponding to the mCherry cDNA, and the loxP flanked hygromycin (hygro) cassette were inserted by HR to the *Oprm1* locus. HR was followed by FRT recombinase treatment in ES cells. (**C**) Double knock-in mice were obtained by crossing homozygote DOP-eGFP and MOP-mCherry mice

13. Fetal Calf Serum (FCS) heat inactivated. **Caution:** *test several batches to determine the best one for your culture conditions.*

14. Glutamax™ (GIBCO, cat. no. 35050061).

15. L-Glutamine (GIBCO, cat. no. 25030081).

16. Penicillin-streptomycin (P/S) (cat. no. 15140122).

17. DNase (Sigma cat. no. DN25).

18. B27 supplement (GIBCO, cat. no. 17504044).

19. Trypan blue solution (Sigma, cat. no. T8154).

20. Paraformaldehyde 32% solution diluted to 4% before use (*see* Sect. 4.2).

21. Phosphate buffer saline (PBS). (Sigma-Aldrich, cat. no. P5493).

22. Cell strainer 70 μm (Falcon, cat. no. 352350).

2.3 Setup and Procedures

1. *Borate buffer:* Dissolve boric acid 1.24 g and sodium tetraborate (borax) 1.9 g in 400 mL MilliQ H_2O. pH should be 8.4. Sterile filter (0.22 μm) before use. **Caution**: *borax is a hazardous substance, manipulate cautiously and eliminate waste according to the safety rules fixed by your institution/government.*

2. *Coverslips sterilization and coating:* Put coverslips in a 100-mm petri dish, sterilize in 70% ethanol for 2 h under gentle agitation, let dry completely under laminar flow, and transfer to culture plates. Rinse once with MilliQ water. Coat coverslips with poly-L-lysine 25 μg/mL final concentration in borate buffer; incubate at 37 °C for 2 h to overnight. Rinse three times with sterile water and prewarm in DMEM medium. Coated coverslips may be prepared several days before use, dried in the laminar flow hood and kept sealed at 4 °C for up to 1 month.

3. *Dissection medium:* Prepare 5 mL of ice cold Hibernate supplemented with 1× P/S and 0.5 nM glutamax per animal and transfer 0.5 mL dissection medium in a 15-mL centrifugation tube per animal (two hippocampi).

4. *Enzyme solution:* Prior to dissection, prepare a fresh solution of papain at 40 U/mL concentration in Hibernate medium, incubate 5 min at 37 °C in a water bath, then keep on ice until use. Prepare 0.5 mL per animal (two hippocampi).

5. *Plating medium:* Prepare 12 mL DMEM medium supplemented with 4.5 g/L glucose + 10% heat inactivated FCS + 2 mM glutamine + Pen/strep) per animal on the day of use.

6. *Growing medium:* Prepare 12 mL Neurobasal medium supplemented with 2% B27, 2 mM glutamax, 0.5 mM glutamine, and 1× P/S per animal on the day of use.

7. *Phosphate-buffered saline (PBS):* Prepare 1 L 1× PBS working solution from 10× stock solution by diluting with MilliQ water. Check the pH and adjust to 7.4 with 1 M HCl or 1 M NaOH solutions if needed. Sterile filter 0.22 μm and keep at 4 °C for up to 6 months.

8. *Fixation solution:* Dilute paraformaldehyde (PFA) 32% solution to 4% final concentration in PBS 0.1 M, adjust pH to 7.4 if needed. Prepare 500 μL per well for use in 24-well plate. Keep up to 5 days at 4 °C and up to 6 months at −20 °C. **Caution**: *PFA is a hazardous highly toxic substance, manipulate under flow hood and eliminate waste according to the safety rules fixed by your institution/government.*

2.4 Dissection and Culture Procedures

2.4.1 Dissection and Cell Dissociation

Decapitate pups. Transfer the head in a 33 mm petri dish with 1.5 mL ice cold dissection medium and isolate the brain. Place the isolated brain in a new 33 mm petri dish with 1.5 mL ice cold dissection medium. Remove the meninges, dissect to isolate the two hippocampi, and keep them in 0.5 mL dissection solution in a 15 mL tube in ice. Add 0.5 mL of papain solution per tube to 0.5 mL dissection medium (1 mL final/2 hippocampi). Place tubes in a water bath at 37 °C for 30 min with gentle shaking every 5–10 min. Five to ten minutes before digestion ends, add DNase at a final concentration of 1 mg/mL.

Remove papain solution by decantation, add 1 mL of Hibernate, and triturate with Pasteur pipet until the tissue is completely dissociated (about 15–20 times up-and-down are sufficient) (*see note 1*). Centrifuge at 1000 g for 5 min at 22 °C. Remove the supernatant, add 1 mL plating medium to resuspend cells (3–5 gentle up-and-down with Pasteur pipet). Filter the cell suspension through a 70 μm-cell strainer to remove any residual aggregates. Count cells by diluting 20 μL of cell suspension in 80 μL of 1:10 Trypan blue solution diluted in PBS. Place 20 μL of this solution in a cell counting chamber and count cells excluding Trypan blue (viable cells) only.

2.4.2 Plating and Feeding

Prewarm poly-L-lysine coated plates in DMEM medium at 37 °C. Remove DMEM and plate cells in 24-well plates at a density of 80,000–100,000 cells per well in a final volume of 500 μL.

Incubate in a humid incubator at 37 °C and 5% CO_2 and allow cells to adhere to the bottom of plates during 1 h. Remove the plating medium by aspiration, rinse once with 500 μL of prewarmed Neurobasal A medium, then add 500 μL of growing medium and return back to the incubator. Let cells grow for at least 12 days with half of the medium replaced every 4–5 days.

3 Processing and Pharmacological Treatments

Pharmacological treatments should be realized between DIV 12 and DIV 15 after plating (*see note 2*). Ligands are added in a volume not exceeding 10% of the culture medium volume. Incubate cells during the appropriate time. At the end of the pharmacological treatment, remove the plate from the incubator and immediately place on ice, carefully aspirate the medium and wash twice with ice cold sterile filtered PBS. Remove PBS and add 500 μL of PFA 4% in ice cold PBS and incubate for 20 min on ice. Remove PFA and rinse twice with ice cold PBS and proceed to immunostaining or keep sealed with Parafilm in 500 μL PBS at 4 °C up to 30 days (*see note 3*).

4   Immunocytofluorescence (ICF)

4.1 Material and Reagents

1. Phosphate buffer saline 0.1 M, pH 7.4.
2. Normal Goat Serum (NGS) (Sigma cat. no. S26).
3. Tween 20 (Euromedex cat. no. 2001-B).
4. Primary and secondary antibodies (see Table 1).
5. ProLong™ Gold Antifade Mounting medium (Molecular Probes cat. no. P36935).
6. DAPI (Sigma cat. no. D9542).
7. Finepoint Forceps (Rubis Switzerland cat.no.1K920).
8. Microscope glass slides.

4.2 Setup and Reagents

1. *PBS Tween 20 solution (PBST):* Add 0.2% (V:V) of Tween 20 solution to 1× PBS solution (*see* Sect. **2.3.7**), mix vigorously to complete dissolution, and keep at 4 °C. Bring at room temperature before use.
2. *Blocking solution:* Add 5% of Normal Goat Serum (NGS) to the PBST solution. Prepare the day of use.

4.3 Method

Incubate fixed cells in 250 μL of blocking solution for 1 h under gentle agitation at room temperature (20–22 °C). Then remove the blocking solution by aspiration and incubate 2 h at RT or overnight at 4 °C with primary antibodies in blocking solution (250 μL/well) under gentle agitation (Table 1). Wash three times in PBST and incubate for 2 h protected from light with specific secondary antibodies diluted in blocking solution (Table 1) (250 μL/well). Wash three times in PBST, incubate 5 min in DAPI

Table 1
Primary and secondary antibodies

Antigen	Antibody	Supplier reference	Dilution
eGFP	Chicken IgY	AVES-GFP1020	1/1000
mCherry	Rabbit IgG	Clontech-632496	1/1000
Chicken IgY	Goat anti-chicken Alexa Fluor 488	Molecular Probes-A11039	1/2000
Rabbit IgG	Goat anti-rabbit Alexa Fluor 594	Molecular Probes-A11012	1/2000

solution (1 μg/mL in PBS) for nuclei staining, and then wash three times in PBS followed by one wash in MilliQ H$_2$O. Remove coverslips from wells with finepoint forceps, let coverslips dry completely at room temperature protected from light, and mount on glass slides with ProLong™ Gold Antifade mounting medium. Keep for up to 1 year at −20° protected from light.

5   Confocal Microscope

Images were acquired with a laser-scanning confocal microscope Leica SP5 using 63×/NA 1.4 oil immersion lens and ×5 numerical zoom. The pinhole was adjusted to 1 airy unit and the gain was adjusted without offset for each filter on a specific scanning plan allowing specific acquisition without saturation. Image acquisition was performed according to Nyquist parameters in *XY* with an average frame of 3 in a sequential scan mode to avoid cross talk between different wavelengths. Z-stacks were obtained by scanning the whole neuron thickness with step of 1 μm in z.

6   Image Analysis with ICY Open Source Software

Confocal images were analyzed with ICY software (http://icy.bio-imageanalysis.org/). Quantification was performed on a single plane extracted from a z-stack. The analysis combined two sequential steps. The first one consists in isolating each neuron to define regions of interest (ROI). The second one involves the detection of the spots in each channel and the determination of the amount of co-localization in each ROI.

6.1   ROI Definition

Each neuron was carefully delineated using the "free-hand area" tool. This initial ROI is filled with the "fill holes in ROI" plugin to define the total cell area (ROI$_{total}$). ROIs were then processed to generate two ROIs corresponding to the cell periphery and the cytoplasm (detailed protocol available online *http://icy.bioimage-analysis.org/*). Based on staining in basal conditions, we estimated that most of the plasma membrane staining was found over an 8 pixels thickness. Therefore, we automatically eroded with the "Erode ROI" plugin the ROI$_{total}$ by 8 pixels and subtracted this new ROI (ROI$_{cyto}$) to ROI$_{total}$ to obtain a ROI corresponding to the cell periphery (ROI$_{peri}$).

6.2   Spots Detection and Co-localization

To detect specific signal in each ROI, we used the "spot detector" plugin which relies on the wavelet transform algorithm [11]. By carefully setting the sensitivity threshold and the scale of objects to detect, it allows detection of spots even in images with low signal to noise ratio. In our conditions, the sensitivity threshold was fixed between 50% and 60% and the scale of objects set at 2 (pixel size 3) for mu and delta receptors. Once parameters were defined, images were processed with the tool "protocol" in Icy which is a graphical interface for automated image processing. Data including the number of spots detected in each channel and ROIs, the number of co-localized objects and the ROI area were automatically collected in excel files. Objects were considered co-localized if the distance of their centroid was equal to or less than 3 pixels. The protocol is available online (NewColocalizer with binary and excel output v1_batch.xml). To obtain histograms, we calculated object densities for each receptor reported to the surface of each ROI. Membrane to cytoplasm density ratios were calculated to illustrate the subcellular distribution of each receptor. The extent of co-localization was calculated according to the following formula for each ROI [% colocalization = 100 × ((colocalized MOP and DOP objects)/($\sum$(detected MOP and DOP objects))].

7   Statistical Analysis

Statistical analyses were performed with Graphpad Prism V7 software (GraphPad, San Diego, CA). Normality of the distributions and homogeneity of the variances were checked before statistical comparison to determine appropriate statistical analysis. In our case, data were not normally distributed and the non-parametric Mann Whitney test was used to compare receptor densities in the plasma membrane and cytoplasm in basal conditions or after agonist treatment. The extent of receptor co-localization was compared using two-way ANOVA with repeated measures followed by post-hoc Sidak's test for multiple comparisons. Basal group was compared to agonist-treated group (first factor) within cytoplasm and plasma membrane localization (second factor).

7.1 Results and Discussion

In this chapter, we have presented an optimized method for monitoring the subcellular distribution of endogenous MOP and DOP receptors. To this aim, we combined the use of a genetically modified mouse line co-expressing functional fluorescently tagged receptors, optimized primary neuronal culture protocol, and automatic quantitative analysis of confocal images with an open source software. Importantly, the image analysis procedure can be easily implemented in any laboratory since data processing does not require extensive mathematical developments or program writing with specialized software.

7.2 Individual Distribution of MOP and DOP Receptors

High magnification confocal images analyzed as single focal plan revealed discontinuous and punctate distributions for DOP-eGFP and MOP-mCherry that were predominantly located at the cell surface in basal conditions (Fig. 2a). Images also revealed a perinuclear cytoplasmic localization of both receptors that likely correspond to receptor stock in the endoplasmic reticulum. These observations were consistent with data from the literature describing a predominant and membrane localization of DOP [10, 12] and MOP [13] but also substantial localization in perikarya [10].

Quantification using the ICY software indicated a higher density of fluorescent objects at the cell surface for both MOP and DOP receptors (Fig. 2B) that was three times higher compared to the cytoplasm (Fig. 2C). Activation by the MOP-DOP agonist CYM51010 led to the appearance of high intensity punctate structures in the cytoplasmic and a dramatic decrease in the plasma membrane labeling of both receptors (Fig. 2A). Accordingly, the ratio corresponding to the density of fluorescent objects at the plasma membrane compared to the cytoplasm dropped dramatically from 3 to 1 for both MOP and DOP receptors which supports internalization of the two receptors in vesicle-like structures (Fig. 2C).

These results are in agreement with previous reports using the density of fluorescence to estimate changes in DOPeGFP subcellular distribution in vivo. In these studies, the ratio of fluorescence density between the plasma membrane and cytoplasm was about 1.5 in basal conditions and significantly decreased following agonist stimulation [10, 14]. We therefore tested our quantification method using images of DOR-eGFP neurons in the hippocampus acquired with similar parameters in confocal microscopy. Using the protocol described above, we found that the density in DOP-eGFP objects under basal conditions was around 3 similar to our results in primary neuronal cultures. We also established that this value corresponded to a ratio in fluorescence density between the plasma membrane and the cytoplasm of about 1.7, similar to previously reported ratios [10]. Moreover, we quantified the ratio of fluorescence density between the plasma membrane and the cytoplasm using the set of images used in Erbs et al. 2016, and calculated a similar increase of about 10% in DOP-eGFP expression at the plasma membrane after chronic morphine treatment in the neurons of the hippocampus [15]. Altogether, these results fully validate the quantification method developed using ICY software.

7.3 Co-localization of MOP and DOP Receptors and Detection of MOP-DOP Heteromers

Our analysis revealed substantial co-localization of MOP and DOP associated signals under basal conditions. In fact, more than 22% of the MOP and DOP objects were co-localized within the plasma membrane, whereas cytoplasmic co-localization was fairly low (around 10%) (Fig. 2D). After selective MOP-DOP activation with the agonist CYM51010, the cytoplasmic co-localization was increased by 87% to reach 18.7%. A twofold decrease in the percentage of MOP-DOP co-localization within the plasma membrane was also observed after CYM51010 treatment (Fig. 2D). These observations support a view in which MOP and DOP

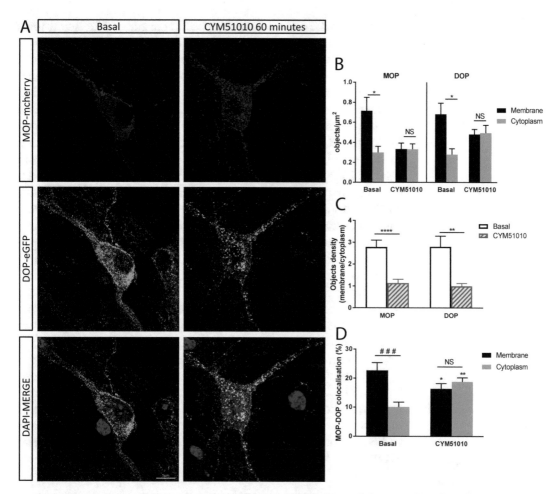

Fig. 2 MOP-DOP heteromer visualization and quantification. (**a**) Representative confocal images illustrating MOP-mCherry and DOP-eGFP co-localization in basal conditions or after treatment during 60 min with the MOP-DOP agonist CYM51010 400 nM. Scale bar: 5 μm. (**b**) Image quantification with ICY software illustrating each receptor distribution. In basal conditions, MOP and DOP densities are significantly higher in the plasma membrane compared to cytoplasm. Mann Whitney test, *p-value <0.05. CYM51010 treatment for 60 min led to changes in receptor subcellular distribution with no more statistical difference in receptor densities between the plasma membrane and cytoplasm (Mann Whitney test, p-value >0.05). (**c**) Changes in receptors distribution. Plasma membrane to cytoplasm ratio of MOP or DOP spots densities in basal conditions or after with 400 nM CYM51010 for 60 min reflects receptor redistribution. Mann Whitney test, **p-value <0.01; ****p-value <0.0001. (**d**) Quantification of MOP-DOP co-localization upon agonist activation with 400 nM CYM51010 for 60 min reveals MOP-DOP co-internalization. Two-way ANOVA with repeated measures, post-hoc Sidak's test. *p-value <0.05 for basal cytoplasm vs CYM51010 cytoplasm; **p-value <0.01 for basal membrane vs CYM51010 membrane. ###p-value <0.001 for basal membrane vs basal cytoplasm; NS: p-value >0.05 for CYM51010 membrane vs CYM51010 cytoplasm

receptors remain associated upon specific activation of the heteromers and undergo common intracellular fate.

It is however important to note that due to the resolution limit of confocal microscopy, MOP-DOP physical interaction has to be assessed by other experimental approaches such as co-localization in electron microscopy or disruption of the physical contact by a transmembrane peptide. The latter has been successfully performed for MOP-DOP receptors in vivo. Indeed, a peptide corresponding to the MOP TM 1 or to the DOP second intracellular loop in fusion with the cell transduction domain of the human immunodeficiency virus (HIV) TAT protein interfered with MOP-DOP co-immunoprecipitation [16–18]. The recently described proximity ligation assay [19] provides high spatial resolution and represents another attractive option to address physical proximity.

In addition, the resolution of the images did not enable to distinguish the pool of receptors associated to the plasma membrane from receptors located in the sub-membrane compartment which represents another limitation of our analysis. Total internal reflection fluorescence microscopy (TIRFM) would be required to differentiate the two compartments and precisely monitor the first steps in receptor internalization.

MOP-mCherry-DOP-eGFP co-localization studies could also be combined with identification of the intracellular compartments using specific antibodies. This would enable for fine mapping of the receptors in the vesicular structures and cellular compartments and would provide detailed information on the intracellular fate of MOP-DOP heteromers.

8   Conclusion

Double knock-in mice represent unique tools to investigate endogenous MOP-DOP neuroanatomical distribution but also to explore functional dynamics in physio-pathological conditions. As described here, the use of double fluorescent knock-in mice combined with confocal imaging and ICY software analyses enables easy quantification of receptor subcellular distribution and co-localization, hence specific MOP-DOP trafficking. Importantly, the field of application of the method described here is not restricted to the study of MOP-DOP heteromers but can be applied to address the heteromerization of any other pair of GPCRs. Moreover, the analysis with the ICY software is not restricted to the study of receptors but is widely applicable to monitor the co-localization of any two objects detected independently and can be applied to any type of high-resolution images.

9   Notes

1. Cell dissociation is a critical step; if the trituration is too gentle, the tissue will not dissociate, if too vigorous cells will break.

2. Because re-expression of DOP-eGFP and MOP-mCherry is only detectable from day in vitro (DIV) 10 in primary neurons, pharmacological treatments should be performed between DIV12 and DIV15 to ensure full expression of the receptors.

3. Paraformaldehyde fixation decreases the fluorescence intensity of eGFP and mCherry and amplification by immunostaining is recommended. The DOP-eGFP construct can also be used for ex vivo real-time imaging of receptor internalization by confocal microscopy [10]. However, due to the low expression level of endogenous MOP receptors and their weak expression at the plasma membrane in basal conditions, real-time monitoring of MOP-mcherry remains presently below the detection limit.

Acknowledgments

The authors would like to acknowledge the financial support of the Fondation pour la Recherche Médicale (LPA20140129364) and CNRS. L. Derouiche was the recipient of an IDEX post-doctoral fellowship of the University of Strasbourg.

Conflict of Interest
The authors declare no conflict of interest.

References

1. Charbogne P, Kieffer BL, Befort K (2014) 15 years of genetic approaches in vivo for addiction research: Opioid receptor and peptide gene knockout in mouse models of drug abuse. Neuropharmacology 76 Pt B:204–217. https://doi.org/10.1016/j.neuropharm.2013.08.028

2. Pasternak GW (2014) Opioids and their receptors: are we there yet? Neuropharmacology 76 Pt B:198–203. https://doi.org/10.1016/j.neuropharm.2013.03.039

3. Civelli O (2008) The orphanin FQ/nociceptin (OFQ/N) system. Results Probl Cell Differ 46:1–25. https://doi.org/10.1007/400_2007_057

4. Ong EW, Cahill CM (2014) Molecular perspectives for mu/delta opioid receptor heteromers as distinct, functional receptors. Cells 3(1):152–179. https://doi.org/10.3390/cells3010152

5. Gendron L, Mittal N, Beaudry H, Walwyn W (2015) Recent advances on the delta opioid receptor: from trafficking to function. Br J Pharmacol 172(2):403–419. https://doi.org/10.1111/bph.12706

6. Gomes I, Ijzerman AP, Ye K, Maillet EL, Devi LA (2011) G protein-coupled receptor heteromerization: a role in allosteric modulation of ligand binding. Mol Pharmacol 79(6):1044–1052. https://doi.org/10.1124/mol.110.070847

7. Fujita W, Gomes I, Devi LA (2014) Revolution in GPCR signalling: opioid receptor heteromers as novel therapeutic targets: IUPHAR review 10. Br J Pharmacol 171(18):4155–4176. https://doi.org/10.1111/bph.12798

8. Massotte D (2015) In vivo opioid receptor heteromerization: where do we stand? Br J Pharmacol 172(2):420–434. https://doi.org/10.1111/bph.12702

162 Lyes Derouiche et al.

9. Erbs E, Faget L, Scherrer G, Matifas A, Filliol D, Vonesch JL, Koch M, Kessler P, Hentsch D, Birling MC, Koutsourakis M, Vasseur L, Veinante P, Kieffer BL, Massotte D (2015) A mu-delta opioid receptor brain atlas reveals neuronal co-occurrence in subcortical networks. Brain Struct Funct 220(2):677–702. https://doi.org/10.1007/s00429-014-0717-9

10. Scherrer G, Tryoen-Toth P, Filliol D, Matifas A, Laustriat D, Cao YQ, Basbaum AI, Dierich A, Vonesh JL, Gaveriaux-Ruff C, Kieffer BL (2006) Knockin mice expressing fluorescent delta-opioid receptors uncover G protein-coupled receptor dynamics in vivo. Proc Natl Acad Sci U S A 103(25):9691–9696. https://doi.org/10.1073/pnas.0603359103

11. Olivo-Marin JC (2002) Extraction of spots in biological images using multiscale products. Pattern Recogn 35(9):1989–1996. https://doi.org/10.1016/S0031-3203(01)00127-3. Pii S0031-3202(01)00127-3

12. Cahill CM, McClellan KA, Morinville A, Hoffert C, Hubatsch D, O'Donnell D, Beaudet A (2001) Immunohistochemical distribution of delta opioid receptors in the rat central nervous system: evidence for somatodendritic labeling and antigen-specific cellular compartmentalization. J Comp Neurol 440(1):65–84

13. Trafton JA, Abbadie C, Marek K, Basbaum AI (2000) Postsynaptic signaling via the [mu]-opioid receptor: responses of dorsal horn neurons to exogenous opioids and noxious stimulation. J Neurosci 20(23):8578–8584

14. Faget L, Erbs E, Le Merrer J, Scherrer G, Matifas A, Benturquia N, Noble F, Decossas M, Koch M, Kessler P, Vonesch JL, Schwab Y, Kieffer BL, Massotte D (2012) In vivo visualization of delta opioid receptors upon physiological activation uncovers a distinct internalization profile. J Neurosci 32(21):7301–7310. https://doi.org/10.1523/JNEUROSCI.0185-12.2012

15. Erbs E, Faget L, Ceredig RA, Matifas A, Vonesch JL, Kieffer BL, Massotte D (2016) Impact of chronic morphine on delta opioid receptor-expressing neurons in the mouse hippocampus. Neuroscience 313:46–56. https://doi.org/10.1016/j.neuroscience.2015.10.022

16. He SQ, Zhang ZN, Guan JS, Liu HR, Zhao B, Wang HB, Li Q, Yang H, Luo J, Li ZY, Wang Q, Lu YJ, Bao L, Zhang X (2011) Facilitation of mu-opioid receptor activity by preventing delta-opioid receptor-mediated codegradation. Neuron 69(1):120–131. https://doi.org/10.1016/j.neuron.2010.12.001

17. Kabli N, Martin N, Fan T, Nguyen T, Hasbi A, Balboni G, O'Dowd BF, George SR (2010) Agonists at the delta-opioid receptor modify the binding of micro-receptor agonists to the micro-delta receptor hetero-oligomer. Br J Pharmacol 161(5):1122–1136. https://doi.org/10.1111/j.1476-5381.2010.00944.x

18. Xie WY, He Y, Yang YR, Li YF, Kang K, Xing BM, Wang Y (2009) Disruption of Cdk5-associated phosphorylation of residue threonine-161 of the delta-opioid receptor: impaired receptor function and attenuated morphine antinociceptive tolerance. J Neurosci 29(11):3551–3564. https://doi.org/10.1523/JNEUROSCI.0415-09.2009

19. Borroto-Escuela DO, Hagman B, Woolfenden M, Pinton L, Jiménez-Beristain A, Oflijan J, Narvaez M, Di Palma M, Feltmann K, Sartini S, Ambrogini P, Ciruela F, Cuppini R, Fuxe K (2016) In situ proximity ligation assay to study and understand the distribution and balance of GPCR homo- and heteroreceptor complexes in the brain. In: Luján R, Ciruela F (eds) Receptor and ion channel detection in the brain. Neuromethods, vol 110. Springer Science+Business Media, New York. https://doi.org/10.1007/978-1-4939-3064-7_9

Chapter 11

Biochemical Characterization of Dopamine D2 Receptor-Associated Protein Complexes Using Co-Immunoprecipitation and Protein Affinity Purification Assays

Ping Su, Frankie H. F. Lee, and Fang Liu

Abstract

Schizophrenia is a devastating mental disorder, affecting almost 1% of the world population, and has a tremendous effect on social and occupational functioning. Dopamine D2 receptors (D2Rs) are the main targets of typical antipsychotic medications for schizophrenia, where they can effectively alleviate the positive symptoms by antagonizing D2Rs. Thus, investigating different modulation of D2R function is important in identifying novel drug targets and therapeutics for better outcomes in schizophrenia. Protein–protein interactions between D2Rs and other proteins are critical in regulating D2R signaling and subsequent downstream physiological functions. Here we described various biochemical methods including co-immunoprecipitation and protein affinity purification assays, that are commonly used to characterize D2R-associated protein complexes. Specifically, we reviewed the D2R–D1R and D2R–DISC1 interactions and discussed their association in the pathophysiology of schizophrenia. This chapter aims to provide systemic guidelines for the standard biochemical techniques in identifying D2R-associated protein–protein interactions, and to investigate the roles of these interactions in the brain.

Key words Protein–protein interaction, Schizophrenia, Dopamine D2 receptor (D2R), Co-immunoprecipitation, Protein affinity purification

1 Introduction

Schizophrenia is a devastating mental disorder, affecting almost 1% of the world population [1], and has a tremendous effect on social and occupational functioning [2, 3]. Dysfunction of the dopamine neurotransmitter system is a predominant theory in the pathophysiology of schizophrenia, especially attributing to the psychotic components. Dopamine D2 receptors (D2Rs) are the main targets of typical antipsychotic medications for schizophrenia, where they can effectively alleviate the positive symptoms by antagonizing D2Rs [4, 5]. Therefore, the effectiveness of antipsychotics depends on the degree of D2R blockade and occupancy. Furthermore,

Kjell Fuxe and Dasiel O. Borroto-Escuela (eds.), *Receptor-Receptor Interactions in the Central Nervous System*, Neuromethods, vol. 140, https://doi.org/10.1007/978-1-4939-8576-0_11, © Springer Science+Business Media, LLC, part of Springer Nature 2018

post-mortem studies, positron emission tomography, and single-photon emission computed tomography have reported elevated levels of D2R mRNA and protein expression in the brains of schizophrenia patients [5, 6].

D2Rs belong to the seven transmembrane G-protein coupled receptor family [7, 8], where they consist of a large third intracellular loop and an extremely short C-terminal tail [8]. D2Rs can couple to the inhibitory G_i protein, resulting in the inhibition of cAMP accumulation [9]. D2Rs can also activate the glycogen synthase kinase 3 (GSK3) signaling pathway that is β-arrestin-dependent, which has been extensively studied for their important role in the pathophysiology of schizophrenia [10–15]. D2Rs interact with other proteins mainly through the third intracellular loop or the C-terminus, and these interactions are crucial in regulating D2R signaling, and subsequent modulation of downstream physiological functions [16–18]. In this chapter, we will be focusing on the D2R–D1R and D2R–DISC1 interactions and discuss their involvement in the brain, especially in relation to schizophrenia.

1.1 D2R-D1R

Dopamine D1 and D2 receptors (D1Rs and D2Rs) are the two main subtypes in the dopamine receptor family. They are classified into D1-like or D2-like subfamily based on receptor structures, pharmacological properties, and sequence homology [8]. D1Rs have a short third intracellular loop and a long C-terminus, whereas D2Rs possess a long third intracellular loop and an extremely short C-terminus [8]. With respect to their downstream signaling pathways, D1Rs couple to $G_{\alpha s}$ and $G_{\alpha olf}$ proteins, which activate adenylyl cyclase, protein kinase A, and phosphatase-1 inhibitor DARPP-32. Besides, D1Rs can couple to $G_{\alpha q}$ protein, resulting in sequential activation of phospholipase C (PLC) and mobilization of intracellular calcium. In contrast, D2Rs inhibit adenylyl cyclase by coupling to $G_{\alpha i}$ and $G_{\alpha o}$ proteins, and can regulate many other downstream effectors, including protein kinases, phospholipase, and ion channels, via the $G_{\beta\gamma}$ subunits of G proteins [9].

Previous studies reported the presence of functional cross-talk between the D1R and D2R, which indicates a possible protein–protein interaction between the two receptors [8, 9, 19]. D1R and D2R co-expression and colocalization have also been reported in neurons of human and rat brains [20]. Studies using fluorescence resonance energy transfer (FRET) have determined that D1R and D2R are localized in close proximity on the cell surface and in the endoplasmic reticulum (ER) [21], providing further evidence for the possibility of a potential D1R-D2R protein complex formation. The existence of D1R-D2R hetero-oligomers was confirmed using co-immunoprecipitation assay in cell lines co-expressing both receptors and in rat brain tissue [20]. Furthermore, GST pull-down assay showed that the residues 257-271 within the third intracellular loop of the D2R long isoform are responsible for mediating the D1R–D2R interaction [22].

The interaction between D1R and D2R can produce a synergistic effect on functions of these two receptors that is distinct from the signal transduction of individual D1R and D2R functions. In the striatum, co-activation of both D1R and D2R leads to a novel PLC-mediated calcium signaling that is different from either the D1R-mediated Gs activation or the D2R-mediated Gi activation. Activating D1R or D2R alone is unable to generate this specific calcium signal [23], but rather results in its desensitization that is not caused by depletion of calcium stores or internalization of the oligomers [23, 24]. Moreover, activation of the D1R-D2R hetero-oligomers increases both total and activated Ca^{2+}/ calmodulin-dependent kinase IIα (CaMKIIα) in a Ca^{2+}-dependent way in the rat nucleus accumbens [23].

Our group has reported that the level of D1R-D2R protein complex was increased in post-mortem brain tissues of patients with major depression detected by co-immunoprecipitation assay. Using GST pull-down and in vitro binding assay, the residues M257-E271 within the D2R long isoform is the fragment mediating the binding to D1R, while the C-terminus of D1R enables it to bind to D2R. Administration of an interfering peptide that disrupts the D1R–D2R interaction in rats significantly reduced immobility in the forced swim test (FST), decreased escape failure in learned helplessness tests, but had no effect on locomotor activity, all of which suggest potential antidepressant effects [22]. Furthermore, using a novel intranasal delivery system with the Pressurized Olfactory Device (POD) where the interfering peptide can be delivered to relevant brain regions of rats, there was a significant antidepressant effect at doses ≥ 1.67 nmol/g in the FST. The clinical relevance for the use of this peptide as a new treatment option for major depression is supported by the preclinical behavioral data and its sustained detection 2 h after administration [25].

The mechanistic action of clozapine, one of the most effective medications in the treatment of schizophrenia, also involves the D1R–D2R interaction [26, 27]. Two binding sites for clozapine at the D1R and D2R were identified, and the formation of D1R-D2R hetero-oligomers markedly changed the binding affinity of D1R [27]. The authors further used FRET to show that a low concentration of clozapine is sufficient in uncoupling the D1R-D2R hetero-oligomers, suggesting that clozapine might antagonize both the D1R and D2R, thus reducing the D1R–D2R interaction [27].

1.2 D2R-DISC1

Disrupted in schizophrenia 1 (*Disc1*) is a strong genetic risk factor for schizophrenia and other psychiatric disorders. It was originally discovered in a large Scottish family where a balanced chromosomal translocation (1:11) (q42.1;q14.3) directly disrupts this gene and co-segregates with schizophrenia, bipolar disorder, and recurrent major depression [28–31]. Structurally, a C-terminal truncated DISC1 protein is formed as a result of this translocation event. By sequencing *Disc1* exons in patients with mental disorders, multiple studies have reported that mutations in *DISC1* are likely involved in the pathophysiology [32–34]. DISC1 functions as a scaffolding protein, where it interacts with many important signaling proteins such as transcription factors and proteins related to cytoskeletal organization [35, 36]. The complex DISC1 interacting network further suggests a strong association between DISC1 and mental illnesses [37–39].

Mutations in the *Disc1* gene and altered DISC1 protein expression have been reported to affect the dopaminergic system in many transgenic animal models. Mice with the *Disc1*-L100P (334T/C) mutation showed histological and behavioral abnormalities, such as enlarged lateral ventricles, abnormal cortical lamination, pre-pulse inhibition deficits (PPI), disrupted latent inhibition, and hyperactivity, all of which are consistent with those observed in human schizophrenia patients [40]. Moreover, these behavioral deficits could be rescued by antipsychotic medications, especially haloperidol [41]. In another transgenic mouse model where a putative dominant-negative DISC1 (DN-DISC1) is expressed under the αCaMKII promotor (exclusive expression in the cortex and hippocampus), the ratio of D2Rs in the striatum vs. cerebellum was consistently higher after administering a D2R antagonist, [^{11}C] raclopride. Similar results were illustrated with [^{3}H] spiperone, another D2R antagonist using autoradiography. Together, these findings indicated that DN-DISC1 mice have more D2Rs available for ligand binding in the striatum. Real-time PCR also showed an increase in mRNA levels of the long, postsynaptic D2R isoform in the striatum [42], consistent with other reports describing a small but significant elevation of striatal D2Rs in untreated schizophrenia patients [43–45]. Behaviorally, DN-DISC1 mice showed significantly greater augmentation in methamphetamine-induced locomotion. In vivo microdialysis also demonstrated a larger increase in the proportion of extracellular dopamine levels in the ventral striatum after methamphetamine injection, agreeing with clinical, pharmacological, and PET studies in humans. Baseline dopamine levels were substantially decreased while the dopamine transporter was increased in the striatum of DN-DISC1 mice [42]. Finally, another group overexpressed full-length human DISC1 in rats and reported a significant increase in dopamine D2 high-affinity receptor levels using [^{3}H] domperidone binding challenged with dopamine [46]. In conclusion, all the present *Disc1* animal models showed some degree of overlapping traits, including: (1) increased locomotion after amphetamine administration and (2) increased dopamine levels in the nucleus accumbens after amphetamine administration [47]. These data provide extensive evidence showing a strong association between DISC1 and the dopaminergic system.

Recently, a direct protein–protein interaction between D2R and DISC1 was identified and confirmed using co-immunoprecipitation in solubilized protein from mouse and rat striatum, as well as in HEK293T cells transfected with D2R and DISC1 [41]. DISC1 was able to be coimmunoprecipitated by anti-D2R antibody, and vice versa using solubilized protein of animal brain tissues or transfected cells. GST pull-down and in vitro binding assays were used to determine the motif within D2R that binds to the N-terminus of DISC1. The binding of DISC1 to D2R results in the activation of D2R-mediated GSK3 signaling. Disruption of this interaction by an interfering peptide rescued the behavior deficits relevant to schizophrenia in mice and rats. Surprisingly, this peptide did not induce catalepsy, which is a strong predictor of extrapyramidal symptoms in humans, indicating a potential novel and effective drug target with minimal side effects for developing therapeutics of schizophrenia.

1.3 Methods Used to Study D2R–D1R and D2R–DISC1 Interactions

There are many methods used in studying protein–protein interactions, including immunohistochemistry/immunocytochemistry, co-immunoprecipitation, fluorescence/bioluminescence resonance energy transfer (FRET/BRET), protein labeling, GST pull-down, yeast two-hybrid system, and in vitro binding assays. Each method has its own specific advantages and disadvantages in determining protein interactions [48]. Immunohistochemistry/immunocytochemistry can provide spatial information of protein localization, and detect colocalization of potential interacting proteins, but this only represents the possibility of an interaction. FRET/BRET, protein labeling, and yeast two-hybrid system require the use of transfected cells or yeast system, which may not completely resemble the in vivo setting and hence unable to confirm the existence of protein interactions. Co-immunoprecipitation assay is a better option in verifying the actual existence of protein interactions in brain tissues, but a disadvantage in this method is the uncertainty of whether the interaction is direct or indirect. In contrast, in vitro binding assay can prove the direct interactions of two proteins. Along with GST pull-down assay, these two methods act in concert in deducing the binding motifs within interacting proteins, which is important for further drug development. To conclude, a combination of different biochemical methods is needed to study the mechanisms and functions of a certain protein–protein interaction in the brain.

As described above, co-immunoprecipitation, GST pull-down, and in vitro binding assays are the main methods to study D2R–D1R and D2R–DISC1 interactions. We will discuss the details of these three methods in this chapter.

2 Materials

2.1 Co-immuno-precipitation and Western Blot

1. Phosphate-buffered saline (PBS): NaCl 8.0 g, $Na_2HPO4 \cdot 12$ H_2O 2.08 g, KCl 0.2 g, KH_2PO_4 0.2 g dissolved in 1 l ultra-pure water, at pH 7.4.

2. Protein A/G plus agarose beads (Santa Cruz Biotechnology).

3. Animal tissue or cell lysis buffer: 150 mM NaCl, 2 mM EDTA, 50 mM Tris-HCl pH 7.4, 0.5% sodium deoxycholate, 1% NP-40, 1% Triton X-100, 0.1% SDS. Add before use: 1 mM (working concentration) phenylmethanesulfonyl fluoride (PMSF), and protease inhibitor cocktail (Sigma-Aldrich) (see **Note 1**).

4. 1× Tris-buffered saline (TBS) (pH 7.4): 10 mM Tris (hydroxy-methyl) aminomethane (Tris), 150 mM NaCl.

5. Co-immunoprecipitation wash buffer: 0.1% (v/v) Triton X-100 (Sigma-Aldrich) prepared in 1× TBS.

6. 2× SDS sample buffer (Bio-Rad): 65.8 mM Tris-HCl, pH 6.8; 2.1% (w/v) SDS; 26.3% (w/v) glycerol; 0.01% (w/v) bromophenol blue. This is used as loading buffer in samples for SDS-PAGE and Western Blot.

7. 10× SDS-running buffer: 30.2 g Tris, 100 ml 10% SDS, and 188 g glycine dissolved in 1 l ultrapure water.

8. 10× transfer buffer: 30.9 g Boric acid, and 7.45 g Ethylenediaminetetraacetic acid (EDTA) dissolved in 1 l ultra-pure water, and modify pH to 8.9.

9. Nitrocellulose membranes or polyvinylidene difluoride (PVDF) membranes (Bio-Rad). Nitrocellulose membranes are soaked directly into transfer buffer before use, while PVDF membranes should be activated with methyl alcohol before soaking into transfer buffer. Methyl alcohol should be used carefully as it is toxic.

10. 1× TBST: 0.05% (v/v) Tween 20 prepared in 1× TBS.

11. Blocking buffer: 5% (w/v) non-fat dry milk prepared in 1× TBST.

12. Primary antibody dilution buffer: 1% (w/v) bovine serum albumin (BSA) (Sigma-Aldrich) and 0.02% (w/v) NaN_3 (extremely toxic, wear personnel protective equipments) prepared in 1× TBST.

13. Secondary antibody dilution buffer: 1× TBST (see **Note 2**).

14. Horseradish peroxidase-conjugated secondary antibody (Sigma-Aldrich) against the primary antibody species.

15. Enhanced cheminoluminescence (ECL) reagents (PerkinElmer) (freshly prepare before use according to manu-

facturer's instructions) and X-ray film (light sensitive, and should be used only in dark room) (Kodak). Nowadays, the ChemiDoc Imaging Systems (Bio-Rad) are often used for Western blot, so X-ray films can be replaced.

16. Tissue homogenizer.

17. Refrigerated centrifuge.

18. Rocking/rotating platform.

19. SDS-PAGE and membrane transfer equipment.

2.2 Cloning of GST-Fusion Protein cDNA Constructs

1. pGEX-4T-3 cloning vector (other pGEX vectors are also available).

2. cDNA of protein of interest.

3. Restriction enzymes suitable for double digestion (New England Biolabs).

4. Taq polymerase with buffer (New England Biolabs) (many other types of polymerase may be used recently).

5. dNTP (New England Biolabs).

6. Customized primers used for polymerase chain reaction (PCR) of genes of interest.

7. 50× TAE buffer (should be prepared in a fume hood): Tris-HCl 242 g, Acetic acid 57.1 ml (may hurt eyes, wear personnel protective equipments), 0.5 M EDTA (pH 8.0) 100 ml are added into ultrapure water to a final volume of 1 l.

8. Agarose gel: 1–3% (w/v) agarose dissolved in 1× TAE buffer, heat with microwave, and add 0.1% (v/v) ethidium bromide (EB) before use. EB is hazardous chemical. Wear personnel protective equipment to avoid skin touch. EB is a highly sensitive visualizing DNA reagent classically used in DNA electrophoresis. Alternatively, safer chemicals such as SYBR Green available from Invitrogen can also be used (see **Note 3** for safely carrying EB procedures).

9. 100× BSA (10 mg/ml, New England BioLabs): 100 time working concentration.

10. T4 DNA ligase with buffer (New England BioLabs).

11. Competent cells: DH5α, BL21 (Invitrogen).

12. Luria broth (LB) medium: 25 g LB powder dissolved in 1 l ultrapure water (add 15 g agar for bacterial growing plates), and autoclave for 45 min (121 °C at 205.8 KPa). Add 100 μg/ml ampicillin in sterilized LB before use.

13. GenElute Plasmid Miniprep Kit (Sigma-Aldrich).

14. 37 °C bacterial incubator.

15. 37 °C bacterial shaker.

2.3 GST Pull-Down Assay	1. Isopropyl β-D-1-thiogalactopyranoside (IPTG) (Bioshop Canada Inc.): IPTG powder dissolved in ultrapure water to a final concentration as 1 M, aliquot and stock at −20 °C.

***2.3 GST
Pull-Down Assay***

1. Isopropyl β-D-1-thiogalactopyranoside (IPTG) (Bioshop Canada Inc.): IPTG powder dissolved in ultrapure water to a final concentration as 1 M, aliquot and stock at −20 °C.

2. Glutathione-Sepharose 4B beads (GE Healthcare).

3. Lysis buffer for bacteria: 1% Triton X-100 dissolved in 1× PBS, add protease inhibitor cocktail prior to use.

4. Glutathione-Sepharose 4B beads wash buffer: 1× PBS.

5. GST-fusion protein elution buffer: 50 mM Tris-HCl, 10 mM reduced glutathione, pH 8.0. It is possible to dispense in 1–10 ml aliquots and store at −20 °C until needed. Avoid more than five freeze/thaw cycles.

6. 2-Mercaptoethanol (BME) (Sigma-Aldrich) (toxic, should be used in a fume hood, and wear personnel protective equipment).

***2.4 In Vitro
Binding Assay***

1. T7 Quick Master Mix (Promega).

2. [^{35}S]-Methionine.

3. Luciferase control DNA.

4. BioMax (Kodak) films.

5. Filter paper (VWR, used for coarse precipitates).

6. Gel dryer.

7. Vacuum pump.

3 Methods

***3.1 Co-immuno-
precipitation Assay***

1. **Preparation of tissue extract**.

   (a) **For animal tissues**: Put animal tissue into either 1.5 or 15 ml centrifuge tubes (acutely dissected tissue is best but frozen tissue kept at −80 °C can also be used). Add lysis buffer into the tubes (1 ml lysis buffer per 100 mg tissue). Homogenize on ice at 8000–10,000 rpm three times for 15 s each when there is not visible pellets in the liquid.

   For cultured cells: remove the medium from the culture plates by suction, and wash the cells with 1× PBS for three times (for non-adherent cells, collect the cells in 1.5 ml tubes, centrifuge and discard the medium, and wash with 1× PBS for three times), and keep the plates on ice. Add lysis buffer into the culture plates (150 μl for 35 mm (diameter) well, 300 μl for 60 mm well, and 500 μl for 100 mm well), and scrape the cells off with cell scrapers, and collect the cells as well as lysis buffer in 1.5 ml centrifuge tubes. Some researchers may prefer to use sonication to lyse the cultured cells, but we do not recommend sonication here for co-immunoprecipitation according to our experience (see **Note 4** for selection of lysis and washing buffers to use in co-immunoprecipitation).

(b) Shake at 4 °C for 1 h (animal tissues) or 30 min (cultured cells) using rocking/rotating platform.

(c) Centrifuge for 10 min at 4 °C, 12,000 g.

(d) Collect the supernatant, which contains the solubilized tissue extract (described here is the method for total protein extracts, however, in some situations, subcellular fractions may be specifically required for co-immunoprecipitation, in which case special fractionation methods are needed to obtain the protein extracts).

(e) Measure the protein concentration with BCA protein assay (Pierce) or Bradford protein assay (Bio-rad).

1. **Co-immunoprecipitation Setup**.

(a) Preparation of protein A/G plus agarose beads (Santa Cruz Biotechnology). Wash 25 µl of protein A/G plus agarose beads with 1 ml cold 1× PBS in a 1.5 ml Eppendorf tube for 5 min using a rocking/rotating platform at 4 °C. Centrifuge at 1000 g in a microfuge for 1 min at 4 °C. Discard supernatant and repeat washing twice. For a large amount of samples, total amount of beads for all the samples can be washed together to make conditions equal between different samples, and the beads can be aliquoted after washing.

(b) Preparation of protein extracts: 500–1000 µg of solubilized protein extract from animal tissues or cultured cells is added into lysis buffer to a final volume of 500–1000 µl in a 1.5 ml Eppendorf tube. Protease inhibitor cocktail should be added to the mixture prior to next step.

(c) Add 2–4 µg primary antibodies or related IgG (as a negative control) to each sample tube (see **Note 5** for selection of primary antibodies).

(d) Add 25 µl of protein A/G plus agarose beads prepared in step **a** to the mixture of antibody and protein extracts. The sequence of adding beads, proteins, and antibody into the final co-immunoprecipitation mixture may be different between different operators, just choose the best working condition for the experiments after performing for several times.

(e) Rotate at 4 °C overnight to allow binding of beads to the primary antibody and antibody to proteins (see **Note 6** for optional preclearing).

(f) Centrifuge at 1000 g for 5 min to pellet the beads and discard the supernatant. In some situation, supernatant may be kept for quality control.

(g) Wash the beads with 1 ml cold tissue lysis buffer at 4 °C. Spin and remove the supernatant. Repeat two more times. For some weak protein–protein interactions, 0.1% Triton X-100 dissolved in 1× PBS can be used as washing buffer.

(h) Following the third wash, discard supernatant. Add 25 μl of 2× SDS sample buffer to beads, and heat at 100 °C for 5 min.

Figure 1 shows a representative example of a co-immunoprecipitation experimental result [41].

3.2 Wet Transfer Western Blotting

1. Prepare SDS-polyacrylamide gels at a concentration of 7.5–15% (see **Note 7**).

2. Load co-immunoprecipitation samples (without the beads) into individual wells of the gel. Run at a constant voltage of 90 V until the dye front reaches the interface between the stacking gel and the separating gel. Then, increase voltage to 120 V and let the dye front run to the bottom of separating gel. The time for running the gel can be modified according to the molecular weight of proteins of interest.

3. Transfer the resolved proteins from gel to a nitrocellulose (or PVDF) membrane using constant current of 400 mA for 2 h with ice cold transfer buffer (see **Note 8**).

4. After transfer, incubate the membrane with 5% non-fat dry milk blocking buffer at room temperature for 1 h.

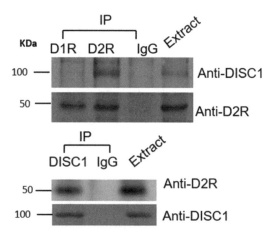

Fig. 1 D2R forms a complex with DISC1. In rat brain striatal lysate, D2R antibody, but not D1R antibody or IgG (negative control), coimmunoprecipitated with DISC1 (top panel); DISC1 antibody coimmunoprecipitated with D2R (bottom panel)

5. Following the blocking step, gently wash the membrane with 1×
TBST and incubate the membrane with the primary antibody
solution at 4 °C overnight with gentle shaking (see **Note 9**).

6. After incubating the primary antibody, wash the membrane
with 1× TBST three times for 5 min each, and incubate with
secondary antibody at room temperature for 2 h with gentle
shaking (see **Note 10**).

7. After incubating with the secondary antibody, wash the membrane with 1× TBST six times for 5 min each, and incubate
with ECL according to the manufacturer's instructions.

8. Finally, the bands are detected with Imaging system or X-ray
films.

3.3 DNA cloning

1. **Design and synthesize primer pairs for target gene
fragments**.

2. **Obtain cDNA encoding the proteins of interest (the cDNA
may be plasmids, DNA fragments, or any types of DNA
containing the cDNA encoding the proteins of interest).**

3. **PCR amplification of target gene fragments**.

   (a) PCR setup:

10× Taq polymerase buffer	5 μl
dNTP mix (2.5 mM each)	4 μl
Template	10–100 ng
Forward primer (10 μM)	0.625 μl
Reversed primer (10 μM)	0.625 μl
Taq DNA polymerase	0.25 μl
Add ultrapure water to a final volume of 50 μl	

   Template: the cDNA encoding proteins of interest without TAT
   tag.

   The amount of template and primers can be modified during
   optimization.

   (b) PCR condition

94 °C	3 min	
94 °C	30 s	30 cycles
58–65 °C	30–60 s	
72 °C	60–90 s (1 kb/min)	
72 °C	10 min	
4 °C	Pause	

The 58–65 °C step can be modified according to the proportion of GC content in the primers, and can be optimized after performing for several times.

(c) Add 6× DNA loading buffer (ThermoScientific) into PCR products, load the mixture into 1–3% DNA agarose gel, and run the gel. Typically, we used the 6× DNA loading buffer but any other concentration should be fine. (As agarose gels are prepared with EB, a hazardous material, wear personnel protective equipment and handle gels safely.)

(d) Check the results with UV light, and cut the bands at the right size (see **Notes 11**).

(e) Extract the DNA with a DNA extraction kit, and store at −20 °C until use (up to 6 months). Typically, the purified DNA is used immediately or within 1 week to obtain the best experimental results.

4. **Digestions of PCR product and cloning vector using two restriction enzymes.**

   (a) Restriction digestion setup

10× NEB buffer	5 μl
Vector or PCR products	1 μg
100× BSA	0.5 μl
Enzyme I (NEB, USA)	1 μl
Enzyme II (NEB, USA)	1 μl
Add ultrapure water to 50 μl	

   (b) Incubate the mixture at 37 °C for 3–4 h (see **Note 12** for the use of restriction enzymes and BSA). Here, restriction enzymes have been selected to obtain a digested DNA with cohesive ends.

   (c) Add 6× DNA loading buffer to digestion reaction, load the mixture onto 1–3% DNA agarose gel, and run the gel (dephosphorylation of the cloning vector can be also done prior to adding DNA loading buffer).

   (d) Check the results with UV light, and cut the bands of the right size.

   (e) Extract the DNA with DNA extraction kit, and store at −20 °C until use (up to 6 months). Typically, the purified DNA is used immediately or within 1 week. However, for this step, long-term storage of the DNA extract is not recommended since the digested double-stranded DNA has cohesive ends.

5. **Ligation of the gene fragments with vector**.

   (a) Ligation reaction setup (see **Notes 13** and **14**).

10× T4 DNA ligase buffer	2 μl
vector: insert	1: 3–10
T4 DNA ligase (NEB)	1 μl
Add ultrapure water to 20 μl	

   (b) Incubate at 4 °C overnight.

6. **DNA transformation of DH5α competent cells**.

   (a) Obtain heat shock competent cells from −80 °C and let them thaw on ice (10–20 min) (see **Note 15** for the use of different competent cells).

   (b) Add the ligation mixture into 100 μl of competent cells, gently mix and allow them to sit on ice for 30 min.

   (c) Heat shock the mixture at 42 °C for 90–120 s, and put on ice immediately for 2 min.

   (d) Add 800 μl LB medium without ampicillin, incubate in shaker at 37 °C with gentle agitation (around 120 rpm) for 40 min (this incubation is necessary for transformation of ligation products, but it may not be needed for transformation of plasmids, since plasmids are easier to incorporate into the bacteria).

   (e) Centrifuge the mixture at 1000 g at room temperature for 2 min and discard 700–800 μl of supernatant. Resuspend the competent cells with the remaining supernatant, transfer and spread evenly onto LB-ampicillin plates. Wait until there is no liquid flowing on the plates, and then put into 37 °C incubator (LB-ampicillin plates do not need to be prewarmed at 37 °C before use. Plates can be left at room temperature 10–20 min before use). Put plates in bacterial incubator to grow colonies overnight (12–16 h). Harvest the colonies once they have reached a sufficient size).

7. **DNA miniprep and sequencing**.

   (a) When the bacterial colonies have reached a suitable amount (depending on the operator), pick single colony with a 10 μl pipette tip (or inoculating ring), and put the tip into a 10 ml bacterial culture tube prefilled with 5 ml LB medium with ampicillin. To allow enough air for the bacteria to grow in the medium, no more than 5 ml LB medium is added in the 10 ml tubes.

   (b) Place the tube into bacterial shaker and incubate until the liquid becomes cloudy (it usually takes an overnight incubation, which is about 12–16 h).

(c) Perform the DNA miniprep with GenElute Plasmid Miniprep Kit.

(d) Send the DNA miniprep samples to a sequencing facility to confirm integrity of DNA sequences following PCR (see **Note 16**. Some confirmation steps can be carried out before sending the samples, such as double restriction digestion and PCR as described in Sect. 3.3, steps **3** and **4**).

3.4 GST Pull-Down Assay

1. **Expression of GST-fusion protein.**

   (a) Transform BL21 competent cells with plasmids. Allow bacteria to grow in 15 ml LB with antibiotics (100 μg/ml) in 50 ml tubes. Incubate bacteria at 37 °C (250 rpm) in a shaking incubator overnight (see **Note 17** for selection of antibiotics).

   (b) Remove the tubes and pour 25 ml of bacterial culture into 225 ml LB with antibiotics (100 μg/ml).

   (c) Allow bacteria to grow at 37 °C in a shaking incubator (250 rpm) for 1.5–2 h. For step **b** and **c**, we use 1:10 dilution of the overnight cultured bacteria, because the OD600 of the bacteria is able to reach 0.6–1.2 in 1.5–2 h according to our experience, which can save some time for the whole experimental procedures. Other dilution ratios can also be taken into account depending on the operator's experience and preference.

   (d) Quantify bacterial growth at an optical density of 600 nm (OD600). OD600 should be between 0.6 and 1.2. If OD600 does not reach the optimal value, continue growing the bacteria and quantify the OD600 every 30 min until it reaches the optimal value.

   (e) Add IPTG to a final concentration of 0.5 mM, and incubate at 30 °C or 28 °C (to inhibit proliferation of bacteria and to enhance the synthesis of GST-fusion proteins), with shaking at 250 rpm for 3 h (see **Note 18**).

   (f) Pour bacterial culture into a 250 ml centrifuge bottle and spin at 1000 *g* for 20 min at 4 °C.

   (g) Discard the supernatant.

   (h) Add 10 ml of bacterial lysis buffer (1% Triton X-100 and protease inhibitor prepared in PBS), resuspend pellet, and transfer to a 15 ml centrifuge tube. Sonicate the suspending pellet for 40 s three times to lyse the bacteria wall.

   (i) Place tubes in a shaker at 4 °C and solubilize bacterial lysate for 1 h.

   (j) Centrifuge at 12,000 *g* at 4 °C for 20 min, and keep the supernatant.

(k) Add 150 µl Glutathione-Sepharose 4B beads to a 1.5 ml Eppendorf tube and wash with 1 ml of 0.1% Triton X-100 prepared in 1× PBS. Place the tube on a rocking/rotating platform at 4 °C and mix for 5 min. Centrifuge at 1000 g for 1 min at 4 °C.

(l) Discard the supernatant and retain the beads.

(m) Mix supernatant (isolated in step **1j**) to beads in a new 15 ml centrifuge tube. Place the tube on a rocking/rotating platform and mix the supernatant and beads for 2 h at 4 °C.

(n) Centrifuge samples at 1000 g for 1–2 min at 4 °C. Discard the supernatant.

(o) Add 10 ml 0.1% Triton X-100 prepared in 1× PBS into beads, and place on a shaker to rotate at 4 °C for 5 min. Centrifuge at 4 °C, 1000 g for 1–2 min, and discard the supernatant. Repeat for three times.

(p) Discard the supernatant as much as possible, and transfer the beads to a 1.5 ml Eppendorf tube.

(q) Add 200 µl Elution buffer into beads. Resuspend beads with the pipette, put on a rocking/rotating platform and mix at room temperature for 20 min.

(r) Centrifuge at 10, 000 g for 20 min at room temperature. Keep the supernatant, and store at −80 °C.

2. **Preparation of Tissue Extract**.

(a) GST pull-down is generally carried out using animal brain tissues, so only the preparation of tissue extract is described in this protocol. Place animal tissue into 1.5 ml or 10 ml centrifuge tubes (acutely dissected tissue is best but frozen tissue kept at −80 °C can also be used). Add lysis buffer into the tubes (1 ml lysis buffer per 100 mg tissue). Homogenize at 8000–10,000 rpm for three times, 15 s each. Put the tube on a rocking/rotating platform at 4 °C and solubilize the lysate for 1 h.

(b) Centrifuge at 10,000 g for 10 min at 4 °C. Transfer the supernatant (solubilized tissue extract) to a new 1.5 ml or 10 ml centrifuge tube.

(c) Measure the protein concentration of tissue extract and GST-fusion protein samples with Bradford protein assay. Since glutathione in the GST-fusion protein elution buffer can react with the BCA working solution, BCA protein assay is not suitable to quantify GST-fusion protein samples.

3. **GST pull-down**.

   (a) Prepare 30 μl of Glutathione-Sepharose 4B beads for GST pull-down as described in Sect. 3.4, steps **1k–1**.

   (b) Add GST-fusion protein (50–100 μg) and solubilized tissue extract (500–1000 μg) to prewashed Glutathione-Sepharose 4B beads. A solution of 0.1% Triton X-100/PBS/protease inhibitor cocktail is added to a final volume of 500 μl (see **Note 19**).

   (c) Incubate at 4 °C overnight using rocking/rotating platform.

   (d) Centrifuge samples at 1000 *g* for 1 min at 4 °C, and discard the supernatant. Add 1 ml of 0.1% Triton X-100/PBS, and put on a rocking/rotating platform at 4 °C for 5 min. Repeat the washing step three times (see **Note 20**).

   (e) Following the last wash, discard supernatant but leave ~30 μl in tube and add 30 μl sample buffer (95% 2× SDS sample buffer + 5% 2-Mercaptoethanol (BME)). Heat tubes at 100 °C for 5 min. Centrifuge the samples at 1000 *g* for 1 min at room temperature. Samples are subjected to SDS-PAGE and Western Blotting.

Figure 2a–d shows a representative example of a GST pull-down experiment result [41].

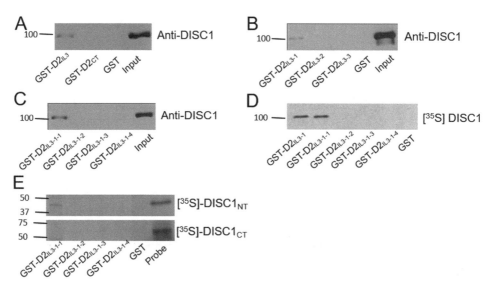

Fig. 2 DISC1 forms a complex with D2R via the K211-T225 region of the IL3 of D2R. (**a**) Western blot showing that GST-D2IL3, but not GST-D2CT, can "pull down" DISC1 from rat striatal tissue. (**b**) Western blot showing that GST-D2IL3-1, but not GST-D2IL3-2 or GST-D2IL3-3, can "pull down" DISC1 from rat striatal tissue. (**c**) Western blot showing that GST-D2IL3-1-1, but not GST-D2IL3-1-2, GST-D2IL3-1-3, or GST-D2IL3-1-4, can "pull down" DISC1 from rat striatal tissue. (**d**) GST-D2IL3-1 and GST-D2IL3-1-1 directly bind to DISC1 in vitro. GST-D2IL3-1, GST-D2IL3-1-1, GST-D2IL3-1-2, GST-D2IL3-1-3, and GST-D2IL3-1-4 were incubated with [35S]-DISC1, and the precipitated proteins were subjected to SDS-PAGE. (**e**) GST-D2IL3-1-1 directly binds to DISC1NT in vitro. GST-D2IL3-1, GST-D2IL3-1-1, GST-D2IL3-1-2, GST-D2IL3-1-3, and GST-D2IL3-1-4 were incubated with [35S]-DISC1NT or [35S]-DISC1CT, and the precipitated proteins were subjected to SDS-PAGE

**3.5 In Vitro
Binding Assay**

1. **Probe Synthesis.**

   (a) **Reaction Setup.**

   T7 Quick Master Mix (Promega; use according to manu-facturer's instructions): 40 µl cDNA (Negative Control: without plasmid DNA; Positive Control: using luciferase control DNA 1 µg): 0.5–1.0 µg (approximately 1 µl).

   [^{35}S]-Methionine (1000 Ci/mM at 10 mCi/ml): 2 µl (if specific activity of the radiolabeled probe is too low, the amount of µCi [^{35}S]-methionine can be increased).

   ddH$_2$O 7 µl.

   Total 50 µl.

   (b) Incubate at 30 °C for 1.5 h (reaction temperature and time can be modified to optimize specific activity of radiola-beled probe).

   (c) Run gel to check the probe.

   The radiolabeled products can be stored at −20 °C for up to 2 months or at −70 °C for up to 6 months. Handling of radioac-tive material should be conducted according to Institutional Radiation Compliance Safety Office.

2. **In vitro binding.**

   (a) 1× PBS (1 ml) + GST-fusion protein (20 µg) + Glutathione-Sepharose 4B beads (10 µl) + Probe synthesized in Sect. 3.5, step 1 (10 µl, stored at −20 °C).

   (b) Incubate at room temperature for 1 h.

   (c) Wash twice with 0.1% Triton X-100 prepared in 1× PBS.

   (d) Add 15 µl of 2× SDS loading buffer.

   (e) Boil samples (Negative Control: GST protein; Positive Control: 1 µl of Probe) at 100 °C for 5 min and load sam-ples onto gels (see **Note 7**).

   (f) Run samples on SDS-PAGE for approximately 1 h.

3. **Gel Handling and Exposure (after running the gel).**

   (a) At the end of the run, put the gel into a plastic box con-tainer in a fume hood. Soak the gel in enhancer solution. The volume of enhancer solution used depends on the size of the gel and the box container. Make sure the gel is com-pletely soaked in solution. Incubate for 20–40 min. Discard enhancer solution into a bottle specific for radioactive liquid waste.

   (b) Wash gel with ultrapure water for three times, 20 min each. The volume of water should be 1.5–2 times of the volume of enhancer solution used in Sect. 3.5, step **3a**.

(c) Cut filter paper to fit the size of gel.

(d) Carefully place gel on the filter paper. Avoid bubble formation in the middle of the gel during this procedure.

(e) Put the gel on a gel dryer. Pour some dry ice around the vacuum pump. Turn on power of gel dryer and vacuum pump.

(f) Let the gel dry for 2 h.

(g) Dried gels are subjected to autoradiography using BioMax (Kodak) film (a 7-day exposure is recommended at room temperature, and the exposure time can be modified depending on the signal).

Figure 2e shows a representative example of results obtained with in vitro binding assay result [41].

4   Notes

1. 100 mM PMSF solution prepared in isopropyl alcohol can be stored at −20 °C for long-term usage, but vortexing is required before use. Wear gloves when using PMSF, which is an extremely toxic substance that can be absorbed through skin.

2. For HRP-conjugated secondary antibody, 1× TBST is the commonly used dilution buffer. However, for some species, such as anti-goat HRP-conjugated secondary antibody, a strong background tends to appear in the membrane when detecting the bands. In this situation, a dilution buffer containing 5% non-fat dry milk prepared in 1× TBST can be used to avoid strong background signal. Furthermore, 5% non-fat dry milk instead of 1% BSA will be an alternative solution in diluting primary antibody for Western blot, which may also help dampen the background signal.

3. **EB Handling**.

   3.1 Safe EB Operating Procedure

   (a) Select a designated EB-working area, and place a clear notice labeling "Ethidium Bromide in use. Toxic!!!".

   (b) Cover the working area with a disposable plastic-backed absorbent pad in case of spills.

   (c) Put everything being used on the pad, and all the containers should be kept closed. Besides, gloves should be placed close to the area for easy access, but not within the working area.

   (d) Procedures without involving EB should be performed outside the area, such as weighing the agarose. Afterwards, all other procedures involving the use of EB should be conducted in the specified area. Gloves should be changed after contact with EB-containing materials. Two layers of gloves should be used to provide additional safety.

(e) Wash your hands when changing gloves. Use a UV light (EB will emit a fluorescence reddish-brown color) to check for contamination in the work area. If decontamination is needed, please follow the methods described below. Be careful when using UV light, especially protect from UV light irradiation, as it may cause skin carcinoma after exposure for a long time.

(f) Use fresh towels and a soap/water solution, or towels soaked in ethanol to wipe the contaminated area or equipment multiple times. Check to make sure if the decontamination process is complete using UV light.

(g) Place fresh paper towels soaked in ethanol over the contaminated surface. Sprinkle activated charcoal on the towels. Use additional towels to remove ethanol/charcoal mixture. Place all clean-up materials into a plastic hazardous waste bag. Check if there is any remaining contamination with UV light and repeat decontamination process if needed.

3.2 Personal Protective Equipment

(a) Before beginning experiments using EB, wear standard nitrile laboratory gloves (two layers are better), and a fully buttoned lab coat with sleeves extending to the wrists. To provide better protection, use tape to hold the sleeves tightly, and make sure the gloves cover the end of the sleeves to avoid any chances of EB contacting skin.

(b) Wear goggles and a face shield as accidental splashes may occur.

(c) Wear UV-blocking eyewear or work behind a UV shielding glass to avoid direct contact with UV light when performing experiments or decontamination processes involving EB. Try to minimize the use UV light as short as possible.

(d) Tyvek sleeves and/or gowns (or other air-tight non-woven textile) should be worn if the arms or torso are exposed to liquid suspensions or dry particles.

3.3 Disposal EB Waste

(a) Trace amounts of EB in gels for electrophoresis gels should not pose a hazard. Higher concentrations, for example, as evident by the color of the gel becoming dark pink or red, should be treated with caution. Less than 0.1% EB can be placed in laboratory trash but gels with more than or equal to 0.1% should be deposited in biohazard box for incineration.

(b) For all materials used in handling EB, including gloves, test tubes, and paper towels, they should be placed in biohazard waste for incineration. Deactivate in bleach before disposal if needed.

(c) In some laboratories, EB solutions may be used to stain the gels. EB solutions should be disposed as follows:
   Aqueous solutions containing <10 μg/ml EB can be drained through the sink. Aqueous solutions containing >10 μg/ml EB should be filtered or deactivated using charcoal filtration or chemical neutralization. Solutions containing heavy metals, organics, cyanides, or sulfides should be disposed as hazardous waste.

4. Since both D2R and D1R are membrane proteins, the type of lysis buffer used for isolation of tissue protein, co-immunoprecipitation, and GST pull-down assays is critical. Selection of lysis and washing buffers should depend on the binding affinity between interacting proteins. Triton X-100 and SDS can be initially considered. Alternatively, commercially available membrane protein isolation kits may also be utilized. In some circumstances, glycerol (10–30%) may be included in the lysis buffer, as it helps preserve the existing protein complexes in the extracts.

5. The specificity of co-immunoprecipitation assays depends primarily on antibodies used for immunoprecipitation. Thus, proper selection of antibodies is essential for success. If the selected antibody targets an epitope that mediates the protein–protein interaction, it will fail to pull down interacting proteins. The basic principle with regard to antibodies used for immunoprecipitation is that they should be of different host species as to those used for Western blot detection. Occasionally, many attempts are required to optimize a protocol that provides ideal results for one co-immunoprecipitation assay.

6. Co-immunoprecipitation experiments using IgG as negative control can allow experimenter to determine the extent of nonspecific immunoprecipitated bands. If nonspecific immunoprecipitation is detected, a preclearing step is recommended to reduce the amount of non-specificity. Prior to adding antibody, solubilized tissue extract is incubated with protein A/G plus agarose beads at 4 °C for 1 h. Centrifuge and transfer the supernatant to a new tube. Add new prewashed protein A/G plus agarose beads and antibodies, and then incubate at 4 °C overnight.

7. The concentration of acrylamide chosen for the gels depends on the molecular weight of proteins in samples. Use a higher concentration if the protein of interest has a low molecular weight, and vice versa for proteins with higher molecular weight. The selection of 10-well or 15-well combs depends on the amount of samples to be loaded on gels.

8. Larger size proteins move slower than smaller ones on SDS-PAGE and wet transfer processes. Thus, larger size proteins may take a longer time during the transfer process. However, the experimenter can change the icebox during transfer to keep the transfer buffer cold for efficient protein transfer.

9. For Western Blotting, highly specific antibodies are critical to obtain reliable results.

10. For some proteins with molecular weights at around 50 kDa, the bands will overlap with the bands of the antibody heavy chain used for immunoprecipitation (the size is also around 50 kDa), which has been a major difficulty in co-immunoprecipitation. It is preferable to choose antibodies from different species used for immunoprecipitation and immunoblotting. Some cross-link immunoprecipitation kits can resolve this issue; however, the interacting proteins will also be cross-linked and become one single protein molecule with altered molecular weights. Nowadays, there are some secondary antibodies or some HRP-conjugated protein binding to the light chains (around 30 kDa) of an antibody and will not recognize the heavy chain of antibodies used for immunoprecipitation. This may provide an alternative solution for detecting proteins with molecular weights at around 50 kDa.

11. After PCR, the time for running the gel is not fixed, but depends on the size of PCR product. It is necessary to check regularly with UV light until the bands are sufficiently separated. The concentration of the gel and the type of DNA ladder used should be adjusted based on the PCR product size. For smaller PCR products, higher concentration of agarose gel should be used, which reduces band diffusion in the gel.

12. It is better to choose enzymes that use the same buffer, and a double digestion system from NEB can be used. Furthermore, all the restriction enzymes used in DNA subcloning should be stored at -20 °C. BSA is necessary in most restriction digestions, but the experimenter will need to decide according to the manufacturer's instructions. 37 °C is normally used for restriction digestions, but temperature may be different with manufacturer's instructions.

13. The vector:insert ratio is essential for a successful ligation reaction, which is normally between 1:3 and 1:10. Different ratios

for one pair of vector and insert can be tested. For smaller fragments (<200 bp), the experimenter can initially choose a ratio close to 1:10, while a ratio close to 1:3 is selected for longer fragments (>1 kb). Several attempts may be needed for successful ligations.

14. T4 ligase buffer is kept frozen in −20 °C, and a salt precipitate is formed upon thawing. Dissolve the salt precipitate by vortexing vigorously or by warming the buffer at 37 °C.

15. Two kinds of competent cell strains are often used in transformation. For recombinant DNA cloning, bacterial strains such as DH5α and TOP10 can be employed, while for expression of fusion proteins, the *E. coli* strain BL21 is usually chosen.

16. Before sequencing the plasmid constructs, digestion and PCR can be used to assess if the insertion is successful and of the correct size. With diagnostic restriction digests, only two bands should be detected in the gel, corresponding to the vector and the insert fragment. With diagnostic PCR, there should be only one band for the insert fragment. Besides, a PCR without template is used as a negative control, and a PCR with the template at the beginning of the cloning acting as a positive control is recommended. For shorter insert fragments, it is better to use PCR, whereas for longer fragments, both digestion and PCR should be conducted.

17. For GST pull-down assays, two antibiotics are commonly used for culturing bacteria: ampicillin and kanamycin. The experimenter should choose the antibiotic according to the antibiotic resistance marker of plasmids used to transform bacteria. For example, expression plasmids encoding GST-tagged proteins typically carry an ampicillin resistance marker.

18. Time and temperature for GST-fusion protein expression can be modified depending on the OD600 value, protein stability, and protein expression levels. Generally speaking for protein stability, larger size proteins are more prone to degradation. Thus, a lower temperature should be used for these proteins to inhibit protease activity of bacteria, but experimenters should sustain the temperature above 18 °C to keep the expression functioning in the bacteria.

19. A final volume of 500 μl is typically used in our GST pull-down assays. If final volume is brought up to 1 ml, the experimenter will need to use more GST-fusion protein and solubilized tissue extract. This may improve detection sensitivity on protein–protein interactions in Western blots.

20. Wash solution recipes for GST pull-down assays depend on the specificity and binding affinity of the interactions. The experimenter can experiment various concentrations of Triton X-100, starting from low to high. Nonspecific bands are fre-

quently detected with the GST alone, in which case a new transformation of BL21 with GST vector should be performed to generate a new purified batch of GST alone for negative control. Freshly transformed BL21 is the ideal choice for GST pull-down assays, because the quality of the bacteria is the best to synthesize GST-fusion protein and avoid nonspecific binding.

References

1. van Os J, Kapur S (2009) Schizophrenia. Lancet 374:635–645
2. Taly A (2013) Novel approaches to drug design for the treatment of schizophrenia. Expert Opin Drug Discov 8:1285–1296
3. Wong AH, Van Tol HH (2003) Schizophrenia: from phenomenology to neurobiology. Neurosci Biobehav Rev 27:269–306
4. Glatt SJ et al (2003) Meta-analysis identifies an association between the dopamine D2 receptor gene and schizophrenia. Mol Psychiatry 8:911–915
5. Seeman P, Kapur S (2000) Schizophrenia: more dopamine, more D2 receptors. Proc Natl Acad Sci U S A 97:7673–7675
6. Roberts DA et al (1994) The abundance of mRNA for dopamine D2 receptor isoforms in brain tissue from controls and schizophrenics. Brain Res Mol Brain Res 25:173–175
7. Bertorello AM et al (1990) Inhibition by dopamine of (Na(+)+K+)ATPase activity in neostriatal neurons through D1 and D2 dopamine receptor synergism. Nature 347:386–388
8. Missale C et al (1998) Dopamine receptors: from structure to function. Physiol Rev 78:189–225
9. Neve KA et al (2004) Dopamine receptor signaling. J Recept Signal Transduct Res 24:165–205
10. Beaulieu JM et al (2007) The Akt-GSK-3 signaling cascade in the actions of dopamine. Trends Pharmacol Sci 28:166–172
11. Beaulieu JM et al (2009) Akt/GSK3 signaling in the action of psychotropic drugs. Annu Rev Pharmacol Toxicol 49:327–347
12. Beaulieu JM et al (2008) A beta-arrestin 2 signaling complex mediates lithium action on behavior. Cell 132:125–136
13. Beaulieu JM et al (2005) An Akt/beta-arrestin 2/PP2A signaling complex mediates dopaminergic neurotransmission and behavior. Cell 122:261–273
14. Beaulieu JM et al (2004) Lithium antagonizes dopamine-dependent behaviors mediated by an AKT/glycogen synthase kinase 3 signaling cascade. Proc Natl Acad Sci U S A 101:5099–5104
15. Beaulieu JM et al (2007) Regulation of Akt signaling by D2 and D3 dopamine receptors in vivo. J Neurosci 27:881–885
16. Su P et al (2015) Protein interactions with dopamine receptors as potential new drug targets for treating schizophrenia. In: Lipina T, Roder J (eds) Drug discovery for schizophrenia, 1st edn. Royal Society of Chemistry, London, pp 202–233
17. Fuxe K et al (2014) Dopamine D2 heteroreceptor complexes and their receptor-receptor interactions in ventral striatum: novel targets for antipsychotic drugs. Prog Brain Res 211:113–139
18. Shioda N et al (2010) Advanced research on dopamine signaling to develop drugs for the treatment of mental disorders: proteins interacting with the third cytoplasmic loop of dopamine D2 and D3 receptors. J Pharmacol Sci 114:25–31
19. Wang M et al (2008) Dopamine receptor interacting proteins (DRIPs) of dopamine D1-like receptors in the central nervous system. Mol Cells 25:149–157
20. Lee SP et al (2004) Dopamine D1 and D2 receptor Co-activation generates a novel phospholipase C-mediated calcium signal. J Biol Chem 279:35671–35678
21. So CH et al (2005) D1 and D2 dopamine receptors form heterooligomers and cointernalize after selective activation of either receptor. Mol Pharmacol 68:568–578
22. Pei L et al (2010) Uncoupling the dopamine D1-D2 receptor complex exerts antidepressant-like effects. Nat Med 16:1393–1395
23. Rashid AJ et al (2007) D1-D2 dopamine receptor heterooligomers with unique pharma-

cology are coupled to rapid activation of Gq/11 in the striatum. Proc Natl Acad Sci U S A 104:654–659

24. So CH et al (2007) Desensitization of the dopamine D1 and D2 receptor hetero-oligomer mediated calcium signal by agonist occupancy of either receptor. Mol Pharmacol 72:450–462

25. Brown V, Liu F (2014) Intranasal delivery of a peptide with antidepressant-like effect. Neuropsychopharmacology 39:2131–2141

26. Dziedzicka-Wasylewska M et al (2008) Mechanism of action of clozapine in the context of dopamine D1-D2 receptor hetero-dimerization—a working hypothesis. Pharmacol Rep 60:581–587

27. Faron-Gorecka A et al (2008) The role of D1-D2 receptor hetero-dimerization in the mechanism of action of clozapine. Eur Neuropsychopharmacol 18:682–691

28. Brandon NJ, Sawa A (2011) Linking neurodevelopmental and synaptic theories of mental illness through DISC1. Nat Rev Neurosci 12:707–722

29. Hodgkinson CA et al (2004) Disrupted in schizophrenia 1 (DISC1): association with schizophrenia, schizoaffective disorder, and bipolar disorder. Am J Hum Genet 75:862–872

30. Porteous DJ et al (2011) DISC1 at 10: connecting psychiatric genetics and neuroscience. Trends Mol Med 17:699–706

31. Taylor MS et al (2003) Evolutionary constraints on the disrupted in schizophrenia locus. Genomics 81:67–77

32. Johnstone M et al (2015) Copy number variations in DISC1 and DISC1-Interacting partners in major mental illness. Mol Neuropsychiatry 1:175–190

33. Luo X et al (2016) Association study of DISC1 genetic variants with the risk of schizophrenia. Psychiatr Genet 26:132–135

34. Thomson PA et al (2014) 708 Common and 2010 rare DISC1 locus variants identified in 1542 subjects: analysis for association with psychiatric disorder and cognitive traits. Mol Psychiatry 19:668–675

35. Duan X et al (2007) Disrupted-In-Schizophrenia 1 regulates integration of newly generated neurons in the adult brain. Cell 130:1146–1158

36. Lee FH et al (2011) Disc1 point mutations in mice affect development of the cerebral cortex. J Neurosci 31:3197–3206

37. Teng S et al (2017) Rare disruptive variants in the DISC1 Interactome and Regulome: association with cognitive ability and schizophrenia. Mol Psychiatry. https://doi.org/10.1038/mp.2017.115

38. Enomoto A et al (2009) Roles of disrupted-in-schizophrenia 1-interacting protein girdin in postnatal development of the dentate gyrus. Neuron 63:774–787

39. Kim JY et al (2012) Interplay between DISC1 and GABA signaling regulates neurogenesis in mice and risk for schizophrenia. Cell 148:1051–1064

40. Clapcote SJ et al (2007) Behavioral phenotypes of Disc1 missense mutations in mice. Neuron 54:387–402

41. Su P et al (2014) A dopamine D2 receptor-DISC1 protein complex may contribute to antipsychotic-like effects. Neuron 84:1302–1316

42. Jaaro-Peled H et al (2013) Subcortical dopaminergic deficits in a DISC1 mutant model: a study in direct reference to human molecular brain imaging. Hum Mol Genet 22:1574–1580

43. Laruelle M (1998) Imaging dopamine transmission in schizophrenia. A review and meta-analysis. Q J Nucl Med 42:211–221

44. Abi-Dargham A (2009) The neurochemistry of schizophrenia : a focus on dopamine and glutamate. In: Charney DS, Nestler EJ (eds) Neurobiology of mental illness, 3rd edn. Oxford University Press, New York, pp 321–328

45. Abi-Dargham A et al (1998) Increased striatal dopamine transmission in schizophrenia: confirmation in a second cohort. Am J Psychiatry 155:761–767

46. Trossbach SV et al (2016) Misassembly of full-length Disrupted-in-Schizophrenia 1 protein is linked to altered dopamine homeostasis and behavioral deficits. Mol Psychiatry 21:1561–1572

47. Dahoun T et al (2017) The impact of Disrupted-in-Schizophrenia 1 (DISC1) on the dopaminergic system: a systematic review. Transl Psychiatry 7:e1015

48. Su P et al (2014) Study of crosstalk between dopamine receptors and ion channels. In: Tiberi M (ed) Dopamine receptor technologies, 1st edn. Humana, New York, pp 277–302

Chapter 12

Methods to Identify the Signature of Trimers Formed by Three G Protein-Coupled Receptors or by Two G Protein-Coupled and One Ionotropic Receptor with Special Emphasis in the Functional Role in the Central Nervous System

Irene Reyes-Resina, Eva Martínez-Pinilla, Dasiel O. Borroto-Escuela, Kjell Fuxe, Gemma Navarro, and Rafael Franco

Abstract

Whereas ionotropic receptors have been considered as functional units consisting of interacting subunits, G protein-coupled receptors were considered as monomeric cell surface receptors. Experimental evidence from the end of the twentieth century has demonstrated that GPCR may also form dimers, trimers, and even high-order oligomers. Novel techniques (BRET, FRET, SRET) to detect interactions between GPCRs have appeared that led to a substantial advancement in the field, i.e., to identify an ever-increasing number of GPCR homo- and heteroreceptor complexes. A main drawback of these techniques is that they cannot be applied to detection of receptor complexes in brain. Fortunately, novel techniques and novel concepts such as the *heteromer signature* may be used now to detect GPCR complexes in specific brain regions and in specific neuronal and/or glial cells. Remarkably those techniques make now possible to detect and give insight into the function of receptors formed by even three GPCRs or by two GPCRs and one ionotropic receptor such as the NMDA glutamate receptor. The central nervous system has been the main target for such a revolution in understanding how cell surface receptors participate in neurotransmission and/or regulate cell signaling and fate.

Key words Cannabinoid CB_1, Cannabinoid CB_2, Dopamine D_1, Histamine H_3, Heteromer signature, GPR55, NMDA receptor, SRET, BRET, FRET, Proximity ligation assay, Bimolecular complementation

1 Introduction

Alzheimer's disease (AD) is one of the most devastating neurodegenerative diseases with enormous personal, social, and economic burden. None of the myriad of therapeutic approaches has been successful. It should be however noted that targeting cell surface neuronal receptors is one of the most studied approaches and, in

Kjell Fuxe and Dasiel O. Borroto-Escuela (eds.), *Receptor-Receptor Interactions in the Central Nervous System*, Neuromethods, vol. 140, https://doi.org/10.1007/978-1-4939-8576-0_12, © Springer Science+Business Media, LLC, part of Springer Nature 2018

fact, one of the two available drugs prescribed for AD patients, memantine, targets N-methyl-D-aspartate (NMDA) ionotropic glutamate receptors.

Ionotropic receptors are main players in neurotransmission in both peripheral and central nervous system. Many years ago, it was demonstrated that these receptors are constituted by different subunits that interact to form a ligand-induced ion channel. As an example, NMDA receptors are tetramers formed by subunits whose nomenclature is NR, and the quaternary structure constitutes a calcium ion channel that opens when glutamate appears in the synaptic cleft (see [1] and references therein).

In contrast to the oligomeric structure of ionotropic channels, functional G protein-coupled receptors (GPCR) were considered as monomeric structures. Recent, cumulative, and solid evidence shows that many GPCRs are expressed in the cell surface and function as dimers, even as higher order oligomeric structures. The number of homodimers, i.e., receptors formed by two identical receptors, and the number of heteromers, i.e., GPCRs formed by different receptors, are steadily increasing and a conservative estimation indicates that 600 dimer/oligomer structures have been already identified (see www.gpcr-hetnet.com/ and references therein).

The occurrence of GPCRs leads to many interesting questions and technological challenges. One is related to the size of the functional units in the cell surface. A relevant example derived from results from our laboratory has demonstrated that adenosine A_1 and A_{2A} receptors form heteromeric structures that function as concentration-sensing devices. The effect at low and high adenosine concentration on glutamate release is opposite and this cannot be explained by GPCR monomers but by A_1-A_{2A} receptor heteromers. A further question that constitutes the focus of this manuscript is how to demonstrate whether GPCRs may interact with other type of cell surface receptors to give rise to novel units that may be relevant for neurotransmission events and targets to combat neurodegenerative diseases.

We have recently provided evidence of macromolecular complexes formed by three receptors: one ionotropic (NMDA) receptor and two GPCRs (dopamine D_1 and histamine H_3) [2]. The reason for investigating such possibility was based on two previous pieces of relevant data. On the one hand, evidence for the formation of seemingly direct interactions between dopamine D_1 and NMDA receptors was provided in two different laboratories [3, 4]. On the other hand, we had shown that dopamine receptors in striatal neurons may form complexes with histamine H_3 receptors and that such complexes qualitatively and quantitatively shape the histaminergic signaling in striatum. In summary, we undertook actions to demonstrate the formation of complexes of three receptors knowing that it is technologically challenging to obtain reliable data. It should be noted that demonstration in natural sources of "trimers" formed by three (A, B, and C) receptors is not possible although some of the already known techniques to detect "dimers" are being explored to be able to detect higher order structures. It is, however, possible to obtain data on trimer formation in heterologous systems and go to a natural source to find for some of the in vitro-obtained specific features. On doing so and on detecting in natural sources dimers of R_1-R_2, R_1-R_3, and R_2-R_3, Occam's razor rule allows establishing (as a working hypothesis) the existence of functional units formed by three different receptors. We here focus on the most recent and relevant techniques that may be used to (1) demonstrate in heterologous systems and to (2) suggest in natural sources the formation of macromolecular complexes formed by three GPCRs or two GPCRs and one ionotropic receptor.

2 Methods

2.1 Co-immuno-precipitation

Co-immunoprecipitation was one of the pioneering techniques to demonstrate interactions (even to propose stoichiometries) between cell surface receptors and cell surface receptor subunits in lymphocytes and other cells of the immune system. The technique was later used to look for protein-protein interaction in any cell type. Especially difficult is to look for interactions in the central nervous system due to the heterogeneity in what concerns different cell types and in the difficulty to obtain human brain samples; that is, whereas human blood and samples of other tissues are available with more or less difficulties, the majority of human neurological samples are only postmortem available and, usually, from patients that have undertaken medications that may have affected the occurrence and function of receptor heteromers. The technique has been widely described and is used today in preliminary assays but not to demonstrate direct receptor-receptor interactions. The reason is that clusters of detergent and proteins may lead to false-positive results, i.e., to suspect direct interaction for proteins that are in the same cluster (even in the same-membrane microdomain) but not directly interacting. Although this technique will not be here addressed, some solid protocols are found in the literature [5–9].

2.2 Energy Transfer Techniques to Detect Dimers

2.2.1 Bioluminescence Resonance Energy Transfer (BRET) Assays

To perform bioluminescence resonance energy transfer (BRET) assays to study interaction between two cell surface receptors, one of them must be fused to the Renilla luciferase (Rluc) and the other one to the variant of a fluorescent protein, for instance, a variant of the green fluorescence protein; yellow fluorescent protein (YFP) is one of the most used (Fig. 1a). It is advisable to try both combinations, i.e., receptor R_1 fused to Rluc with R_2 fused to YFP, and the other way around, R_2 fused to Rluc and R_1 to YFP. Often, one combination works better than the other.

To prepare vectors with appropriate DNA sequences, commercially available plasmids containing Rluc or YFP may be used. Receptor (human, rat, mouse, etc.,) sequences must be inserted without their stop codon. This requires the design of primers and PCR performance and to take into account that Rluc and YFP are fused to the C-termini of GPCRs; that is, they will have a topologically cytosolic orientation. pRLuc-N1 (PerkinElmer, Wellesley, USA) or pEYFP (Clontech, Heidelberg, Germany) may be used. Sense and antisense primers harboring unique restriction sites for two different restriction enzymes will ultimately generate the R-Rluc and R-YFP fusion proteins.

A very common cell model for receptor expression is the HEK-293T cell line, probably the most used cell line of human origin. HEK-293T cells are maintained in a humid atmosphere of 5% CO_2 at 37 °C, and grown in 6-well plates in Dulbecco's modified Eagle's medium (DMEM) (Gibco, Paisley, Scotland, UK) supplemented with 5% (v/v) heat-inactivated fetal bovine serum (FBS) (Invitrogen, Paisley, Scotland, UK), 100 U/ml penicillin/streptomycin, 2 mM L-glutamine, and MEM non-essential amino acid solution (1/100). Transfection may be transient or stable, but the transient one is by far the most used for energy transfer studies, mainly due to the progressive and unequal loss of expression in stable cell clones. Importantly, in all energy transfer assays performed in plate fluorescence readers it is necessary to validate the fusion proteins, i.e., to test whether these fusion proteins are functional as the non-fused receptors. For such purpose, experiments comparing fusion and unfused protein expression in cells must be performed; ERK 1/2 phosphorylation and cAMP-level determination are often used for checking for functional fusion proteins.

Co-transfection may be achieved, as previously described [10], using the polyethylenimine (PEI, Sigma-Aldrich. St. Louis, MO, USA) method with a constant amount of cDNA encoding for R_1-Rluc and increasing amounts of cDNA corresponding to R_2-YFP. Cells are washed 48 h after transfection in HBSS (137 mM NaCl; 5 mM KCl; 0.34 mM Na_2HPO_4; 0.44 mM KH_2PO_4; 1.26 mM $CaCl_2$; 0.4 mM $MgSO_4$; 0.5 mM $MgCl_2$; and 10 mM HEPES pH 7.4) supplemented with 0.1% glucose (w/v), detached by gently pipetting and resuspended in the same buffer. To adjust

Fig. 1 (continued) with BiFC scheme differs from (**A**) in that each half of the BRET acceptor YFP is fused to a GPCR (R_1 and R_2, green and purple, respectively), while Rluc is fused to a third receptor, for instance a subunit (GluN2, NR_2) of an ionotropic NMDA receptor. SRET^2 (**C**) and SRET^1 (**D**) schemes are similar to the BRET scheme, but in this case the fusion proteins are composed of each one of the three GPCRs (R_1, R_2, and R_3, green, purple, and pink, respectively) linked to Rluc, GFP^2, and YFP (**C**) or Rluc, YFP, and DsRed (**D**), respectively. In SRET^2 DeepBlueC is used as Rluc substrate (brown dots), while in SRET^1 coelenterazine H is used (red dots). Panels **C** and **D** bottom: A representation of the excitation (up) and emission (down) spectra of fused proteins, taken from [14]

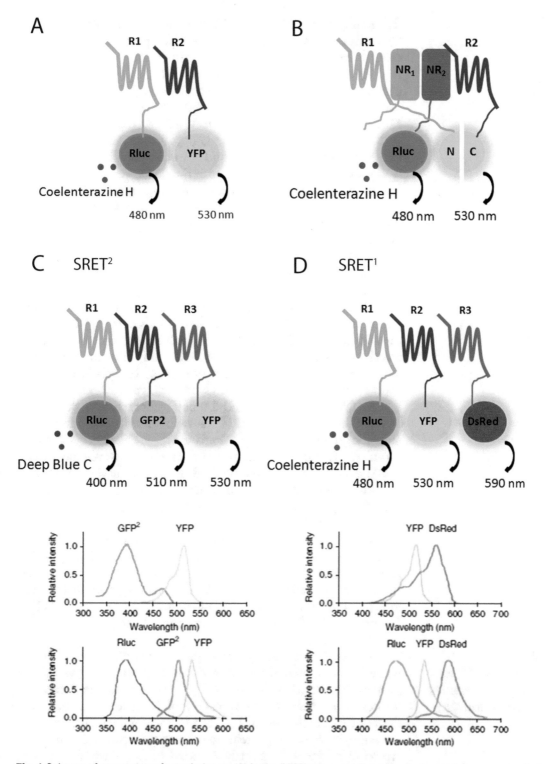

Fig. 1 Scheme of energy transfer techniques. (**A**) In the BRET scheme, interacting fusion proteins composed of GPCRs (R₁ and R₂, green and purple, respectively) linked to Rluc and YFP are shown. The red dots represent the Rluc substrate, coelenterazine H. Numbers indicate the peak wavelength of the emitted light after the addition of coelenterazine H, also represented by the colored halos of Rluc and YFP. (**B**) The BRET combined

the number of cells, a Bradford assay kit (Bio-Rad, Munich, Germany) is used to determine sample protein concentration, using bovine serum albumin dilutions as standards. To quantify R_2-YFP fluorescence expression, cells (20 µg protein) are distributed in 96-well microplates (black plates with a transparent bottom; Porvair, Leatherhead, UK) and fluorescence is read using a Mithras LB 940 (Berthold, Bad Wildbad, Germany) equipped with a high-energy xenon flash lamp, using a 10 nm bandwidth excitation filter at 485 nm, and an emission filter at 530 nm. Receptor-fluorescence expression is the fluorescence of the sample minus the fluorescence of cells expressing only R_1-Rluc; values should be between 5000 and 50,000. To measure BRET signal, the equivalent of 20 µg protein of cell suspension is distributed in 96-well microplates (white plates; Porvair) and 5 µM coelenterazine H (2-4-dehydroxy coelenterazine) (PJK GMBH, Kleinblittersdorf, Germany) is added. Degradation of this substrate by Rluc produces light that emits a wavelength that excites YFP if BRET donor and acceptor are close. Readings are collected using a Mithras LB 940 (Berthold, Bad Wildbad, Germany) 1 min after the addition of coelenterazine H. Signals are detected in the short-wavelength filter at 485 nm (440–500 nm) and the long-wavelength filter at 530 nm (510–590 nm). To quantify R_1-Rluc expression, luminescence readings are performed 10 min after adding 5 µM coelenterazine H. Luminescence should be around 100,000 units. The net BRET is defined as [(long-wavelength emission)/(short-wavelength emission)] – Cf where Cf corresponds to [(long-wavelength emission)/(short-wavelength emission)] the cells expressing only R_1-Rluc. BRET curves are adjusted by using a nonlinear regression equation, assuming a single phase with GraphPad Prism software (San Diego, CA, USA). BRET values are given as milli BRET units (mBU: 1000 × net BRET).

If there is a specific interaction between receptors, a saturation curve should be obtained (values should be higher than 20 mBU to be considered as significant). A negative control is always performed by co-transfecting cDNAs for two receptors that do not interact (one of them fused to Rluc and the other one to YFP). Unspecific BRET gives a linear non-saturable graph.

2.2.2 Detection of Trimers by BRET and Bimolecular Fluorescence Complementation (BiFc)

Bimolecular fluorescence complementation (BiFc) is the technical name for an approach that provides fluorescence only when two proteins interact. Briefly, GPCR R_1-R_2 dimers may be detected if fluorescence arises upon expression of R_1 fused to a half of a fluorescent protein and of R_2 fused to the complementary half of the same fluorescence protein. In other words, two fusion proteins are required for BiFc, namely R_1 fused to the N-terminal half of YFP and R_2 fused to the C-terminal half of YFP. Upon co-expression of fusion proteins and excitation at 485 nm, detection of emission fluorescence at 530 nm indicates a specific (and direct) interaction between R_1 and R_2 receptors. When a BRET design is added by using a receptor fused to Rluc, it is then possible to detect trimers (Fig. 1B). This combination of techniques has been used to detect, *inter alia*, functional calcitonin gene-related peptide (CGRP) receptor [11] and dopamine D_2, cannabinoid CB_1, and adenosine A_{2A} receptor hetero-oligomers [12].

For complementation assays, appropriate vectors must contain the protein/receptor of interest fused to a half of the fluorescent protein. Albeit commercially available, vectors containing the N- and C-terminal truncated versions of, for instance, YFP may be created. For such purpose, full-length YFP is subcloned in the XhoI site of pcDNA3.1 vector (Invitrogen). The N-terminal truncated version of YFP, nYFP (amino acids 1–155), may be made by PCR amplification and cloning into the XhoI site of pcDNA3.1 using the following primers: FnYFP (5′-CCGCTCGAGACCATGGTGAGCAAGG GCGAGGAGC-3′) and RnYFP (5′-CCGTCTAGATCAGGCC ATGATATAGACGTTG-3′). Also, a C-terminal truncated version of YFP, cYFP (amino acids 155–231), may be obtained using the same strategy and the following primers: FcYFP (5′-CGCTCGAGA CCATGGACAAGCAGAAGAACGGC-3′) and RcYFP (5′-CCGT CTAGATTACTTGTACAGCTCGTCCAT-3′).

cDNAs for the three receptors of interest are amplified using primers to suppress the stop codons. Sense and antisense primers harboring unique sites for restriction enzymes are used to subsequently clone each receptor in the following plasmids: R_1 in pRluc-N1 (plasmid containing the Renilla luciferase sequence), R_2 in pcDNA3.1-nYFP (plasmid containing the sequence of the N-terminal part of YFP), and R_3 in pcDNA3.1-cYFP (plasmid containing the sequence of the C-terminal part of YFP). Accordingly, the final vectors are R_1-Rluc, R_2-nYFP, and R_3-cYFP. It should be again noted that all plasmids express luminescent or complementary parts of fluorescent proteins fused at the C-terminal ends of the receptors.

For BRET-BiFc, HEK-293T cells are transiently co-transfected (as indicated above) with a constant amount of cDNA encoding for R_1-Rluc and with increasing equal amounts of cDNAs corresponding to R_2-nYFP and R_3-cYFP. Protein quantification and readings for fluorescence, bioluminescence, and energy transfer are performed as described for BRET. BRET is finally measured as indicated above and results expressed in milli BRET units.

Similarly, Rluc may be divided into two halves, each of them fused to a different receptor. On doing that, tetramer formation may be measured using cells expressing R_1-nRluc, R_2-cRluc, R_3-cYFP, and R_4-nYFP. Technical problems and high levels of cell death and/or low transfection yield may arise from the need to transfect cells with four different cDNAs.

2.2.3 Sequential BRET-FRET (SRET)

Sequential BRET-FRET (SRET) is a biophysical technique that combines BRET and fluorescence resonance energy transfer (FRET) techniques to allow the detection of GPCR trimers at the cell membrane level [13]. It combines a BRET assay, in which energy is transferred from the luciferase Rluc (after oxidation of its substrate, coelenterazine H) to a fluorescent BRET acceptor, with a FRET assay in which the BRET acceptor is a FRET donor. In fact, emission of such FRET donor excites the FRET acceptor. Obviously, the whole process requires that BRET/FRET donor/acceptors are in close proximity.

For a SRET experiment, cells should express a receptor fused to Rluc, a receptor fused to a BRET acceptor (GFP² or YFP), and a receptor fused to a FRET acceptor that fits with the BRET acceptor chosen (YFP in the case of GFP² and DsRed in the case of YFP). Oxidation of the Rluc substrate triggers acceptor excitation and subsequent energy transfer to the FRET acceptor. SRET will only occur between these fusion proteins if the two coupling pairs, Rluc/GFP² and GFP²/YFP or Rluc/YFP and YFP/DsRed, are at a distance of no more than 10 nm. The two aforementioned SRET variants (Rluc-YFP-DsRed and Rluc-GFP²-YFP) are known as SRET[1] (Fig. 1D) and SRET[2] (Fig. 1C), respectively [14]. Both variants are useful for trimer detection. SRET[1] has advantages as no overlapping occurs between YFP and DsRed emission spectra. In contrast, GFP² and YFP emission spectra are quite close in the SRET[2] setup (Fig. 1C, D, bottom). Given the spectral emission of the GFP²/YFP pair, the contribution of GFP² or YFP proteins to the two detection channels is significant. Thus, the spectral signature [15] of these fusion proteins must be determined and considered for SRET and YFP expression evaluation (for details see [14]). The choice of which receptor is fused to which donor/acceptor is important, as the relative orientation of donor and acceptor may lead to negative results in the case of interacting proteins. Therefore, it is advisable to make different combination of receptors and of donor/acceptors for double check on negative results.

As well as in BRET and FRET, single-point determinations may be performed in SRET. Obviously, saturation curves can be done but to obtain a SRET saturation curve is technically more challenging than for BRET. Rluc levels should be (after substrate addition) around 100,000 luminescence units, while fluorescent proteins should render approximately 5000–20,000 fluorescence units. To further improve SRET, preliminary BRET and FRET assays should be performed to adjust transfection ratios of the BRET and FRET pairs. If the donor protein expression is high, then the fluorescence background will be high and SRET signal will be poorly detected. If the donor protein expression is low, the acceptor protein will not be appropriately excited. For saturation curves, cells must be transfected with fixed amounts of the fusion proteins which contain Rluc and YFP or GFP² (BRET acceptor), and with increasing amounts of the fusion protein which contains the FRET acceptor (DsRed or YFP, respectively).

The amount of fusion protein expression must be near within physiological range. If possible, it is also advisable to use confocal microscopy to visualize (by its own fluorescence or by using antibodies anti-Rluc) correct expression and cellular localization of fusion proteins.

Expression vectors are similar to those used in BRET and FRET and the only precaution is to find an appropriate BRET acceptor, because it (in turn) has to become a FRET donor. cDNAs for the human version of GPCRs without their stop codon are obtained by PCR and subcloned to RLuc-containing vector (pRLuc-N1; PerkinElmer, Wellesley, MA), to pGFP^2-N3(h) vector (humanized pGFP^2-N3(h) from Perkin Elmer, Dreieich, Germany), to pEYFP-containing vector (pEYFP-N1; Clontech, Heidelberg, Germany) or to DsRed vector (DsRed-Express from Clontech, Heidelberg, Germany) using sense and antisense primers harboring unique restriction sites for two different restriction enzymes, generating receptor-Rluc, receptor-GFP^2, receptor-YFP, or receptor-DsRed fusion proteins.

For SRET, HEK-293T cells are cultured and transiently transfected with the plasmids encoding for the indicated fusion proteins in each case, and the cell number is controlled as described for BRET. Black plates with a transparent bottom are used for FRET and fluorescence determinations, and white plates with white opaque bottom for Rluc, BRET, and SRET determinations. Using aliquots of transfected cells (20 µg of protein), three trials are performed in parallel:

1. First, acceptor fluorescence is read in a Fluostar Optima Fluorimeter equipped with a high-energy xenon flash lamp, using an excitation filter at 485 nm and 10 nm bandwidth emission filter corresponding to 530 nm (527–536 nm) for R_3-YFP readings and an excitation filter at 400 nm and 10 nm bandwidth emission filter corresponding to 510 nm (506–515 nm) for R_2-GFP^2 readings (for SRET^2). Also, R_2-GFP^2/ R_3-YFP FRET is read exciting R_2-GFP^2 and collecting readings for R_2-GFP^2 and R_3-YFP at the same time. For SRET^1, R_3-DsRed fluorescence is read using an excitation filter at 544 nm and a 10 nm bandwidth emission filter corresponding to 590 nm (586–595 nm), and R_2-YFP fluorescence is read as described above. Also, R_2-YFP/R_3-DsRed FRET is read exciting YFP and simultaneously reading YFP and DsRed fluorescence. Gain settings are identical for all experiments to keep the relative contribution of the fluorophores to the detection channels constant for spectral unmixing. In both cases the second acceptor fluorescence is the fluorescence to be read but subtracting the fluorescence of cells expressing only R_1-Rluc and R_2-GFP^2 (SRET^2 assays) or R_2-YFP (SRET^1 assays).

2. Quantitation of R_1-Rluc expression by determining the luminescence resulting from R_1-Rluc. For these assays, cells are distributed in 96-well microplates (Corning; white plates with

white bottom) and luminescence is determined 10 min after addition of 5 µM coelenterazine H in a Mithras LB 940 multimode reader (Berthold Technologies).

3. SRET^2 measurements in cells distributed in 96-well microplates (Corning; white plates with white bottom) 30 s after adding 5 µM DeepBlueC (Molecular Probes) are performed using a Mithras LB 940 reader with detection filters for short wavelength (400 nm (370–450 nm)) and long wavelength (530 nm (510–590 nm)). For SRET^1 experiments signal is collected after adding 5 µM coelenterazine H, using a Fluostar Optima Fluorimeter using a 590 nm filter (586–595 nm; long wavelength). In this case, we also determined the luminescence resulting from R_1-Rluc 1 min after addition of 5 µM coelenterazine H in a Mithras LB 940 (short wavelength at 485 nm (370–450 nm)). By analogy with BRET, we defined net SRET as ((long-wavelength emission)/(short-wavelength emission)) – Cf, where Cf corresponds to ((long-wavelength emission)/(short-wavelength emission)) in cells expressing R_1-Rluc, R_2-GFP^2, or R_2-YFP and the third receptor not fused to a fluorescence protein.

Results are represented as net SRET versus the ratio R_3-YFP expression/R_1-Rluc expression for SRET^2 or the ratio R_3-DsRed expression/R_1Rluc expression for SRET^1. Adequate controls should also be done. As a positive control, an assay with a fusion protein containing the three Rluc, GFP^2, and YFP or Rluc, YFP, and DsRed proteins fused together, respectively, can be performed. Another possibility is performing an assay in which each of the subunits of a G protein is fused to Rluc, GFP^2, and YFP or to Rluc, YFP, and DsRed, respectively (e.g., G_α-Rluc, G_β-YFP, and G_γ-DsRed). As negative controls, assays in which one of the fusion proteins contains only the fluorescent protein and not the receptors (e.g., R_1-Rluc, R_2-YFP, DsRed), or assays in which one of the fusion proteins is substituted by another one containing a receptor that is known to not interact with the other two, are recommended (this should render a linear graph instead of a saturation curve).

2.2.4 Restraints of Energy Transfer Assays

The main limitation is the need to use heterologous expression systems to test the interactions. It may for instance happen that two receptors that are able for a tight in vitro interaction, will never meet simultaneously in any cell; hence if two proteins susceptible of interaction are never co-expressed in the same cell they will not have the chance for in vivo interaction. Moreover, to our knowledge no transgenic animal has been generated with deletion of the receptor genes and/or expressing transgenes for fusion proteins of receptors with donor/acceptors of FRET or BRET. It is true, however, that it is possible to perform confocal FRET to assess direct interactions between two proteins in preparations of intact cells. This and other more sophisticated imaging approaches are giving insight into different aspects of receptor-receptor interactions. Useful as they are, they do not address yet the intricacies related to oligomers formed by three different proteins, some of which are oligomeric (such as NMDA receptors). For this reason, they will not constitute the main focus of this chapter. While waiting for novel developments, it is possible to undertake approaches that allow evidence for heteromer formation in natural sources.

2.3 Detection of Receptor-Receptor Interactions in Natural Sources

It is worth revisiting one recent review on the evidence of GPCR interactions [16] to know the tools available to detect receptor "dimers" and the challenges to further develop these tools to study "trimers," especially when the signal transduction of the involved receptors is so different as in the case of GPCRs and ionotropic receptors. We here focus on two approaches, one related to the use of tissue slices, i.e., working with living cells, and another related to the use of fixed tissue.

2.3.1 The Heteromer Signature

Identification of GPCR heteromers and their specific function in a specific tissue, as for example central nervous system, is a challenge in the field. By using energy transfer assays, the possibility of receptor-receptor interactions in a heterologous system could be directly demonstrated. To investigate for heteromerization in tissues one may take advantage of the novel ligand-binding properties and intracellular trafficking pathways displayed by the heteromer that are substantially different from those of its constituent receptors. In this way, ligand binding to one receptor can modulate ligand binding to the partner receptor. If this modulation consists of a blockade of the partner receptor signaling, we are facing a phenomenon known as negative cross talk. Often, a selective antagonist of receptor R_1 blocks the signal mediated by R_2 in the R_1-R_2 heteromer; this phenomenon is considered as cross-antagonism. These unique features, acting as a de facto heteromer signature/print, allow unequivocal identification of heteroreceptor complexes. During the last decade, prints have been demonstrated for GPCR heteromers such as those constituted by CB_1R-CB_2R [17], CB_1R-GPR55 [18], CB_2R-GPR55 [19], D_1-H_3 [20, 21], or D_2 and AT_1 angiotensin receptor [22].

The heteromer signature must be first, for instance, detected using ERK1/2 phosphorylation assays in co-transfected cells, and the resulting data may be later used to detect heteromer expression in tissue slices. For this purpose, HEK-293T cells growing in 60 mm diameter plates are transiently transfected with the corresponding R_1 and/or R_2 plasmids using the ramified polyethylenimine (PEI, Sigma-Aldrich) method. Cells are incubated with the corresponding cDNA together with ramified PEI (5 µl of 10 mM PEI for each µg of cDNA) and 150 mM NaCl in serum-starved medium. After 4 h, the medium is substituted by a complete culture medium. For ERK1/2 phosphorylation assays, HEK-293T-transfected cells are cultured in serum-free medium for 16 h before the addition of any further reagent.

For brain slice preparation, rats are anesthetized with 4% isoflurane (2-chloro-2-(difluoromethoxy)-1,1,1-trifluoro-ethane) and decapitated with a guillotine; brains are rapidly removed and placed in ice-cold oxygenated (O_2/CO_2: 95/5%) Krebs-HCO_3^- buffer (124 mM NaCl; 4 mM KCl; 1.25 mM NaH_2PO_4; 1.5 mM $MgCl_2$; 1.5 mM $CaCl_2$; 10 mM glucose; and 26 mM $NaHCO_3$ pH 7.4). Brains are sliced at 4 °C in a brain matrix (Zivic Instruments, Pittsburgh, PA) into 0.5 mm coronal slices and the desired area is dissected. Slices are kept at 4 °C in Krebs-HCO_3^- buffer during the dissection. Each slice is transferred into an incubation tube containing 1 ml of ice-cold Krebs-HCO_3^- buffer. The temperature is raised to 23 °C for 30 min and medium is then replaced by 2 ml Krebs-HCO_3^- buffer (23 °C). The slices are incubated under constant oxygenation (O_2/CO_2: 95/5%) at 30 °C for 4–5 h using an Eppendorf Thermomixer device (5 Prime, Inc., Boulder, CO). The media are replaced by 200 μl of fresh Krebs-HCO_3^- buffer and incubated for 30 min before addition of receptor ligands.

HEK-293T-transfected cells or brain slices are treated with selective agonists and/or antagonists for R_1 and R_2. Then, they are rinsed with ice-cold phosphate-buffered saline (PBS) and lysed by the addition of, respectively, 100 or 150 μl, of ice-cold lysis buffer (50 mM Tris–HCl pH 7.4; 50 mM NaF; 150 mM NaCl; 45 mM β-glycerophosphate; 1% Triton X-100; 0.4 mM $NaVO_4$; and protease inhibitor mixture). The cellular debris is removed by centrifugation at $13,000 \times g$ for 5 min at 4 °C, and the protein is quantified by the bicinchoninic acid method using bovine serum albumin dilutions as standard. To determine the level of ERK1/2 phosphorylation, equivalent amounts of protein (15 μg) are mixed with 6× Laemmli sample buffer; proteins are separated by electrophoresis on a denaturing 10% SDS-polyacrylamide gel and transferred onto nitrocellulose membranes (BioTrace™ NT Nitrocellulose Transfer Membrane, 66485, PALL). After that, Odyssey blocking buffer (LI-COR Biosciences, Lincoln, Nebraska, USA) is added, and the membranes are rocked for 60 min. The membranes are then probed (2–3 h) at room temperature in blocking buffer containing a mixture of a mouse anti-phospho-ERK1/2 (1:1000 dilution; M8159 Sigma-Aldrich), and rabbit anti-ERK1/2 antibodies, which recognize both phosphorylated and unphosphorylated ERK1/2 (1:20,000 dilution; M5670 Sigma-Aldrich). After three washes with TBS/Tween20, membranes are incubated with a mixture of IRDye®CW 800 (anti-mouse) antibody (1:15,000 dilution; 926-32210 LI-COR Biosciences, Lincoln, Nebraska, USA) and IRDye® 680RD (anti-rabbit) antibody (1:15,000 dilution; 926-68071 LI-COR Biosciences, Lincoln, Nebraska, USA) for 1 h at room temperature. Bands are visualized by the Odyssey® Fc Imaging System (LI-COR Biosciences, Lincoln, Nebraska, USA) and their densities quantified using the Image Studio software 1.1 (LI-COR Biosciences, Lincoln, Nebraska, USA). The level of phosphorylated ERK1/2 isoforms is normalized for differences in loading using the total ERK1/2 protein band intensities.

Remarkably, if the increase in pERK1/2 levels due to one receptor agonist is blocked by pretreatment with the partner receptor antagonist (cross-antagonism) in double-transfected cells or in brain slices, the results likely reflect a signature of the R_1-R_2 heteromer. As a control, the effect of the ligands should be tested in cells single transfected with cDNA for each receptor, in untransfected cells, and/or in transgenic animals in which one of the receptors is knocked out.

Detection of trimers in natural sources deserves special attention mainly when the receptor complexes are constituted by different type of receptors, i.e., GPCRs and ionotropic receptors. Despite the technical difficulties, a very recent study demonstrated in the rat and mouse cortex the existence of heteroreceptor complexes formed by dopamine D_1, histamine H_3, and subunits of NMDA glutamate receptors [2]. In a first step, the in vitro identification of heteromers by BRET and BiFc assays using H_3, D_1, and subunits of NMDA receptors fused to Rluc, YFP, and YFP Venus N-terminal fragment (n-YFP) or YFP Venus C-terminal fragment (c-YFP) evidenced that these three receptors are able to form heteromers in HEK-293T cells. In addition, these complexes were also detected, in rat and mouse brain sections and/or slices, by co-immunoprecipitation, by in situ proximity ligation assays (PLA), and by functional interactions via cross-antagonism.

2.3.2 In Situ Proximity Ligation Assay

In situ proximity ligation assay (PLA) has been widely used for typification of cancer, to detect two antigens in the same (tumorous) cell. Quite recently it was applied to the field of receptor-receptor interactions, and more specifically to detect interactions among GPCRs.

The Duolink II in situ PLA detection kit (Duolink® In Situ Detection Reagents Red, DUO92008, developed by Olink Bioscience, Uppsala, Sweden; and now distributed by Sigma-Aldrich as Duolink® using PLA® Technology) is used to detect the presence/absence of receptor-receptor molecular interaction in the sample. More details on the PLA protocol can be found elsewhere [23]. At present, this technique allows the detection of complexes formed by only two receptors (Fig. 2a). In fact, for the detection of GPCR trimers the only possibility is to perform three different PLA assays: to detect the interaction between R_1 and R_2, R_1 and R_3, and R_2 and R_3. If all the three dimers are detected in a high percentage of cells, the formation of the trimer may be assumed.

PLA may be carried out both in cell cultures and in tissue sections. For (rodent brain) section preparation, mouse/rat brains are rapidly removed and fixed with 4% paraformaldehyde solution for 24 h at 4 °C. Brains are then washed in PBS, cryo-preserved in a 30% sucrose solution for 48 h at 4 °C, and stored at −20 °C until sectioning. A freezing cryostat (Leica Jung CM-3000, Leica Microsystems, Mannheim, Germany) is used to cut 30 μm thick sections that are mounted on glass Superfrost™ Plus glass slides (4951PLUS, Thermo Scientific, Waltham, USA). In the case of cells, they should grow on glass coverslips and later be washed with PBS and fixed for 15 min in 4% paraformaldehyde. Samples (brain sections or cells) are washed with PBS containing 20 mM glycine (PBS-Gly) to quench the aldehyde groups, permeabilized with the same buffer containing 0.05% Triton X-100 (5 min treatment for cells and 15 min for brain sections), and finally washed with PBS-Gly. The PLA protocol continues with incubation for 1 h at 37 °C with blocking solution and then with specific antibodies against R_1 and R_2 receptors (usually diluted between 1:50 and 1:500, and which must come from different species; if this is not possible, see next paragraph) overnight at 4 °C. Both antibodies must be previously tested for specificity. After washing with buffer A (150 mM NaCl, 10 mM Tris base, 0.05% Tween-20, pH 7.4) at room temperature, samples are incubated with the plus and minus PLA probes (Duolink II PLA probe plus or minus) for detecting primary antibodies, and with Hoechst for nuclear staining (1/200; B1155, Sigma-Aldrich) for 1 h at 37 °C in a chamber with humid atmosphere. After washing in buffer A at room temperature, samples are incubated with the ligation solution for 1 h at 37 °C, again in chamber with a humid atmosphere. Sections are again washed and subsequently incubated with the amplification solution for 100 min at 37 °C in a humidity chamber. After that, sections are washed in buffer B (100 mM NaCl, 35 mM Tris base, 165 mM Tris–HCl, pH 7.5). Finally, samples are mounted using Mowiol (475904, Calbiochem, Merck KGaA, Darmstadt, Germany). To check for nonspecific labeling, negative controls must be performed by omitting either one of the primary antibodies or one of the probes, or using sections from animals in which one of the receptors is knocked out.

In the case that both primary antibodies come from the same species there is a possibility of directly linking them to the plus and minus oligonucleotides, i.e., with no need of using the so-named plus and minus probes (Fig. 2B). In fact, we may create our own proximity probes by covalent attachment of the 5′ end of oligonucleotides to our affinity-purified antibodies with a kit available from Sigma-Aldrich. To create PLA probes one of the primary antibodies is conjugated with a PLUS oligonucleotide (Duolink® In Situ Probemaker PLUS DUO92009, Sigma-Aldrich) and the other one with a MINUS oligonucleotide (Duolink® In Situ

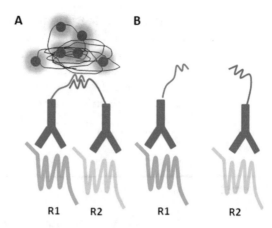

Fig. 2 Scheme of the PLA technique. Specific antibodies (blue and pink) linked to the plus and minus probes are bound to the GPCRs (R_1 and R_2, green and yellow, respectively). If the GPCRs are in close proximity the plus and minus probes will hybridize and the DNA amplification will take place (**A**), yielding red fluorescent signal under the microscope. If the GPCRs are not forming a heteromer, there will be no PLA signal (**B**). PLA images may be observed for inter alia papers related to work with the following heteromers: CB_1R-GPR55 [18], CB_2R-GPR55 [19], CB_1R-CB_2R [24], D_1R-D_2R [25], D_1R-D_3R [26] or D_2R-AT_1R [22]

Probemaker MINUS DUO92010, Sigma-Aldrich). Thus, 2 µl of conjugation buffer is mixed with 20 µl of the primary antibody. Then the mix is added to the vial which contains the lyophilized probe and incubated overnight at room temperature. After that, 2 µl of stop reagent are added and the mix is left for 30 min at room temperature. Finally, 24 µl of storage solution are added, and the probes are kept at 4 °C.

Samples are observed in a Leica SP2 confocal microscope (Leica Microsystems, Mannheim, Germany) equipped with an apochromatic 63× oil-immersion objective (N.A. 1.4). For each field of view a stack of two channels (1 per staining) and several Z stacks (approximately 6 for cell cultures and 20 for brain sections) with a step size of 1 µm are acquired. To ensure appropriate visualization of the labeled elements and to avoid false-positive results, the emission following excitation with the laser at 561 nm is filtered through a band-pass filter of 576–648 nm and color coded in red. Finally, a long-pass filter of UV laser at 406–501 nm is used to visualize the emission from the laser at 364 nm and color coded in blue.

Duolink Image tool software (DUO90806, Sigma-Olink) is used to measure the number of PLA-positive signals (red dots) per cell and the total number of cells (blue nucleus) in the sample field in images taken from the fluorescence microscope. Then, the software allows calculation of the number of cells containing the PLA red fluorescent label (in % versus total cells) and the ratio r (mean number of red spots/cell) in cells containing spots. Upon proper protocol development, the images have high resolution; that is, the results are quite reliable. Examples of PLA images obtained using brain sections from rodent or primates are found in the following reports for the indicated receptor pairs: CB_1R-GPR55 [18], CB_2R-GPR55 [19], CB_1R-CB_2R [24], D_1R-D_2R [25], D_1R-D_3R [26] or D_2R-AT_1R [22].

Acknowledgements

This work was supported by BFU-64405-R grant from the Spanish Ministry of Industry and Competitiveness (it may contain FEDER funds).

References

1. Chen PE, Wyllie DJA (2006) Pharmacological insights obtained from structure-function studies of ionotropic glutamate receptors. Br J Pharmacol 147:839–853

2. Rodríguez-Ruiz M et al (2017) Heteroreceptor complexes formed by dopamine D1, histamine H3, and N-methyl-D-aspartate glutamate receptors as targets to prevent neuronal death in Alzheimer's disease. Mol Neurobiol 54:4537–4550

3. Wang M, Wong AH, Liu F (2012) Interactions between NMDA and dopamine receptors: a potential therapeutic target. Brain Res 1476:154–163

4. Fiorentini C, Missale C (2004) Oligomeric assembly of dopamine D1 and glutamate NMDA receptors: molecular mechanisms and functional implications. Biochem Soc Trans 32:1025–1028

5. Truitt KE, Hicks CM, Imboden JB (1994) Stimulation of CD28 triggers an association between CD28 and phosphatidylinositol 3-kinase in Jurkat T cells. J Exp Med 179:1071–1076

6. Rodgers W, Crise B, Rose JK (1994) Signals determining protein tyrosine kinase and glycosyl-phosphatidylinositol-anchored protein targeting to a glycolipid-enriched membrane fraction. Mol Cell Biol 14:5384–5391

7. van der Geer P (2014) in. Methods Enzymol 541:35–47

8. Kim YJ et al (1993) Novel T cell antigen 4-1BB associates with the protein tyrosine kinase p56lck1. J Immunol 151:1255–1262

9. Fraser JD, Goldsmith MA, Weiss A (1989) Ligand-induced association between the T-cell antigen receptor and two glycoproteins. Proc Natl Acad Sci U S A 86:7133–7137

10. Navarro G et al (2012) NCS-1 associates with adenosine A2A receptors and modulates receptor function. Front Mol Neurosci 5:53

11. Héroux M, Hogue M, Lemieux S, Bouvier M (2007) Functional calcitonin gene-related peptide receptors are formed by the asymmetric assembly of a calcitonin receptor-like receptor homo-oligomer and a monomer of receptor activity-modifying protein-1. J Biol Chem 282:31610–31620

12. Navarro G et al (2008) Detection of heteromers formed by cannabinoid CB1, dopamine D2, and adenosine A2A G-protein-coupled receptors by combining bimolecular fluorescence complementation and bioluminescence energy transfer. ScientificWorldJournal 8: 1088–1097

13. Navarro G et al (2013) Detection of receptor heteromers involving dopamine receptors by the sequential BRET-FRET technology. Methods Mol Biol 964:95–105

14. Carriba P et al (2008) Detection of heteromerization of more than two proteins by sequential BRET-FRET. Nat Methods 5:727–733

15. Zimmermann T, Rietdorf J, Girod A, Georget V, Pepperkok R (2002) Spectral imaging and linear un-mixing enables improved FRET efficiency with a novel GFP2-YFP FRET pair. FEBS Lett 531:245–249

16. Franco R, Martínez-Pinilla E, Lanciego JL, Navarro G (2016) Basic pharmacological and structural evidence for Class A G-protein-coupled receptor heteromerization. Front Pharmacol 7:76

17. Callén L et al (2012) Cannabinoid receptors CB1 and CB2 form functional heteromers in brain. J Biol Chem 287:20851–20865

18. Martínez-Pinilla E et al (2014) CB1 and GPR55 receptors are co-expressed and form heteromers in rat and monkey striatum. Exp Neurol 261:44–52

19. Balenga NA et al (2014) Heteromerization of GPR55 and cannabinoid CB2 receptors modulates signalling. Br J Pharmacol 171:5387–5406

20. Ferrada C et al (2009) Marked changes in signal transduction upon heteromerization of dopamine D1 and histamine H3 receptors. Br J Pharmacol 157:64–75

21. Moreno E et al (2011) Dopamine D1-histamine H3 receptor heteromers provide a selective link to MAPK signaling in GABAergic neurons of the direct striatal pathway. J Biol Chem 286:5846–5854

22. Martínez-Pinilla E et al (2015) Dopamine D2 and angiotensin II type 1 receptors form functional heteromers in rat striatum. Biochem Pharmacol 96:131–142

23. Borroto-Escuela DO et al (2016) Receptor and ion channel detection in the brain. Humana, New York, pp 109–124

24. Navarro G et al (2018) Receptor-heteromer mediated regulation of endocannabinoid signaling in activated microglia. Role of CB1 and CB2 receptors and relevance for Alzheimer's disease and levodopa-induced dyskinesia. Brain Behav Immun 67:139–151. https://doi.org/10.1016/j.bbi.2017.08.015

25. Rico AJ et al (2017) Neurochemical evidence supporting dopamine D1-D2 receptor heteromers in the striatum of the long-tailed macaque: changes following dopaminergic manipulation. Brain Struct Funct 222:1767–1784

26. Farré D et al (2014) Stronger dopamine D1 receptor-mediated neurotransmission in dyskinesia. Mol Neurobiol 52:1408–1420. https://doi.org/10.1007/s12035-014-8936-x

Chapter 13

Bimolecular Fluorescence Complementation Methodology to Study G Protein-Coupled Receptor Dimerization in Living Cells

Doungkamol Alongkronrusmee, Val J. Watts, and Richard M. van Rijn

Abstract

Proteins, such as G protein-coupled receptors (GPCRs), can interact with each other to form dimeric or higher order oligomeric complexes with novel pharmacological properties. GPCRs play a crucial role in numerous physiological processes and diseases, and much research has been performed to prove the existence of GPCR heterodimerization and to investigate the physiological role of the heterodimers. GPCRs are targeted by roughly 25% of all FDA-approved drugs, but heterodimers may represent an untapped additional source of novel drug targets. However, study of GPCR heteromers is not trivial, with most methods having distinct strengths and weaknesses. One method to study GPCR dimerization in living cells is through bimolecular fluorescence complementation (BiFC). The BiFC technique is based on the complementation of two nonfluorescent fragments of a fluorescent protein that is facilitated by fusing the fragments to two interacting proteins. The advantage of BiFC over alternative resonance energy transfer techniques is a high signal-to-noise ratio due to its strong intrinsic fluorescence without exogenous fluorogenic or chromogenic agents required. Here we provide a detailed description of protocols to measure dimerization-induced BiFC in a low-throughput, high-resolution approach using confocal microscopy and in a medium-throughput, low-resolution approach using an automated cell imaging multimode plate reader (Biotek Cytation 3). In this chapter, we use mu and delta opioid receptor heterodimerization to provide a step-by-step BiFC protocol; however, the protocol can be adapted for use with other receptors as well as other confocal or automated microscopes.

Key words Bimolecular fluorescence complementation, G protein-coupled receptor, Dimerization, Screening, Confocal microscopy

1 Introduction

G protein-coupled receptors (GPCRs) perform a broad range of essential physiological actions, and many FDA-approved drugs target GPCRs to overcome aberrant, defective signaling or to counter maladaptive processes. GPCRs not only may signal as monomers but are also able to interact with themselves or other GPCRs through homo- or heteromerization, respectively [1, 2]. To be designated as a heteromer, the receptor heteromer should co-localize within the

Kjell Fuxe and Dasiel O. Borroto-Escuela (eds.), *Receptor-Receptor Interactions in the Central Nervous System*, Neuromethods, vol. 140, https://doi.org/10.1007/978-1-4939-8576-0_13, © Springer Science+Business Media, LLC, part of Springer Nature 2018

same cellular compartment or demonstrate a physical association in native tissue or primary cells. Also, at least two of the following three criteria should be met: (1) evidence for physical interaction between both subunits in native tissue; (2) the function of the heteromer must be distinct from the individual receptors of which it consist; and (3) the functional property for the heteromer could be modified in the absence of either one of the subunits [3, 4]. However, providing evidence for the existence of heteromers, even in vitro, is not trivial. Many assays have been developed to study GPCR dimerization, each with its own unique strength and weakness [5]. There are various bimolecular techniques to visualize the GPCR interactions in living cells including fluorescence (or Förster) resonance energy transfer (FRET) [6], bioluminescence resonance energy transfer (BRET) [7], proximity ligation assay [8], and bimolecular fluorescence complementation (BiFC) [9]. This chapter describes the use of BiFC which is a fluorescence imaging technique widely used in live cells, plants, and animals [10–16].

BiFC is based on the complementation of two nonfluorescent protein fragments of a fluorescent protein. The two fragments are brought into close proximity by the interactions of their fusion partners. The functional chromophore is reconstituted, and a fluorescent signal can be measured (Fig. 13.1). In case that there is no interaction between the two fusion proteins, little to no fluorescence should be detected [9]. The BiFC assay has high sensitivity because the fluorescent signal can be detected even when a small fraction of complex is formed and even with a low expression level of fusion proteins. There are several fluorescent proteins that can be used in BiFC studies such as eGFP, eCFP, eYFP, or DsRed monomer [10]. However, the N-terminal (VN) and C-terminal fragments (VC) of YFP-derivative Venus fluorescent protein are widely used in mammalian cells [17] because Venus has less sensitivity to environmental factors compared with other fluorescent proteins [17, 18]. The advantages of BiFC over FRET and BRET are that it is simple to measure using the single excitation and emission of the parent protein, results in high signal-to-noise ratio, and does not require complex analysis. The irreversible nature of BiFC helps to detect transient or weak interactions in a high-content screening. However, its application for visualization of the site of interaction and dynamic interactions should be carefully considered [11, 19]. BiFC can be combined with FRET or BRET to investigate trimerization by co-expressing a third protein attached to a proper energy transfer donor or even tetramerization by splitting the second fluorescent protein or Renilla luciferase enzyme in half [20, 21]. BiFC also allows direct visualization of the subcellular localizations of the formed complexes when detected by a confocal microscope [22], while an automated digital microscope provides rapid and quantitative BiFC measurements that are susceptible to high-content screening.

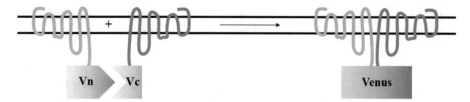

Fig. 13.1 Schematic overview of biomolecular fluorescence complementation (BiFC) assay: Venus yellow fluorescent protein can be separated into two fragments (VN155 and VC155) that can be fused to opioid receptors and when the opioid receptor form heteromers the fluorescent protein fragments will be in close enough proximity to reassemble into a functional Venus fluorescent protein

The BiFC efficiency is used to assess the specificity and strength of the interaction. It is necessary to include a negative control in BiFC experiments to detect the amount of fluorescent signal in the absence of an interaction. Preferably, the negative control is a mutant protein that disrupts the heterodimer interface or noninteracting protein. For the transfection, three plasmids are required: (1) protein "A" fused to VN, (2) protein "B" fused to VC, and (3) a fluorescent protein with different spectral characteristics from Venus, such as mCherry or Cerulean fluorescent protein. The latter plasmid is used as an internal control to normalize BiFC signals and to prevent differences in transfection efficiencies across conditions [23]. In the case of the GPCR heterodimerization, we utilize a membrane-bound protein fused to mCherry [24].

In this chapter, we use mu (MOR) and delta (DOR) opioid receptors which have been reported by different research groups to undergo heterodimerization in vitro and in vivo [25–28] as examples to illustrate step-by-step protocols of BiFC assay using the Nikon A1 confocal microscope and the BioTek Cytation 3 cell imaging multimode reader. We quantify and compare between the BiFC efficiencies of the MOR and DOR interactions in neuronal Cath.a-differentiated (CAD) cells measured by the confocal microscope and the automated digital microscope of the Cytation.

2 Detection of GPCR Interactions in Living Cells Using BiFC

2.1 Materials

2.0.1 Construction of GPCR-BiFC Fusion Vectors

1. Plasmid vectors containing BiFC fragment sequences of Venus: pBiFC vectors and their sequences can be acquired from Addgene (https://www.addgene.org/Chang-Deng_Hu/).

2. DNA encoding the receptors of interest: MOR and DOR are used as examples in this protocol.

3. Plasmid Maxi Kit (Qiagen) or equivalent.

4. Additional reagents and equipment for PCR reaction and DNA sequencing.

2.1.1 Transfection
of GPCR-BiFC Fusion
Vectors into Cells

1. BiFC fusion vectors expressing the receptors of interest: MOR-VC155 and DOR-VN155.

2. Plasmid vector containing a membrane-localized version of mCherry: memCherry is used for aiding in normalization of BiFC signals because it is expressed alongside the BiFC-tagged GPCRs on the cell surface. Membrane targeting of mCherry can be achieved by fusing the mCherry to the N-terminal portion of neuromodulin/GAP43 (Mem vectors, Clontech). The Mem-mCherry was generously provided by Dr. Berlot, Geisinger Medical Center [24].

3. X-tremeGENE 9 (06365787001, Roche) or other transfection reagents.

4. Four chamber 35 mm glass-bottom dish (Cellvis, Mountain View, CA, USA) for a confocal microscopic detection or 96-well black clear-bottom plate (Costar), poly-D-lysine (Sigma) coated for use with an automated digital microscopic detection.

5. Cell system for expressing BiFC constructs: Neuronal Cath.a-- differentiated (CAD) cell line (generously provided by Dr. Chikaraishi, Duke University) or other mammalian cell lines such as HEK293, COS-1, or HeLa cells. The CAD cell is of CNS neuronal origin and is proper to study the GPCR dimerization relevant to the nervous system. It has previously been used to study for example adenosine and dopamine receptor heterodimerization [22].

6. Growth medium: Dulbecco's modified Eagle medium (DMEM) (11995-065, Life Technologies) supplemented with 10% fetal bovine serum (FBS) (F0926, Sigma) for growing CAD cells prepared by adding 50 mL of FBS into 450 mL of DMEM.

7. Serum-free medium: Opti-MEM (31985-070, Life Technologies).

8. Live-cell imaging medium: FluoroBrite DMEM Medium (A1896701, Life Technologies) for enhancing the signal-to-noise ratio of fluorophores and helping to preserve cell health during live-cell imaging.

    (a) Instead of FluoroBrite DMEM, HBSS (14025-092, Life Technologies) + 1 mg/mL glucose can be used.

9. CO_2 tissue culture incubator.

10. Cell culture hood.

2.2 Laser-Scanning Confocal Microscopic Detection

2.2.1 Laser-Scanning Confocal Imaging

1. Nikon Eclipse Ti-E inverted microscope (Nikon Instruments, Melville, NY, USA).

2. Nikon A1 confocal microscope system (Nikon Instruments, Melville, NY, USA).

3. Mercury arc lamp.

4. Multi-argon laser: Use the 561.5 nm and 514 nm laser lines for the excitation of mCherry and Venus, respectively.

5. Primary dichroic mirror: Use the 405/488/561 and 400–457/514 primary dichroic mirrors for the detection of mCherry and Venus, respectively.

6. Band-pass emission filter: The 595/50 nm and 585/65 nm emission filters for the detection of mCherry and Venus, respectively.

7. Plan Apo VC 20× lens (Nikon Instruments, Melville, NY, USA) for BiFC efficiency.

8. Stage-top heating incubator and digital gas mixer (Tokai Hit Co., Shizuoka-ken, Japan).

2.2.2 Image Analysis

1. Desktop computer.

2. NIS-Elements Advanced Research (Version 4.20.01) analysis software (Nikon Instruments, Melville, NY, USA).

3. Microsoft Excel.

2.3 Automated Digital Microscopic Detection

1. Cytation 3 Cell Imaging Multi-Mode Reader (BioTek Instruments, Inc., Winooski, VT).

2. Bottom cube contains a high-powered LED and focusing lens.

2.3.1 Automated Digital Imaging

3. Top cube contains the excitation filter, dichroic mirror, and emission filter. Use the 586/647 nm and 500/542 nm excitation/emission filters for the detection of Texas Red and YFP, respectively.

4. 20× Microscope objectives.

5. Grayscale CCD camera with a Sony chip for image collection: The 16-bit camera records over 65,500 different intensity values per pixel in the image.

6. CO_2/O_2 gas controller.

7. Temperature control ranges from 4 °C over ambient to 45 °C.

2.3.2 Image Analysis

1. Desktop computer.

2. BioTek Gen5 software.

3. Microsoft Excel.

3 Methods

3.1 Construction of GPCR-BiFC Fusion Vectors

1. Using plasmids containing the cDNAs of the GPCR perform a PCR reaction with oligonucleotides that should omit the stop codon by removal or mutation and should contain unique restriction sites, to insert the GPCR-coding sequence in the pBIFC-VN155-I152L or pBiFC-VC155 vector (https://www.addgene.org/Chang-Deng_Hu/). The GPCR is inserted

in frame with the VC or VN fragment and the receptor and Venus fragment are separated by an optimal linker sequence. Purify plasmid cDNA using Qiagen Plasmid Maxi Kit, and verify the constructs by DNA sequencing.

2. To verify the expression and localization of BiFC-tagged GPCRs, use radioligand binding, enzyme-linked immunosorbent assay, flow cytometry, immunofluorescence, or other appropriate assays to determine if the expression level and localization of the receptor are altered by the addition of the fluorescent fragment.

3. To verify the functionality of BiFC-tagged GPCRs: Measure downstream signaling such as cAMP production, ERK phosphorylation, or calcium mobilization (for a G_q-coupled receptor) to determine if the functionality of the receptor is retained after the addition of the fluorescent fragment.

3.2 Transfection of GPCR-BiFC Fusion Vectors into Cells

1. Suspend cells at an appropriate density to allow cells to grow to 60–80% confluency over 24 h. For confocal microscopy, suspend 40,000–45,000 CAD cells in 600 μL of growth medium, and plate into one chamber of a four-chamber 35 mm glass-bottom dish. For automated digital microscopy, suspend 8000–10,000 CAD cells in 100 μL of growth medium, per well of a 96-well black clear-bottom plate, poly-D-lysine coated.

   Dilute poly-D-lysine in sterile tissue culture grade water to a final 50 μg/mL working concentration. Plates are coated for a minimum of 1 h at room temperature. Remove poly-D-lysine solution by aspiration and rinse with sterile PBS. Allow to dry at least 2 h before introducing cells and medium.

2. Incubate cells at 37 °C and 5% CO_2 for 16–24 h or until cells reach a confluency of 60–80% which is optimal for transfection and imaging.

3. Prior to transfection, prepare the plasmid DNA solution using sterile TE (Tris/EDTA) buffer or sterile water so that the amount of DNA required for each transfection is 1 μL.

4. Once cells have reached 60–80% confluency, prepare one 1.5 mL Eppendorf tube per transfection inside the cell culture hood. Dilute X-tremeGENE9 with serum-free medium Opti-MEM to a concentration of 3 μL reagent/100 μL medium, and gently mix the solution by flicking the tube.

5. Add 1 μg of total DNA to 100 μL of diluted X-tremeGENE 9/Opti-MEM solution. Mix the transfection solution by gently flicking the side of the 1.5 mL Eppendorf tube. Make sure that the solution is completely at the bottom of the tube.

Since the normalization plasmids are relatively small in size and robustly expressed, the amount of normalization plasmid and BiFC-tagged GPCR plasmid DNAs should be optimized to achieve best transfection efficiency, and it should be kept similar across biological replicates. The normalization plasmids should be expressed in more than 50–70% of the cells. Here we use a ratio of DOR-VN:MOR-VC:memCherry = 1:1:0.1.

6. Incubate the transfection reagent:DNA complex for 15 min in the cell culture hood at room temperature to allow the DNAs to complex with the X-tremeGENE 9 transfection reagent. A shorter or longer incubation may decrease the transfection efficiency.

7. Remove the culture vessel from the incubator. Removal of growth medium is not necessary; however, replacing the growth medium with Opti-MEM or other reduced-serum medium will limit further cell growth. Slowly add the transfection complex to the CAD cells in the designated well in a drop-wise manner. Mix by tilting plate to ensure even distribution over the entire plate surface. Once the transfection complex has been added to the cells, there is no need to replace with fresh medium.

8. Following transfection, incubate cells at 37 °C and 5% CO_2 for 18–24 h, and image the transfected cells. The optimal incubation time will vary depending on the transfected vector construct, cell line, cell medium, and cell density. The transfection procedure should be performed according to the instructions of the manufacturer's transfection reagent.

9. To reduce background fluorescence, the cell culture medium could be exchanged before imaging with FluoroBrite DMEM or other live-cell imaging medium.

3.3 Laser-Scanning Confocal Microscopic Detection

3.3.1 Laser-Scanning Confocal Imaging

1. Place the culture vessel containing the transfected CAD cells in a 37 °C, 5% CO_2, stage-top heating incubator and digital gas mixer.

If a stage-top heating incubator and digital gas mixer are unavailable, seal the culture vessel with Parafilm to maintain CO_2 levels and promptly complete imaging.

2. Open the NIS-Elements analysis software and arrange control panels as desired such as Ti Pad for microscope controls and A1 Compact Graphical User Interface (GUI).

3. To visualize the mCherry signal, the 561.5 nm laser line for excitation, 405/488/561 primary dichroic mirror, and 595/50 nm band-pass emission filter are used on the first light path configuration (A1 Compact GUI).

4. To visualize the complemented Venus signal, the 514 nm laser line for excitation, 400–457/514 primary dichroic mirror, and 585/65 nm band-pass emission filter are used on the second light path configuration (A1 Compact GUI).

5. Click on the E100 under the Light Path (Ti Pad) and select the Eyeport icon (A1 Compact GUI). The Remove Interlock icon should automatically turn to red (A1 Compact GUI).

6. Open the shutter on the mercury arc lamp and select the mCherry fluorescence filter cube (Ti Pad).

7. For quantification of BiFC efficiency, use an objective lens of 20×, and view transfected cells via the binocular Eyeport of the Nikon Eclipse Ti-E inverted microscope.

8. Focus on the CAD cells via the internal control mCherry signal using the course knob on the microscope.

9. After focusing, click on the L100 under the Light Path (Ti Pad). This converts the microscope to the confocal mode. The light is directed to the confocal scan box. Then, deselect the Eyeport and Remove Interlock icons (A1 Compact GUI). The Remove Interlock should turn from red to grey. This indicates that the safety is removed, and the laser light can get through the scan head.

10. Set up scanning parameters in A1 Compact GUI by selecting an image size at a 512 × 512 pixel for the faster frame rate with the pinhole at 1.0 AU. Since the size of the pinhole at 1.0 AU depends on the excitation wavelength, the pinhole size should be individually selected and remain consistent for each channel throughout the rest of the imaging process.

11. Begin scanning and focus on the z-plane to find the brightest Venus signal using the adjustment knob on the microscope base or the stage control. When using the joystick, toggle through fine or extra-fine for optimizing coordinates.

12. Optimize the laser power, pixel dwell time, and detector gain (A1 Compact GUI) that produce the greatest mCherry and Venus signals by hovering over the image of the positive interaction pair and the negative control. The imaging settings can be different for each channel. Record the HV and offset values used for the capture since these values will be standardized throughout all the captured images. In this protocol, 5% laser power, a detector gain of less than 100, and 4.8 μs pixel dwell are used.

The laser power, pixel dwell time, and detector gain should be coordinated with each other. The goal is to avoid over- and undersaturation, which can be detected using the Pixel Saturation Indicator. Higher laser power or pixel dwell time may cause photobleaching, which can be resolved by increasing the detector gain or the pinhole size more than 1.0 AU. While enhancing the image brightness, overdoing it may result in reduction of the resolution.

13. When the viewing field is optimized at an image size of 512 × 512 pixels, locate the area of interest where CAD cells express the internal control.

14. Increase an image size to 1024 × 1024 pixels (A1 Compact GUI). The pixel dwell time will automatically be adjusted to 6.2 μs, causing the image to become brighter. If there is an oversaturation, reduce the laser power and capture a new image. Also, capture separate images of mCherry and Venus with the pinhole at 1.0 AU. Save both images within the same nd2 file.

15. Acquire more images for each experimental and control sample by repeating the steps above in a new area of interest. It is highly recommended that large-cell populations approximately 50–100 cells should be used to quantify the BiFC efficiency, and to prevent cell-to-cell variations in expression levels from transient transfections.

3.3.2 Image Analysis

1. Open an nd2 image file containing both mCherry and Venus signals in the NIS-Elements analysis software.

2. Subtract the respective background values from both the mCherry and Venus signals. Select Turn Background ROI On/Off. Draw one or more background regions of interest (ROI) in areas devoid of cells, and select Subtract Background using Background ROI.

3. Separate the mCherry and Venus images by dragging the mCherry image tab to an empty area on the screen. Repeat this step with the Venus image and close the original nd2 file.

The captured images have dimensions of 1024 × 1024 total pixels. Each pixel has a pixel value which represents the brightness of the pixel. The intensity of a 12-bit grayscale image ranges from 0 (black) to 4096 (white). The distribution of pixel intensities within the image is analyzed by the histogram, which is a valuable tool to aid in deciding the threshold value when converting a grayscale image to a binary one.

4. A plasma membrane binary layer is created using signals from the mCherry. Select Define Threshold under Binary tab. Use the histogram to define the minimum and maximum values for the range of pixel intensities. Here we set up a range of the intensity from 500 to 1500 to exclude autofluorescence signals from untransfected cells in the analysis. The threshold is kept constant throughout the experiment.

5. To optimize the binary layer, use the smoothing and cleaning function at levels ranging from 1× to 16×, or use binary functions under the Binary Toolbar [right-click on an empty area of the screen > Analysis Controls > Binary Toolbar].

   (a) To add a plasma membrane that has been excluded from the binary layer, use Auto-Detect or Draw functions. Scroll down to decrease the area and right-click to finalize the selection.

   (b) To remove a plasma membrane that cannot be distinguished from the cytoplasm, use the Delete function.

   (c) To decrease the size of ROI, use the Erode function.

   (d) To divide grouped cells, use the Separate function.

6. To collect the mean fluorescence intensity, select Object Features under the Measure menu, and add Mean Intensity to the selected measurements. Right-click on the Measurements tab and select Load Default.

   (a) To collect mCherry intensities, select Perform Measurement under the Measure menu. Export the data of all mCherry intensities in the Measurements window to Microsoft Excel.

   (b) To collect Venus intensities, Copy Binary from the mCherry image and Paste Binary into the corresponding Venus image. Clear the mCherry intensities from the Measurements window and select Perform Measurement. Export the data of all Venus intensities to Microsoft Excel.

7. In Microsoft Excel, match mCherry values with Venus values that have been obtained from the same plasma membrane using the object ID values. Calculate the ratio of the average Venus signal over the average mCherry signal for each cell. The ratio corresponds to the relative complementation efficiency for the combination of BiFC-tagged GPCRs.

8. Repeat the above steps for all captured images, and generate a Microsoft Excel data sheet for each sample such as the positive interaction pair or the negative control.

9. Calculate the median Venus-mCherry ratio from the positive interaction pair and from the negative control. Median values are used because fluorescence intensities usually fall into a non-Gaussian distribution [9]. Calculate the BiFC efficiency by dividing the median Venus-mCherry ratio of the positive interaction pair by that of the negative control (see Table 13.1 for an example of data analysis).

10. Calculate the average BiFC efficiency and standard deviation from at least three independent experiments.

Table 13.1

Example of data analysis for calculating BiFC efficiency from images acquired using confocal microscopy

Object/cell ID	Mean intensity of Venus	Mean intensity of mCherry	Venus/mCherry
Positive control			
1	291.2	1416.32	0.21
2	192.45	1330.45	0.14
...	...	...	...
50–100	168.78	1017.95	0.17
Median			0.17
Negative control			
1	14.78	1029.67	0.01
2	19.44	899.57	0.02
...	...	...	...
50–100	30.93	1761.21	0.02
Median			0.02

BiFC efficiency = Median of positive control/Median of negative control = 9.48

11. Data interpretation: If the BiFC efficiency of the BiFC-tagged GPCR is significantly different from that of the negative control or is higher than 1 [29], the observed fluorescence signals are likely to represent a positive or specific receptor-receptor interaction. If the BiFC efficiency is similar, the fluorescence signals may represent nonspecific complementation (Fig. 13.2).

3.4 Automated Digital Microscopic Detection

3.4.1 Automated Digital Imaging

1. Push the carrier eject button on the front of the reader to load the plate.

2. Open the Gen5 software and create a new protocol by selecting Standard Protocol.

3. Select Procedure under File menu for setting the Read, Temperature, and Plate Type.

4. Under the Read function, select Image as the Detection Method, with Endpoint/Kinetic and Filters selected by default.

5. In the Read step, select 20× objective for imaging. By default, the entire plate is selected for imaging; however, use the Full Plate button on the top right of the window to select a subset of well to be measured. The channels for imaging are selected as follows:

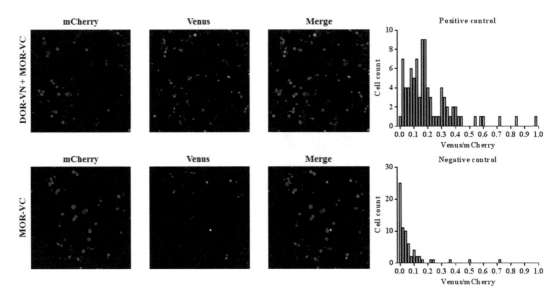

Fig. 13.2 Confocal microscopy-assisted BiFC detection as measurement of MOR-DOR heterodimerization. (Top) Heterodimerization between delta opioid receptor-VN and mu opioid receptor-VC produces clear BiFC. (Bottom) Absence of BiFC when transfecting only MOR-VC in CAD cells. The histograms from a positive and negative control are also shown. The median is less in the negative control. A 20× objective lens is used for observing a large number of cells and providing a general glimpse of cellular localization

> (a) To visualize the mCherry signal, the 586 nm excitation filter, 647 nm emission filter, and Top Optics Position are used on the Texas Red channel.
>
> (b) To visualize the complemented Venus signal, the 500 nm excitation filter, 542 nm emission filter, and Top Optics Position are used on the YFP channel.

6. Under the Temperature function, select Incubator On and adjust the temperature to 37 °C. Use the Preheat before continuing with next step.

7. Under the Plate Type function, select 96-well plate and Use lid.

8. Once the protocol has been created, open a new experiment by using an existing protocol. Experiment mode can collect images from multiple wells and can be defined to capture images only in wells whose fluorescence intensity values are above the defined threshold, decreasing the amount of collected data.

9. Select the Procedure under the Protocol, and double-click the Read in the Description window to validate the protocol. Optimize the numbers of capture images per well by deselecting Single Image set and defining the target area. The image can be offset from the center of the well by adjusting the Horizontal and Vertical offset from the center of well.

10. Open a manual mode session to optimize the Exposure including LED intensity, Integration time, and Camera gain by hovering over the image of the positive interaction pair and the negative control. Increasing LED intensity, integration time, and camera gain individually will increase the brightness of the image. The imaging settings can be different for each channel. Record all the parameter values used for the capture since these values will be standardized throughout all the captured images. In this protocol, we use LED intensity of 6100 ms of Integration time, and Camera gain of less than 20.

11. Focus on the z-plane to find the brightest Venus signal. Perform a FM Scan which displays where the plane of optimal sample focus is based on a focus metric ratio. In case that there are differences between the Texas Red and YFP positions, adjust this Offset in the Options function.

12. After optimizing coordinates, begin image acquisition by saving the Experiment and selecting the Read in the Toolbar.

3.4.2 Image Analysis

1. When reviewing an image, adjust brightness and contrast per channel. Adjusting brightness and contrast does not affect image analysis; however, it improves image viewing on a computer monitor. All images in an experiment will have the similar brightness and contrast settings, allowing legitimate cross comparison among all images.

2. Click Analyze and select I want to set up a new Image Analysis data reduction step.

3. To improve the efficacy of image analysis, these options are available via the Analysis Settings link in the Cellular Analysis window. Three attributes of the image can be optimized: image flatness, image noise, and elevated background levels.

4. In the Analysis settings under the Cellular Analysis, correct for local background variations by using Background flattening size specified in the Rolling Ball diameter box. This makes the background across the image more uniform and improves the accuracy in the cellular analysis. If the rolling ball diameter value is too small, the background correction may be too aggressive. If the rolling ball diameter value is too big, it will be hard to differentiate the background from the cells. In this protocol, we use 80 μm (Rolling Ball diameter) of the background flattening size.

5. Determine the Image smoothing strength to decrease the image noise so that the background is less variable, and the border identification is more accurate. In this protocol, we use three Cycles of 3 × 3 average filter of the Image smoothing strength.

6. Evaluate background on % of lowest pixels. The background level is typically elevated by some residual fluorescent stain or other factors. This parameter will raise the threshold value and remove the lowest background intensity, improving the accuracy of the object identification. In this protocol, we evaluate background on 10% of lowest pixels.

7. After the pre-analysis step, cellular analysis of images requires the coordinated optimization of multiple analysis parameters, such as intensity threshold and minimum and maximum object size, in order to obtain a correct cell count. Changing the threshold intensity setting and minimum and maximum object size results in changes in the number, shape, and size of the intended objects.

8. To optimize a Threshold, use the View Line Profile tool to trace several representative regions that represent the dynamic range of the samples including the positive interaction partner and the negative control. The same threshold value is applied to all images so that data can be compared across images. In this protocol, we use 1500 of parameter threshold to set the minimum pixel intensity value for the mCherry.

9. Determine the approximate size of the counted objects by using the zoom feature in conjunction with the scale bar across different imaged objects. This eliminates unwanted debris that may cause fluorescence. Select Bright objects on a dark background, Split touching objects, and Include edge objects. Click Add step. In this protocol, we set 15 μm of the minimum object size and 60 μm of the maximum object size.

10. Once the optimal analysis parameters have been determined, cellular analysis of the entire plate can be performed by clicking I want to edit an existing Image Analysis step and Apply changes.

11. By default, all channels return intensity measurements. The results are displayed in the right pane. Export the data of all mCherry and Venus intensities to Microsoft Excel.

12. In Microsoft Excel, match mCherry values with Venus values that have been obtained from the same object ID values. Calculate the ratio of the average Venus signal over the average mCherry signal for each cell.

13. Calculate the BiFC efficiency by dividing the median Venus-mCherry ratio of the positive interaction pair by that of the negative control.

14. Calculate the average BiFC efficiency and standard deviation from at least three independent experiments.

15. Data interpretation: If the BiFC efficiency of the BiFC-tagged GPCR is higher than 1 [29], the observed fluorescence signals

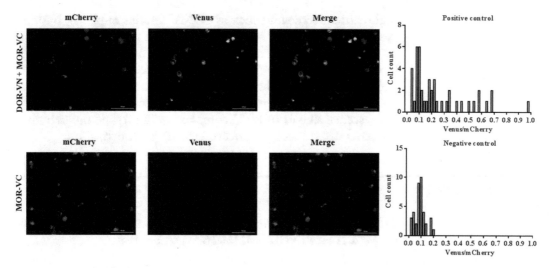

Fig. 13.3 Automated digital microscopy-assisted BiFC detection as measurement of MOR-DOR heterodimerization. (Top) Heterodimerization between delta opioid receptor-VN and mu opioid receptor-VC produces clear BiFC. (Bottom) Absence of BiFC when transfecting only MOR-VC in CAD cells. The histograms from a positive and negative control are also shown. The median is less in the negative control. A 20× objective lens is used for observing a large number of cells and providing a general glimpse of cellular localization

are likely to represent a positive or specific receptor-receptor interaction. If the BiFC efficiency is similar, the fluorescence signals may represent nonspecific complementation (Fig. 13.3).

16. For a rough estimate for the BiFC efficiency, select the Result Options and add Object sum integral to a subpopulation analysis. This will measure the total intensity change, taking area into account. View the results of an experiment in Gen5's main workspace using the Plate View. Use the drop-down list for Data to display the Object sum integral of Venus signals. Use the Quick Export feature to instantly export the current view to Excel.

4 Summary

The BiFC assay provides a simple, yet powerful method for detection of GPCR dimerization in living cells. The potential of BiFC has demonstrated its application to screen for novel interaction partners of the protein of interest, validate suspected interaction partners, and characterize a dimer interface and/or sequence motif in a given protein pair interaction. The fluorescence of the heteromers can be detected using fluorescence microscopes, which enables determination of the subcellular localization of the complex with high spatial resolution. Recent advances in automated digital microscopes also allow BiFC assays to be applied in high-content screening such as cDNA libraries in plants, budding yeast, and mammalian cells. To increase the reliability and versatility of BiFC assays, validation of the proper controls and other combinatorial methods as well as further development of a reversible BiFC system and additional fluorescent proteins with enhanced signal-to-noise ratio are needed. The quantitative assessment of fluorescence signals of BiFC complexes will provide opportunities for novel drug discovery involved in GPCR dimerization.

Acknowledgements

This work was supported by funding from the National Institute on Mental Health (R33MH101673) to Dr. Watts.

References

1. Maggio R, Novi F, Scarselli M, Corsini GU (2005) The impact of G-protein-coupled receptor hetero-oligomerization on function and pharmacology. FEBS J 272(12): 2939–2946

2. Fuxe K, Canals M, Torvinen M, Marcellino D, Terasmaa A, Genedani S, Leo G, Guidolin D, Diaz-Cabiale Z, Rivera A, Lundstrom L, Langel U, Narvaez J, Tanganelli S, Lluis C, Ferre S, Woods A, Franco R, Agnati LF (2007) Intramembrane receptor-receptor interactions: a novel principle in molecular medicine. J Neural Transm (Vienna) 114(1):49–75

3. Gomes I, Ayoub MA, Fujita W, Jaeger WC, Pfleger KD, Devi LA (2016) G protein-coupled receptor Heteromers. Annu Rev Pharmacol Toxicol 56:403–425. https://doi.org/10.1146/annurev-pharmtox-011613-135952

4. Pin JP, Neubig R, Bouvier M, Devi L, Filizola M, Javitch JA, Lohse MJ, Milligan G, Palczewski K, Parmentier M, Spedding M (2007) International Union of Basic and Clinical Pharmacology. LXVII. Recommendations for the recognition and nomenclature of G protein-coupled receptor heteromultimers. Pharmacol Rev 59(1):5–13. https://doi.org/10.1124/pr.59.1.5

5. Vidi PA, Ejendal KF, Przybyla JA, Watts VJ (2011) Fluorescent protein complementation assays: new tools to study G protein-coupled receptor oligomerization and GPCR-mediated signaling. Mol Cell Endocrinol 331(2):185–193. https://doi.org/10.1016/j.mce.2010.07.011

6. Truong K, Ikura M (2001) The use of FRET imaging microscopy to detect protein-protein interactions and protein conformational changes in vivo. Curr Opin Struct Biol 11(5):573–578

7. Pfleger KD, Eidne KA (2006) Illuminating insights into protein-protein interactions using bioluminescence resonance energy transfer (BRET). Nat Methods 3(3):165–174

8. Koch S, Helbing I, Bohmer SA, Hayashi M, Claesson-Welsh L, Soderberg O, Bohmer FD (2016) In situ proximity ligation assay (in situ PLA) to assess PTP-protein interactions. Methods Mol Biol 1447:217–242. https://doi.org/10.1007/978-1-4939-3746-2_13

9. Hu CD, Chinenov Y, Kerppola TK (2002) Visualization of interactions among bZIP and Rel family proteins in living cells using bimolecular fluorescence complementation. Mol Cell 9(4):789–798

10. Kodama Y, Hu CD (2012) Bimolecular fluorescence complementation (BiFC): a 5-year update and future perspectives. BioTechniques 53(5):285–298

11. Shyu YJ, Hu CD (2008) Fluorescence complementation: an emerging tool for biological research. Trends Biotechnol 26(11):622–630. https://doi.org/10.1016/j.tibtech.2008.07.006

12. Walter M, Chaban C, Schutze K, Batistic O, Weckermann K, Nake C, Blazevic D, Grefen C, Schumacher K, Oecking C, Harter K, Kudla J (2004) Visualization of protein interactions in living plant cells using bimolecular fluores-

cence complementation. Plant J 40(3):428–438. https://doi.org/10.1111/j.1365-313X.2004.02219.x

13. Sung MK, Huh WK (2007) Bimolecular fluorescence complementation analysis system for in vivo detection of protein-protein interaction in Saccharomyces cerevisiae. Yeast 24(9):767–775

14. Kerppola TK (2008) Bimolecular fluorescence complementation (BiFC) analysis as a probe of protein interactions in living cells. Annu Rev Biophys 37:465–487

15. Shyu YJ, Suarez CD, Hu CD (2008) Visualization of ternary complexes in living cells by using a BiFC-based FRET assay. Nat Protoc 3(11):1693–1702

16. Duffraisse M, Hudry B, Merabet S (2014) Bimolecular fluorescence complementation (BiFC) in live drosophila embryos. Methods Mol Biol 1196:307–318. https://doi.org/10.1007/978-1-4939-1242-1_19

17. Shyu YJ, Liu H, Deng X, Hu CD (2006) Identification of new fluorescent protein fragments for bimolecular fluorescence complementation analysis under physiological conditions. BioTechniques 40(1):61–66

18. Kodama Y, Hu CD (2010) An improved bimolecular fluorescence complementation assay with a high signal-to-noise ratio. BioTechniques 49(5):793–805

19. Miller KE, Kim Y, Huh WK, Park HO (2015) Bimolecular fluorescence complementation (BiFC) analysis: advances and recent applications for genome-wide interaction studies. J Mol Biol 427(11):2039–2055. https://doi.org/10.1016/j.jmb.2015.03.005

20. Ciruela F, Vilardaga JP, Fernandez-Duenas V (2010) Lighting up multiprotein complexes: lessons from GPCR oligomerization. Trends Biotechnol 28(8):407–415. https://doi.org/10.1016/j.tibtech.2010.05.002

21. Vidi PA, Chen J, Irudayaraj JM, Watts VJ (2008) Adenosine a(2A) receptors assemble into higher-order oligomers at the plasma membrane. FEBS Lett 582(29):3985–3990. https://doi.org/10.1016/j.febslet.2008.09.062

22. Vidi PA, Chemel BR, Hu CD, Watts VJ (2008) Ligand-dependent oligomerization of dopamine D(2) and adenosine a(2A) receptors in living neuronal cells. Mol Pharmacol 74(3):544–551. https://doi.org/10.1124/mol.108.047472

23. Kodama Y, Hu CD (2013) Bimolecular fluorescence complementation (BiFC) analysis of protein-protein interaction. how to calculate signal-to-noise ratio Methods Cell Biol 113:107–121. https://doi.org/10.1016/B978-0-12-407239-8.00006-9

24. Yost EA, Mervine SM, Sabo JL, Hynes TR, Berlot CH (2007) Live cell analysis of G protein beta5 complex formation, function, and targeting. Mol Pharmacol 72(4):812–825. https://doi.org/10.1124/mol.107.038075

25. Gomes I, Gupta A, Filipovska J, Szeto HH, Pintar JE, Devi LA (2004) A role for heterodimerization of mu and delta opiate receptors in enhancing morphine analgesia. Proc Natl Acad Sci U S A 101(14):5135–5139. https://doi.org/10.1073/pnas.0307601101

26. Milan-Lobo L, Enquist J, van Rijn RM, Whistler JL (2013) Anti-analgesic effect of the mu/delta opioid receptor heteromer revealed by ligand-biased antagonism. PLoS One 8(3):e58362

27. O'Dowd BF, Ji X, O'Dowd PB, Nguyen T, George SR (2012) Disruption of the mu-delta opioid receptor heteromer. Biochem Biophys Res Commun 422(4):556–560

28. He SQ, Zhang ZN, Guan JS, Liu HR, Zhao B, Wang HB, Li Q, Yang H, Luo J, Li ZY, Wang Q, Lu YJ, Bao L, Zhang X (2011) Facilitation of mu-opioid receptor activity by preventing delta-opioid receptor-mediated codegradation. Neuron 69(1):120–131

29. Vidi PA, Przybyla JA, Hu CD, Watts VJ (2010) Visualization of G protein-coupled receptor (GPCR) interactions in living cells using bimolecular fluorescence complementation (BiFC). Current protocols in neuroscience Chapter 5:Unit 5 29

Chapter 14

Detection and Quantitative Analysis of Dynamic GPCRs Interactions Using Flow Cytometry-Based FRET

Barbara Chruścicka, Shauna E. Wallace Fitzsimons, Clémentine M. Druelle, Timothy G. Dinan, and Harriët Schellekens

Abstract

Heterodimerization of specific G protein-coupled receptor (GPCR) protomers is associated with increased receptor signaling diversity and exhibits unique biochemical, functional, and pharmacological properties. Evidence for the formation of heteroreceptor complexes has been demonstrated in vitro using cellular models and biochemical assays and ex vivo using brain slices and primary cell cultures. Since mechanisms that lead to brain pathologies such as depression, anxiety, addiction, and schizophrenia involve GPCR signaling, the distinct pharmacological profiles of GPCR assemblies may serve as new target for the development of novel therapeutic strategies with enhanced specificity. Therefore, development and standardization of novel methods for detection and analysis of dimer pairs both in recombinant systems and in native tissue is warranted. This chapter describes a step-by-step protocol for detecting and quantifying dynamic receptor–receptor interactions in living cells using flow cytometry-based fluorescence (Förster) resonance energy transfer (fcFRET). This method has significant potential to identify novel GPCR dimers within the central nervous system while simultaneously allowing analysis of the dynamic nature of these receptor interactions, which is poised to contribute significantly to the field of GPCR neuropsychopharmacology across brain diseases.

Key words Fluorescent resonance energy transfer, Flow cytometry, G protein-coupled receptors, Heteroreceptor complexes, Receptor–receptor interactions, Dimerization

---

1 Introduction

G protein-coupled receptors (GPCRs) comprise a large superfamily of seven-transmembrane (7TM) domain proteins, which are major signaling mediators for a variety of endogenous hormones and neurotransmitters involved in diverse physiological functions ranging from glucose metabolism and appetite regulation to immune response and neurotransmission [1, 2]. GPCRs mediate their downstream signaling, following binding of their respective endogenous ligands, which leads to changes in receptor conformation and further intracellular signaling through interaction with

Kjell Fuxe and Dasiel O. Borroto-Escuela (eds.), *Receptor-Receptor Interactions in the Central Nervous System*, Neuromethods, vol. 140, https://doi.org/10.1007/978-1-4939-8576-0_14, © Springer Science+Business Media, LLC, part of Springer Nature 2018

intracellular signaling molecules such as heterotrimeric G proteins, GPCR kinases, and arrestins. Subsequently, modulation of downstream enzymes such as adenylate cyclase, phospholipase C, or extracellular-signal-regulated kinases leads to specific cellular responses [3–5].

Initially, GPCRs were thought to exist and function exclusively as monomeric signaling units [6]. However, oligomerization of GPCRs, in which protomers of the same or different families combine to generate homo- or heterodimers, as well as higher-order multimers, is becoming increasingly accepted as a fundamental process in GPCR signaling [7–10]. These higher-order complexes have been suggested to exhibit unique biochemical and functional properties, compared to their monomeric counterparts [11]. Heterodimerization of specific protomers has been shown to affect ligand binding to the receptors [12, 13], alter G protein subunit coupling, and influence intracellular signaling [14], as well as ligand-mediated internalization of the receptors [15, 16]. Interestingly, GPCR heterodimers are able to activate intracellular signaling cascades that cannot be triggered by each of the individual receptors alone [17]. Furthermore, GPCR heterodimers can potentiate or attenuate their downstream signaling by modulating the expression, maturation, and trafficking of the protomers from the endoplasmic reticulum (ER) to the cell membrane, changing their subcellular localization [18]. Since GPCRs are considered to be major drug targets, the distinct pharmacological profiles of GPCR assemblies are likely to serve as novel mechanisms, important for the development of more specific pharmacological strategies to modulate cell response, and regulate a plethora of physiological processes [2]. Formation of heteroreceptor complexes and altered functionality has also been demonstrated in the brain [12, 19–21]. What is more important is that dysfunctions of heteroreceptor complexes may be involved in many brain pathologies [22, 23]. For instance, modulation of neuronal signaling via dopamine 2 receptor (D2R) heterocomplexes with adenosine 2A, serotonin 2A, NMDA, and metabotropic glutamate 5 receptors found in striatum has been implicated in the pathophysiology of schizophrenia and may provide new possibilities for the treatment of positive, negative, and cognitive symptoms of schizophrenia [7, 24, 25]. Moreover, heterodimers of D2R with adenosine receptor are implicated in cocaine associated addiction [26], while more recently established positive allosteric interactions between D2R and oxytocin receptor seem to be essential for social behavior [27]. The formation and functioning of serotonin 1A heteroreceptor complexes with galanin 2, serotonin 7 and 2A receptors, located mainly in the raphe-hippocampal system, may serve as new targets for drug development for conditions such as major depression and anxiety [28, 29].

Due to the physiological importance of GPCR receptor interactions in the central nervous system, development of novel methods for the identification, detection, and analysis of dimer pairs has seen a surge in research interest. Physical receptor–receptor interactions have been investigated using a plethora of in vitro and in vivo approaches [30]. Co-immunoprecipitation followed by western blotting has been considered for years as the reference method to demonstrate a physical interaction. Nevertheless, this technique requires the use of non-physiological buffers and detergents to lyse and solubilize tissue, which causes disruption to the natural cells environment, and in consequence preclude the analysis of subcellular localization rising to artifacts. In addition, biophysical proximity assays, based on resonance energy transfer (RET) between two chromophores, allow investigations into physical interaction in intact living cells. Bioluminescence resonance energy transfer (BRET) compared to fluorescence (Förster) based RET (FRET) is characterized by a more specific signal that is easier to quantify. However, this method doesn't allow for microscopic observation and analysis of protein–protein interactions at the subcellular level [31, 32]. The advantage of FRET-based techniques (FRET, TR-FRET, TIRF-FRET, FRET spectroscopy) is that they provide this type of additional information concerning the physiology of receptor–receptor interactions in living cells. In this chapter, we describe the protocol for the quantitative analysis of dynamic receptor–receptor interactions through FRET measurements of fluorescently tagged GPCR pairs using flow cytometry (fcFRET).

2 The Principle of fcFRET Methodology

FRET is a physical phenomenon known since 1946 [33], and from then onwards has become a powerful tool to study protein–protein interactions [34–36]. FRET is based on a non-radiative transfer of energy from an excited fluorophore (donor) to another fluorophore (acceptor), leading to a reduced emission of energy from the donor (donor quenching) and an increased emission energy from the acceptor (FRET signal) [37]. The efficiency of energy transfer depends on the distance between donor and acceptor and occurs only within the range of 1–10 nm (10–100 Å). Therefore, FRET-based biosensors can work as spectroscopic rulers, used for monitoring the proximity changes at the molecular level (Fig. 1) [38]. High performance of FRET requires an emission spectrum of the donor that overlaps significantly with the excitation spectrum of the acceptor [39]. Three main types of FRET biosensors are small organic dyes, quantum dots, and fluorescent proteins (FPs). However, only FPs have the ability to label sensing domains without the need for antibodies (compared to quantum dots), and allow the analysis of FRET in intact living cells (compared to

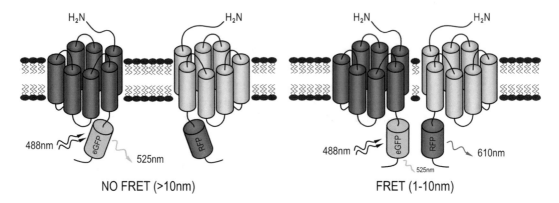

Fig. 1 Principle of fluorescence resonance energy transfer (FRET). FRET between fluorescent proteins (FPs) attached to the carboxy-terminus of two interacting GPCRs. Upon excitation with light at 488 nm, the eGFP emits light at 525 nm (left panel). When the distance between two FPs is between 1 and 10 nm, the excited eGFP (donor) transfers energy to the RFP (acceptor) via a dipole-dipole resonance energy transfer mechanism, causing the RFP to emit light at 610 nm, which is measured as the FRET signal (right panel)

organic dyes) (Note 1). The choice of FRET pair depends, beyond their spectroscopic features, on the instrument used for FRET measurements. FRET can be detected by all instruments capable of recording fluorescence emission, like spectrofluorometers, microscopes, and flow cytometers. In contrast to microscopy-based FRET, flow cytometry analysis allows the evaluation of interactions in large population of cells in a short period of time, providing statistically more robust and reliable data. Compared to spectrofluorometers, flow cytometry makes it possible to classify the population of cells and elucidate the difference in FRET efficiency with other cellular parameters [40, 41]. On top of that, fcFRET allows for quantitative analysis of ligand-mediated changes of FRET signal in heterologous expression systems (in vitro) and in cells endogenously expressing receptors of interest (ex vivo) (Note 2). In conclusion, the fcFRET-based proximity assay provides a unique opportunity to combine the quantitative measurement of physical GPCR interactions with the ability to analyze dynamic changes following exposure to ligands and novel neurotherapeutics.

3   Materials

3.1   Generation of Human GPCRs Expressing Cell Lines

3.0.1   Cells, Cell Culture Media and Buffers

1. Cell lines: Human Embryonic Kidney (HEK) cell lines 293A and 293T-17 were purchased from Invitrogen and American Type Culture Collection (ATCC), respectively. Alternatively, any cells that can be prepared as a suspension can be used for fcFRET.

2. Complete culture media: Dulbecco's Modified Eagle's Medium (DMEM)—high glucose (Sigma, #D5796-500ML) supple-

mented with 1% MEM Non-Essential Amino Acids solution (NEAA) (Gibco, #11140035) and 10% heat-inactivated Fetal Bovine Serum (FBS) (Sigma, #F7524) is stored at 4 °C and pre-warmed to 37 °C before use.

3. Cell dissociation reagent: 0.25% Trypsin-EDTA solution (Sigma, #T4049) is stored at 4 °C, pre-warmed before use.

4. Washing buffer: Phosphate-buffered saline (PBS) without calcium and magnesium, pH = 7.4, (Gibco, #10010015) is stored in room temperature (RT).

3.1.1 Development of a Stable Cell Line Expressing GPCR-A Tagged with eGFP

1. Expression vector: Plasmids containing the open reading frame (ORF) coding for the receptor of interest fused with eGFP (e.g., pCMV-GPCR-A-eGFP) and resistance markers (e.g., G-418) are commercially available (OriGene, GeneCopoeia, cDNA Resource Center) or can be cloned.

2. Transfection reagent: Lipofectamine LTX Reagent with PLUS Reagent (Invitrogen, #15338100) is stored at 4 °C and pre-warmed to RT before transfection.

3. Selection antibiotic: G-418 (Calbiochem, #345812) is stored at 4 °C.

4. Fluorescence enrichment: Flow Associated Cell Sorter (FACS) is used to sort cells expressing GPCR-A-eGFP (FACSAriaII, BD Biosciences). An epifluorescence microscope (Olympus IX70) and a flow cytometer (FACSCalibur, BD Biosciences) are used to monitor expression level of GPCR-A-eGFP in cells.

3.1.2 Lentiviral Transduction of Cells to Transiently Co-express GPCR-B Tagged with RFP

1. Viral plasmids: Lentiviral expression vector containing sequence of the receptor under investigation fused with RFP (pHR-GPCR-B-tagRFP), packaging plasmid (e.g., pCMV ΔR8.91), envelope plasmid (pMD.D-VSV.G) [16].

2. Transfection reagent: Transfection reagent prepared prior to use contains 2.5 mM calcium chloride (Sigma, #C1016) and N,N-bis (2-hydroxyethyl)-2-aminoethanesulfonic acid sodium salt solution (BES). BES solution is prepared as a 2× concentrated stock by adding 50 mM BES (Sigma, #B2891), 280 mM NaCl, and 1.5 mM Na_2HPO_4. Solution is adjusted to pH = 7.0 and stored in 4 °C.

3. Packaging media: DMEM supplemented with 1% NEAA and 2% heat-inactivated FBS.

4. Packaged virus harvesting: 0.45 μm filters (Sartorius, #16555).

5. Transduction Media: DMEM supplemented with 1% NEAA, 2% heat-inactivated FBS, and 8 μg/ml of polybrene (Sigma, #H9268).

**3.2 Sample
Preparation for fcFRET**

1. Washing buffer: PBS without calcium and magnesium, pH = 7.4 (Gibco, #10010015).

2. Suspension buffer: Suspension buffer is prepared prior to use by adding 20× concentrated EDTA stock (Sigma, #E5134) to the washing buffer. 40 mM EDTA stock is dissolved in H_2O and stored at 4 °C.

3. Single-cell suspension: 100 μm nylon mesh cell strainers (VWR, #10199–658).

4. Flow cytometry tubes: 5 ml round-bottom polystyrene tubes (Corning, #352054).

**3.3 Flow Cytometer
and Software
Configuration
for fcFRET Analysis**

1. The measurement of cell-by-cell FRET can be performed with the use of any flow cytometer equipped with two lasers to allow for separate excitation of donor and acceptor biosensors. fcFRET analysis between eGFP (GPCR-A-eGFP) as a donor and tagRFP (GPCR-B-tagRFP) as an acceptor is performed with the use of blue (488 nm) and yellow/green (561 nm) lasers (Fig. 2). eGFP is excited at 488 nm from blue laser and detected with a 525/50 nm bandpass filter, whereas TagRFP is excited at 561 nm from yellow/green laser and detected with a 610/20 nm bandpass filter. FRET signal between eGFP and TagRFP is measured by excitation at 488 nm from blue laser and detection with a 610/20 nm bandpass filter located on the same laser (excitation of the donor and measurement of the

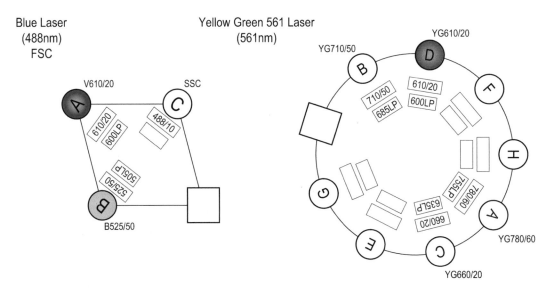

Fig. 2 LSR II configuration used for measurement of FRET between eGFP as a donor and tagRFP as an acceptor. eGFP is excited at 488 nm from blue laser and detected with a 525/50 nm bandpass filter (B), whereas tagRFP is excited at 561 nm from yellow/green laser and detected with a 610/20 nm bandpass filter (D). FRET signal between eGFP and tagRFP is measured by excitation at 488 nm from blue laser and detection with a 610/20 nm bandpass filter (A)

acceptor emission). For the proper separation of eGFP fluorescence and FRET emission from blue laser, a 505 Long Pass (LP) dichroic mirror (DM) should be used [42, 43].

2. To obtain the proper analysis of cell-by-cell FRET signal, data should be analyzed with flow cytometric data analysis programs such as FlowJo (FlowJo, LLC), BD FACSDiva (BD Biosciences), or FCS 6 Express Cytometry (De Novo Software).

4 Methods

4.1 Generation of Cells Co-expressing Fluorescently Tagged GPCR Pairs

4.3.1 Cell Culture

HEK293A and HEK293T-17 cells are cultured in complete culture media and maintained at 37 °C and 5% CO_2 in a humidified atmosphere to a confluence of >85%, after which the cells are washed with the use of washing buffer, dissociated with cell dissociation reagent, and passaged to a lower density. Cells should be passaged according to a strict schedule in order to ensure reproducible behavior, optimal health and GPCR expression levels.

4.1.1 Generation of Stable Human GPCR-A Expressing Cell Line

HEK293A cells are stably transfected with the plasmid construct of the receptor A (GPCR-A) C-terminally tagged with eGFP using transfection protocol, according to manufacturer's instructions. Cells with the highest expression of the GPCR-A-eGFP fusion protein are further selected using FACS, following by monoclonal expansion on 96-well plates. The monoclonal HEK293A-GPCR-A-eGFP cell line is then maintained in complete media supplemented with 400 ng/μl G-418 (selection media). Stability of the GPCR-A-eGFP fusion protein expression is monitored with the use of epifluorescent microscopy and flow cytometry. Functionality of GPCR-A is validated by measuring agonist-mediated signaling in a cellular-based assay.

4.1.2 Obtaining Cells with the Transient Co-expression of GPCR-B Using Lentiviral Transduction

The DNA fragment containing the sequence of GPCR-B is cloned into a HIV-based, replication deficient, lentiviral expression plasmid pHR-SIN-BX-tagRFP, suitable for the second generation lentiviral packaging system. HIV-based lentiviral particles are produced using HEK293T-17 cell line, by transient transfection of the cloned expression construct, pHR-GPCR-B-tagRFP; the packaging construct, pCMV ΔR8.91; and the envelope construct, pMD.G-VSV-G at a concentration of 1.9 μg, 1.5 μg, and 1 μg, respectively. Transfection reagent is used to increase efficiency of plasmid packaging. Twenty-four to thirty hours post transfection media is changed to packaging media. Forty-eight hours post media change the supernatant containing pHR-GPCR-B-tagRFP virus is removed and filtered. HEK293A cells with the stable expression of GPCR-A-eGFP can now be transduced with the pHR-GPCR-B-tagRFP packaged virus diluted in transduction media. The dilution of virus for transduction as well as incubation time is optimized for each cloned expression construct. The efficiency of transduction is monitored with the use of fluorescent microscope and flow cytometry (Note 3). Functionality of the transiently expressed GPCR-B is validated by measuring agonist-mediated signaling in a cellular-based assay.

4.2 Flow Cytometry Measurements

4.2.1 Sample Preparation

Media is aspirated and cells are washed twice with washing buffer. To avoid possible effects of the trypsin and/or EDTA on distribution and functionality of the receptors, we recommend to suspend cells mechanically with the use of washing buffer. Cell suspension is then centrifuged for 4 min at $200 \times g$, at room temperature. Supernatant is discarded and cell pellet is suspended in 400 μl of suspension buffer. Prior to analysis, cells should be passed through a cell strainer with 100 μm nylon mesh and collected in flow cytometry tubes.

In order to assess FRET between stably expressed GPCR-A-eGFP and transiently co-expressed GPCR-B-tagRFP receptors using flow cytometry, several control samples must be included in the experimental protocol:

Sample 1: HEK293A (control for fluorescence background correction).
Sample 2: HEK293A-GPCR-A-eGFP (control for compensation).
Sample 3: HEK293A-Lv-GPCR-B-tagRFP (control for compensation).
Sample 4: HEK293A-GPCR-A-eGFP-Lv-tagRFP (nonspecific FRET signal).
Sample 5: HEK293A-GPCR-A-eGFP-Lv-GPCR-B-tagRFP (FRET efficiency).

As a positive control, the same set of samples with known molecular interaction pair of receptors can be used. If it's known that one of the receptors under investigation forms homodimers (GPCR-A), the positive control sample set can be as follows (sample 1, 2, and 4 are the same as above):

Sample 3′: HEK293A-Lv-GPCR-A-tagRFP (control for compensation).
Sample 5′: HEK293A-GPCR-A-eGFP-Lv-GPCR-A-tagRFP (FRET efficiency).

4.2.2 Flow Cytometer Settings and Data Acquisition

Non-transfected HEK293A cells (sample 1) are firstly used for initial instrument setup. Forward Scatter versus Side Scatter (FSC vs SSC) plot is used to identify cells of interest based on their size and granularity. Sample 1 is also used for setting up photomultiplier tubes (PMT) voltage in order to correct eGFP and tagRFP fluorescence background. In the next step, cells expressing donor or acceptor construct only (sample 2 and 3, respectively) are used to fine tune PMT settings. Sample 2 and 3 are also necessary to perform the proper compensation for spectral bleed through and cross-excitation, in particular for eGFP emission in the tagRFP-FRET detector and tagRFP emission caused by excitation of tagRFP from the blue laser [42, 44]. Next, cells co-expressing the donor construct with the empty acceptor construct (sample 4) are analyzed for nonspecific FRET signal, following by analysis of cells with the expression of the two receptors under investigation (sample 5). At least 10^4 cells for each sample need to be recorded.

4.2.3 Analysis of FRET Signal

Non-transfected cells (sample 1) are first used to differentiate cells based on their size and granulation, according to forward and side scattering plot (FSC/SSC). This step allows to eliminate doublets, dead cells, and debris from further analysis. Additionally, the viability dye can be used to definitely eliminate dead cells. Next, two-dimensional dot-plot of eGFP fluorescence against tagRFP is constructed to gate cell populations with co-expression of both GPCRs under investigation (samples 1, 2, and 3 are used for the proper gate placement). Those cells are next used to create two-dimensional dot-plot of eGFP fluorescence against FRET signal. The gate for FRET positive signal is further corrected by using cells co-expressing donor with control acceptor constructs (sample 4) (Note 4).

An example of fcFRET analysis between the 5-HT2C-VSV and GHS-R1a receptors is demonstrated in Fig. 3. Following lentiviral transduction of HEK293A WT and HEK293A-5-HT2C-VSV cells with the lentiviral GHS-R1a-tagRFP vector, 61.6% and 52.2% of cells were analyzed as positive for tagRFP expression (sample 3 and 5, respectively). HEK293A-5-HT2C-VSV cells transduced with the control-tagRFP vector gave equal high transduction efficiency (61.5%, sample 4). Subsequent analysis of FRET signal was performed on the gated population of the successfully transduced cells and showed an increase in FRET signal from 1.9% in control sample (sample 4) to 30.26% in cells co-expressing both receptors (sample 5) (FRET vs eGFP plots). No tagRFP or FRET signal in HEK293A WT or HEK293A-5-HT2C-VSV cells was observed (sample 1 and 2).

5 Notes

1. *The choice of fluorescent protein pair*

Choice of optimal FRET pair should be considered not only based on spectroscopic features of individual FP, but also depending on the biological question and the compatibility of the fluorescent pairs with the selected FRET signal measurement method. Selected and validated FP pairs for flow cytometry-based FRET are depicted in Table 1.

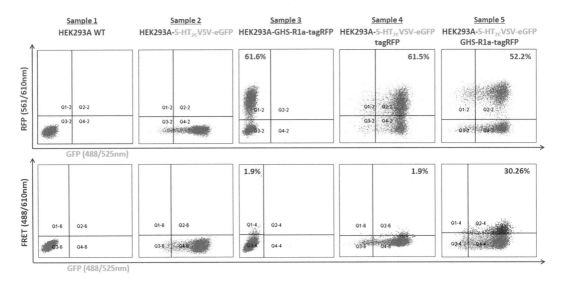

Fig. 3 fcFRET between 5-HT2C-VSV/GHS-R1a receptor pair is analyzed using the LSRII flow cytometer. HEK293A cells stably expressing the partially edited 5-HT2C isoform were transiently transduced with lentiviral vectors expressing control-tagRFP or GHS-R1a-tagRFP. Percentages indicate levels of tagRFP expression (tagRFP vs eGFP plots) or FRET levels (FRET vs eGFP plots). Adapted with permission from Schellekens et al., 2015. "Ghrelin's Orexigenic Effect Is Modulated via a Serotonin 2C Receptor Interaction." ACS Chemical Neuroscience 6 (7): 1186–97. doi: https://doi.org/10.1021/cn500318q. Copyright, 2015, American Chemical Society

Tailored FPs are now possible due to improved optical properties that have resulted in highly sensitive FRET biosensor pairs and improved FRET efficiency. One of the most common biosensor pairs used for fcFRET are GFPs and RFPs, in particular their brighter derivatives (eGFP and tagRFP), which have been used successfully to demonstrate GPCR interactions [43]. The eGFP/tagRFP combination demonstrates greater excitation wavelengths, larger emission peak separation, reduced auto-fluorescence and phototoxicity, compared to CFPs-YFPs pairs [45]. While the eGFP/tagRFP pair has been validated extensively and has many advantages, additional FPs with better optic properties, greater dynamic range and photostability are warranted to allow the fcFRET method to reach its full potential [46].

2. *Antibodies as an alternative for tagged fluorescent proteins*

FPs fused with receptors of interest are commonly used to investigate receptor–receptor interactions. However, in some cases binding of FPs to the N- or C-terminus of the receptors can influence their maturation, structure, subcellular localization and subsequently cause changes in function. Moreover, detection of receptors tagged with FPs requires heterogeneous expression systems and is not specific for membrane-bound GPCRs. This can be overcome by using fluorescence-conjugated antibodies targeting

Table 1
Fluorophore pairs suitable for fcFRET analysis

Donor $\lambda_{Ex}/\lambda_{Em}$ (nm)	Acceptor $\lambda_{Ex}/\lambda_{Em}$ (nm)	Biological process/cellular parameter	References
CFP (435/485)	YFP (514/527)	Molecule–molecule interaction Protein aggregation	[51–53]
	mCitrin (516/529)	Conformational changes	[54]
GFP (488/510)	RFP (555/584)	Receptor–receptor interaction Intramolecular conformational changes	[43, 55]
	Alexa-555 (555/580)	Enzymatic activity	[56]
	mOrange (548/562)	Molecule–molecule interaction	[57]
Alexa-555 (555/580)	Alexa-647 (650/665)	Subunit receptor spatial arrangement	[58]
mCerulean (433/475)	mVenus (515/528)	Molecule–molecule interaction	[59]
	mCitrin (516/529)	Enzymatic activity	[60]
Alexa-488 (490/525)	Alexa-594 (590/617)	Molecule–molecule interaction	[61]
	Alexa-546 (556/573)	Molecule–molecule interaction Receptor–receptor interaction	[62, 63]
PE (488/578)	Cy5 (650/670)	Molecule–molecule interaction Receptor–receptor interaction	[64, 65]
	Alexa-647 (650/665)	Receptor variant clusters	[66]

$\lambda_{Ex}/\lambda_{Em}$—values for maximum excitation/emission of fluorescent biosensor

receptors of interest in cells with endogenous GPCRs expression and used to confirm the occurrence of membrane receptor–receptor interactions (without permeabilization) [47]. Furthermore, dynamic changes in membrane receptor–receptor interactions induced by specific treatments can be studied after sample fixation. In this case, staining and fixation have to be optimized for each antibody. Also, the relative expression of the donor and the acceptor receptors have to be determined in order to use the appropriate concentration of respective antibody (i.e., antibody concentration must be high enough to label all receptor binding sites available at the cell membrane). If fluorescence-conjugated antibodies are not commercially available, a multiple step staining must be performed. In this case higher spatial separation of two fluorophores is observed, which rises the risk of false negative results. Even though fixed samples can be stored at 4 °C until measurement, fluorescence intensity decreases over time. Also, time required for labeling needs to be considered. Another major limitation of antibody labeling for receptors interaction analysis is the consequence of increasing protein structure size (with a single antibody adding 10 nm, this being doubled with a secondary antibody), although nanobodies are being developed to tackle such issues [48].

3. *Stable or transient receptor expression*

Monoclonal cell lines with stable receptors expression have the advantage over transient expression because of the same or very similar receptor expression level maintained over time, leading to a consistency in FRET signal between experiments [49]. However, the establishment of a monoclonal stable cell line is time consuming, while lentiviral transduction-mediated transient expression allows analysis of a wide range of receptors in a much shorter time frame. Furthermore, lentiviral expression is a highly efficient method for transient expression of genes of interest in neuronal cell [50]. This allows for the analysis of dynamic GPCR interactions in more physiologically relevant cell models.

4. *Alternative gating strategy*

In this book chapter, cells with the expression of donor construct (GPCR-A-eGFP) versus acceptor construct (GPCR-B-tagRFP) are used to determine the gating parameters of the population of cells with co-expression of both GPCRs and create a two-dimensional dot-plot for analysis of FRET signal. Gating of FRET signal is also corrected using cells co-expressing donor construct with control acceptor construct (GPCR-A-eGFP and control-tagRFP). Alternatively, cells expressing only the donor FP (eGFP) or acceptor FP (tagRFP), without target receptors, can be used to determine the gating parameters of cells co-expressing both FPs. Gating for FRET positive signal is then corrected using cells co-expressing donor with acceptor FP constructs (eGFP and tagRFP). This gating strategy can be implemented especially for the initial high throughput screening of new protein–protein interactions [51].

Acknowledgements

This research was funded in part by Science Foundation Ireland in the form of a Research Center grant (SFI/12/RC/2273) to APC Microbiome Ireland.

References

1. Sunahara RK, Insel PA (2016) The molecular pharmacology of G protein signaling then and now: a tribute to Alfred G. Gilman. Mol Pharmacol 89:585–592. https://doi.org/10.1124/mol.116.104216

2. Congreve M, Marshall F (2010) The impact of GPCR structures on pharmacology and structure-based drug design. Br J Pharmacol 159:986–996. https://doi.org/10.1111/j.1476-5381.2009.00476.x

3. Tesmer JJG (2016) Hitchhiking on the heptahelical highway: structure and function of 7TM receptor complexes. Nat Rev Mol Cell Biol 17:439–450. https://doi.org/10.1038/nrm.2016.36

4. Venkatakrishnan A, Flock T, Prado DE, Oates ME, Gough J, Madan Babu M (2014) Structured and disordered facets of the GPCR fold. Curr Opin Struct Biol 27:129–137. https://doi.org/10.1016/j.sbi.2014.08.002

5. Ward RJ, Milligan G (2014) Structural and biophysical characterisation of G protein-coupled receptor ligand binding using resonance energy transfer and fluorescent labelling techniques. Biochim Biophys Acta Biomembr 1838:3–14. https://doi.org/10.1016/j.bbamem.2013.04.007

6. Kenakin T (2004) Principles: receptor theory in pharmacology. Trends Pharmacol Sci 25:186–192. https://doi.org/10.1016/j.tips.2004.02.012

7. Borroto-Escuela DO, Fuxe K (2017) Diversity and bias through dopamine D2R heteroreceptor complexes. Curr Opin Pharmacol 32:16–22. https://doi.org/10.1016/j.coph.2016.10.004

8. Parmentier M (2015) GPCRs: heterodimer-specific signaling. Nat Chem Biol 11:244–245. https://doi.org/10.1038/nchembio.1772

9. Terrillon S, Bouvier M (2004) Roles of G-protein-coupled receptor dimerization. EMBO Rep 5:30–34. https://doi.org/10.1038/sj.embor.7400052

10. Hébert TE, Bouvier M (1998) Structural and functional aspects of G protein-coupled receptor oligomerization. Biochem Cell Biol 76:1–11. https://doi.org/10.1139/o98-012

11. Ferré S, Baler R, Bouvier M, Caron MG, Devi LA, Durroux T, Fuxe K, George SR, Javitch JA, Lohse MJ, Mackie K, Milligan G, Pfleger KDG, Pin J-P, Volkow ND, Waldhoer M, Woods AS, Franco R (2009) Building a new conceptual framework for receptor heteromers. Nat Chem Biol 5:131–134. https://doi.org/10.1038/nchembio0309-131

12. Romero-Fernandez W, Borroto-Escuela DO, Agnati LF, Fuxe K (2013) Evidence for the existence of dopamine D2-oxytocin receptor heteromers in the ventral and dorsal striatum with facilitatory receptor-receptor interactions. Mol Psychiatry 18:849–850. https://doi.org/10.1038/mp.2012.103

13. Siddiquee K, Hampton J, McAnally D, May L, Smith L (2013) The apelin receptor inhibits the angiotensin II type 1 receptor via allosteric trans-inhibition. Br J Pharmacol 168:1104–1117. https://doi.org/10.1111/j.1476-5381.2012.02192.x

14. Moreno Delgado D, Møller TC, Ster J, Giraldo J, Maurel D, Rovira X, Scholler P, Zwier JM, Perroy J, Durroux T, Trinquet E, Prezeau L, Rondard P, Pin J-P (2017) Pharmacological evidence for a metabotropic glutamate receptor heterodimer in neuronal cells. Elife 6:pii: e25233. https://doi.org/10.7554/eLife.25233

15. Jordan BA (2001) Oligomerization of opioid receptors with beta 2-adrenergic receptors: a role in trafficking and mitogen-activated protein kinase activation. Proc Natl Acad Sci U S A 98:343–348. https://doi.org/10.1073/pnas.011384898

16. Schellekens H, van Oeffelen WEPA, Dinan TG, Cryan JF (2013) Promiscuous dimerization of the growth hormone secretagogue receptor (GHS-R1a) attenuates ghrelin-mediated signaling. J Biol Chem 288:181–191. https://doi.org/10.1074/jbc.M112.382473

17. Bellot M, Galandrin S, Boularan C, Matthies HJ, Despas F, Denis C, Javitch J, Mazères S, Sanni SJ, Pons V, Seguelas M-H, Hansen JL, Pathak A, Galli A, Sénard J-M, Galés C (2015) Dual agonist occupancy of AT1-R-α2C-AR heterodimers results in atypical Gs-PKA signaling. Nat Chem Biol 11:271–279. https://doi.org/10.1038/nchembio.1766

18. Terrillon S, Durroux T, Mouillac B, Breit A, Ayoub MA, Taulan M, Jockers R, Barberis C, Bouvier M (2003) Oxytocin and vasopressin V1a and V2 receptors form constitutive homo- and heterodimers during biosynthesis. Mol Endocrinol 17:677–691. https://doi.org/10.1210/me.2002-0222

19. Rashid AJ, So CH, Kong MMC, Furtak T, El-Ghundi M, Cheng R, O'Dowd BF, George SR (2007) D1-D2 dopamine receptor heterooligomers with unique pharmacology are

coupled to rapid activation of Gq/11 in the striatum. Proc Natl Acad Sci U S A 104:654–659. https://doi.org/10.1073/pnas.0604049104

20. Kern A, Mavrikaki M, Ullrich C, Albarran-Zeckler R, Brantley AF, Smith RG (2015) Hippocampal dopamine/DRD1 signaling dependent on the ghrelin receptor. Cell 163:1176–1190. https://doi.org/10.1016/j.cell.2015.10.062

21. Borroto-Escuela DO, Romero-Fernandez W, Tarakanov AO, Marcellino D, Ciruela F, Agnati LF, Fuxe K (2010) Dopamine D2 and 5-hydroxytryptamine 5-HT2A receptors assemble into functionally interacting heteromers. Biochem Biophys Res Commun 401:605–610. https://doi.org/10.1016/j.bbrc.2010.09.110

22. Borroto-Escuela DO, Carlsson J, Ambrogini P, Narváez M, Wydra K, Tarakanov AO, Li X, Millón C, Ferraro L, Cuppini R, Tanganelli S, Liu F, Filip M, Diaz-Cabiale Z, Fuxe K (2017) Understanding the role of GPCR heteroreceptor complexes in modulating the brain networks in health and disease. Front Cell Neurosci 11:1–20. https://doi.org/10.3389/fncel.2017.00037

23. Schellekens H, Dinan TG, Cryan JF (2013) Taking two to tango: a role for ghrelin receptor heterodimerization in stress and reward. Front Neurosci 7:148. https://doi.org/10.3389/fnins.2013.00148

24. Fribourg M, Moreno JL, Holloway T, Provasi D, Baki L, Mahajan R, Park G, Adney SK, Hatcher C, Eltit JM, Ruta JD, Albizu L, Li Z, Umali A, Shim J, Fabiato A, MacKerell AD, Brezina V, Sealfon SC, Filizola M, González-Maeso J, Logothetis DE (2011) Decoding the signaling of a GPCR heteromeric complex reveals a unifying mechanism of action of antipsychotic drugs. Cell 147:1011–1023. https://doi.org/10.1016/j.cell.2011.09.055

25. Cabello N, Gandía J, Bertarelli DCG, Watanabe M, Lluís C, Franco R, Ferré S, Luján R, Ciruela F (2009) Metabotropic glutamate type 5, dopamine D2 and adenosine A2a receptors form higher-order oligomers in living cells. J Neurochem 109:1497–1507. https://doi.org/10.1111/j.1471-4159.2009.06078.x

26. Pintsuk J, Borroto-Escuela DO, Lai TKY, Liu F, Fuxe K (2016) Alterations in ventral and dorsal striatal allosteric A2AR-D2R receptor-receptor interactions after amphetamine challenge: Relevance for schizophrenia. Life Sci 167:92–97. https://doi.org/10.1016/j.lfs.2016.10.027

27. de la Mora MP, Pérez-Carrera D, Crespo-Ramírez M, Tarakanov A, Fuxe K, Borroto-Escuela DO (2016) Signaling in dopamine D2 receptor-oxytocin receptor heterocomplexes and its relevance for the anxiolytic effects of dopamine and oxytocin interactions in the amygdala of the rat. Biochim Biophys Acta 1862:2075–2085. https://doi.org/10.1016/j.bbadis.2016.07.004

28. Millón C, Flores-Burgess A, Narváez M, Borroto-Escuela DO, Santín L, Gago B, Narváez JA, Fuxe K, Díaz-Cabiale Z (2016) Galanin (1–15) enhances the antidepressant effects of the 5-HT1A receptor agonist 8-OH-DPAT: involvement of the raphe-hippocampal 5-HT neuron system. Brain Struct Funct 221:4491–4504. https://doi.org/10.1007/s00429-015-1180-y

29. Naumenko VS, Popova NK, Lacivita E, Leopoldo M, Ponimaskin EG (2014) Interplay between serotonin 5-HT1A and 5-HT7 receptors in depressive disorders. CNS Neurosci Ther 20:582–590. https://doi.org/10.1111/cns.12247

30. Guo H, An S, Ward RJ, Yang Y, Liu Y, Guo X-X, Hao Q, Xu T-R (2017) Methods used to study the oligomeric structure of G protein-coupled receptors. Biosci Rep 37:BSR20160547. https://doi.org/10.1042/BSR20160547

31. Boute N, Jockers R, Issad T (2002) The use of resonance energy transfer in high-throughput screening: BRET versus FRET. Trends Pharmacol Sci 23:351–354. https://doi.org/10.1016/S0165-6147(02)02062-X

32. Busnelli M, Mauri M, Parenti M, Chini B (2013) Analysis of GPCR dimerization using acceptor photobleaching resonance energy transfer techniques, 1st ed. Methods Enzymol 521:311–327. https://doi.org/10.1016/B978-0-12-391862-8.00017-X

33. Forster T (1946) Energiewanderung und Fluoreszenz. Naturwissenschaften 33:166–175. https://doi.org/10.1007/BF00585226

34. Fernández-Dueñas V, Llorente J, Gandía J, Borroto-Escuela DO, Agnati LF, Tasca CI, Fuxe K, Ciruela F (2012) Fluorescence resonance energy transfer-based technologies in the study of protein–protein interactions at the cell surface. Methods 57:467–472. https://doi.org/10.1016/j.ymeth.2012.05.007

35. Ciruela F (2008) Fluorescence-based methods in the study of protein-protein interactions in living cells. Curr Opin Biotechnol 19:338–343. https://doi.org/10.1016/j.copbio.2008.06.003

36. Piston DW, Kremers G-J (2007) Fluorescent protein FRET: the good, the bad and the ugly. Trends Biochem Sci 32:407–414. https://doi.org/10.1016/j.tibs.2007.08.003

37. Masters BR (2014) Paths to Förster's resonance energy transfer (FRET) theory. Eur Phys

J H 39:87–139. https://doi.org/10.1140/epjh/e2013-40007-9

38. Stryer L (1978) Fluorescence energy transfer as a spectroscopic ruler. Annu Rev Biochem 47:819–846. https://doi.org/10.1146/annurev.bi.47.070178.004131

39. Shrestha D, Jenei A, Nagy P, Vereb G, Szöllősi J (2015) Understanding FRET as a research tool for cellular studies. Int J Mol Sci 16:6718–6756. https://doi.org/10.3390/ijms16046718

40. Chan F-M, Holmes K (2004) Flow cytometric analysis of fluorescence resonance energy transfer. Flow Cytom Protoc 263:281–292. https://doi.org/10.1385/1-59259-773-4:281

41. Nagy P, Vereb G, Damjanovich S, Mátyus L, Szöllősi J (2006) Measuring FRET in flow cytometry and microscopy. Curr Protoc Cytom Chapter 12:Unit12.8. doi: https://doi.org/10.1002/0471142956.cy1208s38

42. Chan FK-M, Siegel RM, Zacharias D, Swofford R, Holmes KL, Tsien RY, Lenardo MJ (2001) Fluorescence resonance energy transfer analysis of cell surface receptor interactions and signaling using spectral variants of the green fluorescent protein. Cytometry 44:361–368. https://doi.org/10.1002/1097-0320(20010801)44:4<361::AID-CYTO1128>3.0.CO;2-3

43. Schellekens H, De Francesco PN, Kandil D, Theeuwes WF, McCarthy T, Van Oeffelen WEPA, Perelló M, Giblin L, Dinan TG, Cryan JF (2015) Ghrelin's orexigenic effect is modulated via a serotonin 2C receptor interaction. ACS Chem Neurosci 6:1186–1197. https://doi.org/10.1021/cn500318q

44. Chan FK-M, Holmes KL (2004) Flow cytometric analysis of fluorescence resonance energy transfer: a tool for high-throughput screening of molecular interactions in living cells. In: Flow cytometry protocol. Humana, New Jersey, pp 281–292

45. Bajar BT, Wang ES, Zhang S, Lin MZ, Chu J (2016) A guide to fluorescent protein FRET pairs. Sensors (Switzerland) 16:1–24. https://doi.org/10.3390/s16091488

46. Lam AJ, St-Pierre F, Gong Y, Marshall JD, Cranfill PJ, Baird MA, McKeown MR, Wiedenmann J, Davidson MW, Schnitzer MJ, Tsien RY, Lin MZ (2012) Improving FRET dynamic range with bright green and red fluorescent proteins. Nat Methods 9:1005–1012. https://doi.org/10.1038/nmeth.2171

47. Borroto-Escuela DO, Hagman B, Woolfenden M, Pinton L, Jiménez-Beristain A, Oflijan J, Narvaez M, Di Palma M, Feltmann K, Sartini S, Ambrogini P, Ciruela F, Cuppini R, Fuxe K (2016) In Situ proximity ligation assay to study and understand the distribution and balance of GPCR homo- and heteroreceptor complexes in the brain. In: Luján R, Ciruela F (eds) Receptor and ion channel detection in the brain: methods and protocols. Springer, New York, pp 109–124

48. Strack R (2015) Protein labeling in cells. Nat Methods 13:33–33. https://doi.org/10.1038/nmeth.3702

49. Chruścicka B, Burnat G, Brański P, Chorobik P, Lenda T, Marciniak M, Pilc A (2015) Tetracycline-based system for controlled inducible expression of group III metabotropic glutamate receptors. J Biomol Screen 20:350–358. https://doi.org/10.1177/1087057114559183

50. Naldini L (1998) Lentiviruses as gene transfer agents for delivery to non-dividing cells. Curr Opin Biotechnol 9:457–463. https://doi.org/10.1016/S0958-1669(98)80029-3

51. Banning C, Votteler J, Hoffmann D, Koppensteiner H, Warmer M, Reimer R, Kirchhoff F, Schubert U, Hauber J, Schindler M (2010) A flow cytometry-based FRET assay to identify and analyse protein-protein interactions in living cells. PLoS One 5:e9344. https://doi.org/10.1371/journal.pone.0009344

52. Hagen N, Bayer K, Rösch K, Schindler M (2014) The intraviral protein interaction network of hepatitis C virus. Mol Cell Proteomics 13:1676–1689. https://doi.org/10.1074/mcp.M113.036301

53. Holmes BB, Furman JL, Mahan TE, Yamasaki TR, Mirbaha H, Eades WC, Belaygorod L, Cairns NJ, Holtzman DM, Diamond MI (2014) Proteopathic tau seeding predicts tauopathy in vivo. Proc Natl Acad Sci U S A 111:E4376–E4385. https://doi.org/10.1073/pnas.1411649111

54. Abankwa D, Hanzal-Bayer M, Ariotti N, Plowman SJ, Gorfe AA, Parton RG, McCammon JA, Hancock JF (2008) A novel switch region regulates H-ras membrane orientation and signal output. EMBO J 27:727–735. https://doi.org/10.1038/emboj.2008.10

55. Kara E, Marks JD, Fan Z, Klickstein JA, Roe AD, Krogh KA, Wegmann S, Maesako M, Luo CC, Mylvaganam R, Berezovska O, Hudry E, Hyman BT (2017) Isoform and cell type-specific structure of Apolipoprotein E lipoparticles as revealed by a novel Forster Resonance Energy Transfer assay. J Biol Chem 292:14720–14729. https://doi.org/10.1074/jbc.M117.784264

56. Suzuki M, Tanaka S, Ito Y, Inoue M, Sakai T, Nishigaki K (2012) Simple and tunable Förster resonance energy transfer-based bioprobes for

high-throughput monitoring of caspase-3 activation in living cells by using flow cytometry. Biochim Biophys Acta Mol Cell Res 1823:215–226. https://doi.org/10.1016/j. bbamcr.2011.07.006

57. Larbret F, Dubois N, Brau F, Guillemot E, Mahiddine K, Tartare-Deckert S, Verhasselt V, Deckert M (2013) Technical advance: actin CytoFRET, a novel FRET flow cytometry method for detection of actin dynamics in resting and activated T cell. J Leukoc Biol 94:531–539. https://doi.org/10.1189/ jlb.0113022

58. Botzolakis EJ, Gurba KN, Lagrange AH, Feng HJ, Stanic AK, Hu N, Macdonald RL (2016) Comparison of γ-aminobutyric acid, type A (GABAA), receptor αβγ and αβδ expression using flow cytometry and electrophysiology: Evidence for alternative subunit stoichiometries and arrangements. J Biol Chem 291:20440–20461. https://doi.org/10.1074/jbc. M115.698860

59. Hassinen A, Pujol FM, Kokkonen N, Pieters C, Kihlström M, Korhonen K, Kellokumpu S (2011) Functional organization of Golgi N- and O-glycosylation pathways involves pH-dependent complex formation that is impaired in cancer cells. J Biol Chem 286:38329–38340. https://doi.org/10.1074/jbc.M111.277681

60. Gaber R, Majerle A, Jerala R, Benčina M (2013) Noninvasive high-throughput single-cell analysis of HIV protease activity using ratiometric flow cytometry. Sensors (Basel) 13:16330–16346. https://doi.org/10.3390/ s131216330

61. Bunaciu RP, Jensen HA, MacDonald RJ, La Tocha DH, Varner JD, Yen A (2015) 6-Formylindolo(3,2-b)carbazole(FICZ) modulates the signalsome responsible for RA-induced differentiation of HL-60 myeloblastic leukemia cells. PLoS One 10:e0135668. https://doi.org/10.1371/journal. pone.0135668

62. Damjanovich L, Volkó J, Forgács A, Hohenberger W, Bene L (2012) Crohn's disease alters MHC-rafts in CD4 + T-cells. Cytometry A 81(A):149–164. https://doi. org/10.1002/cyto.a.21173

63. Nagy P (2008) Quantitative characterization of the large-scale association of ErbB1 and ErbB2 by flow cytometric homo-FRET measurements. Biophys J 95:2086–2096. https://doi. org/10.1529/biophysj.108.133371

64. Marconi M, Ascione B, Ciarlo L, Vona R, Garofalo T, Sorice M, Gianni AM, Locatelli SL, Carlo-Stella C, Malorni W, Matarrese P (2013) Constitutive localization of DR4 in lipid rafts is mandatory for TRAIL-induced apoptosis in B-cell hematologic malignancies. Cell Death Dis 4:e863. https://doi.org/10.1038/ cddis.2013.389

65. Brockhoff G, Heiss P, Schlegel J, Hofstaedter F, Knuechel R (2001) Epidermal growth factor receptor, c-erbB2 and c-erbB3 receptor interaction, and related cell cycle kinetics of SK-BR-3 and BT474 breast carcinoma cells. Cytometry 44:338–348. https://doi.org/10.1002/1097-0320(20010801)44:4<338::AID-CYTO1125>3.0.CO;2-V. [pii]

66. Darbandi-Tehrani K, Hermand P, Carvalho S, Dorgham K, Couvineau A, Lacapere J-J, Combadiere C, Deterre P (2010) Subtle conformational changes between CX3CR1 genetic variants as revealed by resonance energy transfer assays. FASEB J 24:4585–4598. https:// doi.org/10.1096/fj.10-156612

Chapter 15

Assessing GPCR Dimerization in Living Cells: Comparison of the NanoBiT Assay with Related Bioluminescence- and Fluorescence-Based Approaches

Elise Wouters, Lakshmi Vasudevan, Francisco Ciruela, Deepak K. Saini, Christophe Stove, and Kathleen Van Craenenbroeck

Abstract

G protein-coupled receptors (GPCRs) modulate cellular signaling pathways, including differentiation, proliferation, hormonal regulation, and neuronal activity. Therefore, it is not surprising that almost 50% of the drugs available in the pharmaceutical market target GPCRs. Recently, an emerging body of evidence has proven the formation of GPCR dimers and even higher order oligomers. For neurodegenerative diseases, such as Parkinson's or Alzheimer's disease, it is crucial to characterize these receptor–receptor interactions in the brain to elucidate their role in neuronal disease-relevant processes. As a first step, a robust *in cellulo* assay is essential to identify and characterize specific GPCR–GPCR interactions. In the past 20 years, considerable efforts have been directed towards the development of GPCR dimerization screening assays to evaluate these receptor–receptor interactions in living cells. Interestingly, most of the approaches employ noninvasive fluorescence- and luminescence-based assays. Here, we present an efficient strategy to study GPCR dimerization dynamics, namely a protein complementation assay (PCA) based on the reconstitution of a luminescent protein, the NanoLuciferase (NL). Thus, GPCRs of interest are fused to complementary NL fragments which upon GPCR dimerization may reconstitute to a functional reporter, of which activity can be measured. The experimental procedure takes 2–4 days to complete, depending on the cell type and complexity of the experimental setup. In contrast to alternative protein complementation assays (also described in this book chapter), this method can also be implemented to analyze the kinetics of ligand-dependent modulation of dimerization, broadening its application potential. Additionally, high throughput screenings can also be performed, which is highly relevant given the growing interest and effort to identify small molecule drugs that can target disease-relevant dimers (or even selectively alter GPCR dimer function).

Key words G protein-coupled receptor, Dimerization, Protein complementation assay, NanoBiT, Split VENUS, Split *Renilla* luciferase

1 Introduction

G protein-coupled receptors (GPCRs) constitute the largest family of plasma membrane receptors. GPCRs consist of seven transmembrane domains which upon ligand activation trigger different signaling pathways. Nowadays, almost 50% of marketed drugs

Kjell Fuxe and Dasiel O. Borroto-Escuela (eds.), *Receptor-Receptor Interactions in the Central Nervous System*, Neuromethods, vol. 140,
https://doi.org/10.1007/978-1-4939-8576-0_15, © Springer Science+Business Media, LLC, part of Springer Nature 2018

treating a variety of short- and long-term illnesses, including conditions affecting the central nervous system (CNS), target GPCRs [1]. Interestingly, until now pharmaceutical industry has primarily focused on GPCR monomers as a target for drug development. However, an emerging body of evidence has demonstrated the formation of several GPCR dimers and even higher order oligomers [2–8]. Therefore, it has been recently suggested that certain GPCR dimers may represent operative targets for clinically effective small molecule drugs. Within this mindset, Fuxe et al. [9] proposed the idea of "multi-therapy," which implies the combined use of several ligands to target different GPCRs. This may decrease the effective dose necessary to obtain the desired effect, thereby reducing the risk of side effects. In the same context, "bivalent ligands," which consist of two ligands linked by a spacer, represent another pharmaceutical tool that could lead to more potent and selective compounds [10]. Overall, this field is empowered by a remarkable evolution in the knowledge of structural biology and functional signaling of multiple GPCRs. Although notable milestones have been reached, signaling pathways of GPCRs are more complex than first envisioned, possibly due to di- and/or oligomerization.

A widely accepted evidence for the existence of GPCR dimers is the dimer formation of the Class C GABA receptors [11]. These are obligate dimers, as $GABA_{B1}$ receptor is necessary for ligand binding while $GABA_{B2}$ receptor ensures efficient cell trafficking and signal transduction. Importantly, both protomers are not functional when expressed alone. In addition to obligate GPCR dimers observed in Class C receptors, an increasing amount of evidence suggests that there is also—transient—dimer formation for GPCRs of Class A [2–5, 12]. To investigate these potential interactions between GPCRs, multiple biochemical and biophysical techniques have been applied, such as co-immunoprecipitation (Co-IP) [6], fluorescence or bioluminescence resonance energy transfer (FRET or BRET, respectively) [13–15], and fluorescence- and bioluminescence-based or β-galactosidase protein complementation assays [16–20].

GPCR dimers have been studied with protein complementation assays (PCAs), of which bimolecular fluorescence and luminescence complementation (BiFC and BiLC, respectively) assays have been applied most. This approach relies on the ability of two nonfluorescent or non-luminescent protein fragments to restore their fluorescence or luminescence, respectively, when fused to and co-expressed with respective partners. BiFC is easy to monitor since it does not require a substrate, and visualization of the localization of the interaction within the cell is achieved by fluorescence detection. Several fluorescent proteins have been successfully optimized and applied in BiFC assays, such as GFP, YFP, VENUS, and mCherry [21]. On the other hand, BiLC uses a split version of a luminescent protein, like *Renilla* or Firefly luciferase (*R*Luc or FLuc, respectively) or nanoluciferase (NL). These approaches require addition of a substrate, i.e., coelenterazine *h* or furimazine. One of the more recently optimized luminescence complementation assays is the "NanoLuciferase Binary Technology" or NanoBiT, commercialized by Promega (Fig. 15.1) [22–24]. The main difference between the existing luminescent protein fragments based on *Renilla*/Firefly luciferase-based PCAs and nanoluciferase is that the NanoBiT subunits consist of two dissimilar fragments in size, namely a small 1.3 kDa subunit (Small BiT; SmBiT) and a large 18 kDa subunit (Large BiT; LgBiT). Both subunits have been thoroughly characterized, showing the advantage of superior bright luminescence upon reconstitution of the functional nanoluciferase. Importantly, the SmBiT sequence has been modified to provide a low affinity to prevent spontaneous reconstitution (K_D = 190 µM), thus allowing the study of protein–protein interaction (PPI) kinetics (i.e., association/dissociation).

Finally, it is of paramount importance to highlight that whatever PCA (i.e., BiFC and BiLC) is used to demonstrate the interaction of interest, adequate controls are essential to avoid false positives. In addition, designing flexible linkers and exploring appropriate expression systems is required to ensure the robustness of the assay. In this chapter, we describe the use of different PCAs based on VENUS, *Renilla* luciferase, and nanoluciferase (see Table 1) to study homodimerization of a prototypical GPCR, the dopamine D_2 receptor (D_2R). Overall, this chapter aims at providing an overview of the pros and cons of the PCAs used in the study of GPCR oligomerization. These findings can be extended into other research areas where PPIs constitute the main focus.

2   Materials

2.1   Plasmids

1. GPCR-VN and GPCR-VC plasmids. The fluorescent protein VENUS is split in two fragments, namely the N-terminal part (VN155 (aa 1–154)) and the C-terminal part (VC155 (aa 155–238)) (see **Note 1**). The candidate GPCR can be cloned into the backbones, N-terminally of VN or VC, with a short linker between the receptor and the split fragment of VENUS, normally a leucine and glutamate residue. The expression of the genes is driven by a CMV promoter. The addition of a VN control plasmid is optional but strongly recommended (see **Note 2**).

2. GPCR-*R*Luc1 and GPCR-*R*Luc2 plasmids. For the expression of a candidate GPCR fused to a split fragment of *Renilla* luciferase, the *R*Luc1 (aa 1–229) and *R*Luc2 (aa 230–311) constructs can be used as backbone plasmids. The linker between the receptor and *R*Luc1/*R*Luc2 consists of 24 amino

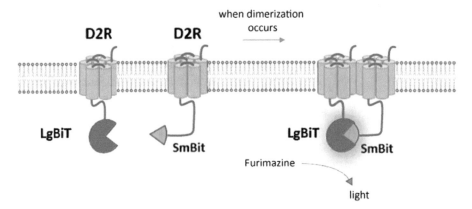

Fig. 15.1 Schematic representation of the NanoBiT system. Candidate GPCRs (i.e., D₂R) are fused to SmBiT or LgBiT. Upon GPCR dimerization, the NanoBiT fragments come into close proximity resulting in NL reconstitution, which can convert the Furimazine substrate leading to the emission of light

Table 1
Overview of the fluorescent and luminescent proteins, already in use for complementation assays

Reporter protein	Readout	Signal-to-noise ratio	Sensor size (kDa)
Renilla luciferase	Luminescence	Low	36
VENUS	Fluorescence	Low	27
NanoBiT	Luminescence	High	19

acids (ATGLDLELKASNSAVDGTAGPVAT). The expression is driven by a CMV promoter. The addition of *R*Luc1 control plasmid is optional but strongly recommended (see **Note 2**).

3. GPCR-LgBiT and GPCR-SmBiT plasmids. The NanoBiT constructs (Promega) express the small (SmBiT) or large (LgBiT) fragment of the NL, wherein the candidate GPCR can be cloned N-terminally of (preferably both) NL fragments (see **Note 3**). Furthermore, a flexible GS linker of 15–21 amino acids (see **Note 4**) between the receptor and the fragment of the NL is present. These constructs contain a HSV-TK promoter, which is not a very strong promoter and thereby limits overexpression.

4. Plasmids constitutively expressing yellow fluorescent protein (i.e., YFP or VENUS) or cyan fluorescent protein (CFP).

2.2 Cells

1. Human embryonic kidney 293T (HEK-293T) cells or any other cell line that can give a high transfection efficiency and easily drives the expression of chimeric GPCRs. Although not

always possible, it is good to use a cell line which does not show endogenous expression of GPCRs which can interfere with the dimerization assay.

2.3 Reagents

1. Dulbecco's Modified Eagle's Medium (DMEM) supplemented with GlutaMAX supplemented with 10% heat inactivated fetal bovine serum (FBS), 100 mg/ml streptomycin, and 100 IU/l penicillin. Serum free medium as well as 2% FBS DMEM medium will also be required. The medium depends on the cell line used.

2. Hank's Balanced Salt Solution (HBSS).

3. Phosphate Buffered Saline (PBS).

4. Transient mammalian cell transfection reagent Polyethylenimine (PEI).

5. Coelenterazine *h* when using *R*Luc or The Nano-Glo Live Cell reagent, a non-lytic detection reagent containing the cell permeable furimazine substrate (Promega) when using the nanoluciferase.

6. Ligands (i.e., agonist, antagonist) for the candidate receptor.

2.4 Equipment

1. 75 cm^2 flasks for maintenance and growth of the cell line.

2. Cell culture facility including a laminar hood and incubator set to 5% CO_2 and 37 °C.

3. 6-well cell culture plates.

4. 96-well plates (black with a flat bottom for fluorescence-based assays, white with a flat bottom for luminescence-based assays).

5. Fluorescence microscope to check transfection efficiency.

6. Luciferase and/or fluorescence microplate reader, e.g., ClarioSTAR (BMG), Glomax 96 (Promega), or TriStar (Berthold Technologies).

3 Methods

1. Design the appropriate (sequence-verified) expression plasmids for the fluorescence-based (GPCR-VN and GPCR-VC) or luminescence-based (GPCR-*R*Luc1 and GPCR-*R*Luc2 or GPCR-SmBiT and GPCR-LgBiT) assays. Proper controls need to be included (see **Note 2**).

2. Prepare and purify the required amount of DNA for the assays and dilute the concentration to 100 ng/µl for the transfection.

3. Seed 2.5×10^5 HEK293T cells in 2 ml DMEM with 10% FBS in each well of a 6-well plate. Place the plate in the incubator.

4. Refresh with 1.8 ml of DMEM +2% FBS the next day.

5. Prepare a DNA mixture containing a combination of candidate GPCR plasmids, which need to be evaluated for dimerization. Prepare at least two replicates for each condition (i.e., for 2× a 6-well), so a proper number of transfected cells for the assay will be obtained. To test for dimerization, first use a 1:1 ratio of both GPCR plasmids, i.e., 200 ng each (see **Note 5**). Furthermore, co-transfect a plasmid coding YFP or CFP (or any other suitable fluorescent protein) (100 ng/well) for normalization of the data (see **Note 6**). Add a mock vector, e.g., pcDNA3 (Invitrogen), to have a final DNA concentration of 2 μg for each DNA mixture.

6. Dilute the PEI transfection reagent (stock: 1 μg/μl) 20× in serum free DMEM.

7. Pipet the PEI/serum free DMEM mixture onto the DNA mixture (1:1 volume ratio), vortex and incubate for 10 min.

8. Transfer 200 μl of the DNA transfection mix onto the cells.

9. After 4–5 h, refresh the medium with 2 ml DMEM supplemented with 10% FBS.

10. Twenty-four hours after transfection (see **Note 7**), it is advisable to check the transfection efficiency (e.g., YFP/CFP expression) with the fluorescence microscope.

11. Additional step if applicable: To test the occurrence of ligand-dependent dimerization, stimulate/block one of the receptors with the appropriate ligand (e.g., 1–10 μM) in 2 ml of DMEM +2% FBS. Incubation times can be varied between 4 and 16 h prior to the readout of the assay.

12. Forty-eight hours after transfection (see **Note 8**), aspirate the medium and wash the cells twice with 1 ml of pre-heated PBS.

13. Scrape the cells with 500 μl of pre-heated HBSS and transfer them into a microcentrifuge tube.

14. Centrifuge for 5 min at 3000 rpm.

15. Aspirate the buffer and add 400 μl of fresh pre-heated HBSS (see **Note 9**). Transfer 100 μl of the cell suspension in a 96-well plate. It is advisable to run at least three replicates per condition to obtain sufficient recordings for statistical analysis.

16. For the fluorescence-based PCA, samples need to be transferred in a 96-well black plate. Firstly, measure the VENUS signal and afterwards the CFP signal with the microplate reader (see **Note 10**).

17. For the luminescence-based PCA, the samples need to be transferred into a 96-well white plate. Firstly, measure the YFP signal with the microplate reader. Afterwards, add 25 μl of coelenterazine h (40× dilution into HBSS, see **Note 11**) in the

case of the *R*Luc-based PCA. Add 25 μl of the Nano-Glo Live Cell reagent, containing the cell permeable furimazine substrate (see **Note 12**), in case of the NanoBiT PCA. Avoid light exposure of the substrate at all times! Measure the luminescence with an endpoint measurement with the microplate reader. For an overview of the excitation and emission wavelengths of the fluorescent and luminescent complementation proteins, see **Note 13**.

18. Calculate the mean, standard deviation, and standard error of the mean using an Excel spreadsheet or GraphPad Prism 5 (GraphPad Software, San Diego California, USA). Plot a bar graph to visualize the data. Examples of D_2R homodimerization with the fluorescence-based PCA using VENUS fragments (Fig. 15.2) (see **Note 14**), the luminescence-based PCA using *R*Luc fragments (Fig. 15.3) (see **Note 14**) as well as the NanoBIT PCA (Fig. 15.3) (see **Note 15**) are shown.

4   Notes

1. Plasmids of the fluorescent protein VENUS split at other positions are also commercially available, i.e., VN173 (aa 1–173) and VC155 (aa 155–238) (Addgene, Javitch group) [25] but also here a low signal-to-noise ratio was obtained.

2. The fragments of a fluorescent or luminescent protein can show a high affinity for each other, which can be the driving force for reconstitution of the protein, resulting in emission of a signal. Therefore, it is highly recommended to include proper controls in your assay, i.e., a non-interacting partner

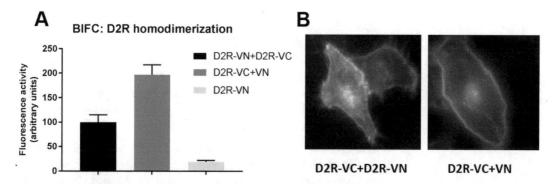

Fig. 15.2 Bimolecular fluorescence complementation (BiFC) with D_2R fused to VENUS fragments (VN155 and VC155). (**a**) D_2R constructs were transiently transfected (200 ng each) in HEK293T cells. Fluorescence activity shows the reconstituted VENUS signal, normalized for the CFP signal. Endpoint measurement of the fluorescent signal was conducted with the ClarioSTAR. Each value represents the mean ± SD from three independent experiments. (**b**) D_2R constructs were transiently transfected (50 ng each) in HeLa cells. Microscopy results confirm the high amount of self-assembly of the split VENUS fragments

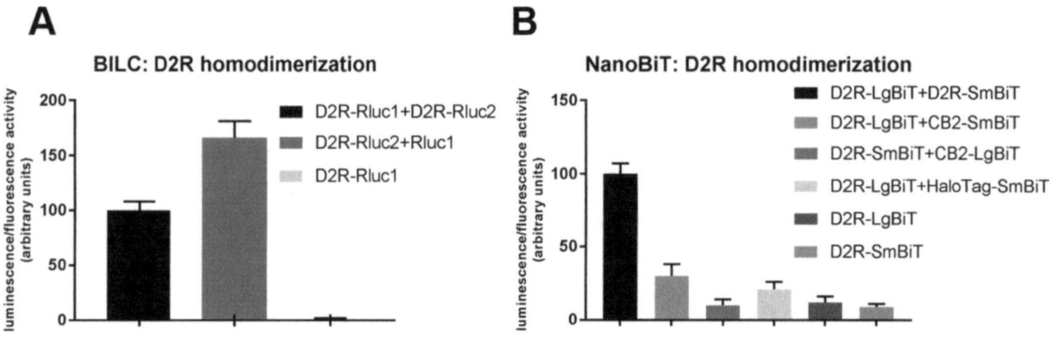

Fig. 15.3 Bimolecular luminescence complementation (BiLC) with D_2R fused to *Renilla* luciferase fragments (*R*Luc1 and *R*Luc2) (**a**) or NL fragments (LgBiT and SmBiT) (**b**). (**a** and **b**) D_2R constructs were transiently transfected (200 ng) in HEK293T cells. VENUS measurements were included for normalization of the data. The ClarioSTAR was used for endpoint measurements of the luminescent signal. Each value represents the mean ± SD from three independent experiments

(another receptor or a mutated version of the receptor at the site of interaction) or the fluorescent or luminescent fragment not fused to a protein of interest (POI). To avoid misleading results, always check if the commercially available plasmids will actually encode a protein (e.g., presence of a start codon, correct reading frame, etc.).

3. When assessing heterodimerization instead of homodimerization with the NanoBiT assay, it is highly recommended to test the different constructs (i.e., GPCR fused to LgBiT or SmBiT), since a specific combination always may be favored, i.e., GPCR1-SmBiT and GPCR2-LgBiT or the other way around, depending on the GPCR dimer of interest.
4. Depending on the restriction site which is chosen within the multiple cloning site of the NanoBiT plasmid, a different linker length will be obtained (see Table 15.2). In this setup with the D_2R, the *Eco*RI restriction site was chosen, so an 18-amino acid linker is present between the D_2R and LgBiT as well as between D_2R and SmBiT. More details on the linker length in the different NanoBiT plasmids (protein of interest fused to N- as well as C-terminally) can be found in the manufacturer's manual (Promega).
5. Different GPCR:GPCR ratios need to be validated if one or both of the GPCRs show different expression levels. An indication of the expression levels can be obtained by running a western blot and quantifying the bands with an open-source program, ImageJ. Ideally, the same antibody should be used to visualize both receptors, e.g., by using N-terminally tagged proteins.
6. It is advisable to co-transfect the same amount of DNA of a plasmid expressing YFP or any other fluorescent protein

Table 15.2
Overview of available linker length—and the restriction site needed—in the NanoBiT plasmids when the POI is fused N-terminally of LgBiT or SmBiT

Fusion protein	Restriction enzyme
POI-GAQGNS-GSSGGGGSGGGGSSG-LgBiT	*Sac*I
POI-GNS-GSSGGGGSGGGGSSG-LgBiT	*Eco*RI
POI-GSSGGGGSGGGGSSG-LgBiT	*Xho*I

(e.g., 100 ng) in all conditions. This allows normalization of the data and better comparison. Important: when running a fluorescence-based assay with the VENUS fragments, CFP (or mCherry) should be used as a control for normalization. Since there is a small overlap between the emission peak of CFP and the excitation peak of VENUS, it is advisable to first measure the VENUS fluorescence and afterwards the CFP fluorescence signal, to avoid cross-excitation.

7. At this time point, it is also possible to reseed the transfected cells immediately in black or white plates. In this case number 12–15 can be left out of the protocol [24].

8. To obtain more robust experimental results, it is recommended to generate stable cell lines once the ideal combination of GPCR plasmids of interest is known. This way, the experimental setup can be reduced to 2 instead of 4 days.

9. This volume is adjustable, depending on the number of replicates and the assay setup.

10. Excitation and emission wavelengths of YFP and CFP are shown in Table 15.3. A bandwidth of 20 nm is recommended.

11. A stock solution of coelenterazine h of 1 mM can be stored at −20 °C. Dilute 40× in HBSS to obtain a final concentration of 25 µM in each well.

12. Dilute the Nano-Glo Live Cell reagent 20× using Nano-Glo LCS Dilution buffer, just prior to measurements.

13. Excitation and emission wavelengths of the complemented proteins are shown in Table 15.4.

14. For the fluorescence-based complementation assay with the VENUS fragments (see Fig. 15.2a) and the luminescence-based complementation assay with the *Renilla* luciferase fragments (see Fig. 15.3), a low signal-to-noise ratio was observed due to a high affinity of the VN and VC or the *R*Luc1 and

Table 15.3
Overview of the excitation and emission maxima of the fluorescent proteins for normalization of the data

Fluorescent protein	Excitation wavelength (nm)	Emission wavelength (nm)
CFP	434	477
YFP/VENUS	497	540

Table 15.4
An overview of the excitation and emission maxima of the reviewed complementation proteins

Protein fragments	Excitation wavelength (nm)	Emission wavelength (nm)
VENUS (VN155, VC155)	515	528
Renilla luciferase (*R*Luc1, *R*Luc2)	–	480
NanoBiT (SmBiT, LgBiT)	–	460

RLuc2 fragments for one another, whereby an interaction of the fusion partners (in this case: D_2R) is not required. This self-assembly phenomenon of the fluorescence-based complementation assay was confirmed by live cell imaging (see Fig. 15.2b) [26].

15. With the NanoBiT assay, a high signal-to-noise is observed due to a low amount of self-assembly of the protein fragments. The cannabinoid receptor 2 (CB2) is an appropriate non-interacting partner for D_2R in this assay. In addition, ligand-induced modulation of D_2R dimerization can be studied. Preliminary data show that following incubation with D_2R ligands, the level of dimerization can be altered (Wouters et al., unpublished).

Acknowledgment

This work was supported by IWT/SBO 140028, Bijzonder Onderzoeksfonds BOF.DCV.20—BOF15/DOS/021 of Ghent University and the FWO G021715N.

References

1. Xu EH, Xiao R (2012) A new era for GPCR research: structures, biology and drugs discovery. Acta Pharmacol Sin 33:289–290
2. Fernàndez-Duenãs V et al (2012) Molecular determinants of A2AR-D2R allosterism: role of the intracellular loop 3 of the D2R. J Neurochem 123(3):373–384
3. Wang M et al (2010) Schizophrenia, amphetamine-induced sensitized state and acute amphetamine exposure all show a common alteration: increased dopamine D2 receptor dimerization. Mol Brain 3:25
4. Perreault ML, Hasbi A, O'Dowd BF, George SR (2011) The dopamine d1-d2 receptor heteromer in striatal medium spiny neurons: evidence for a third distinct neuronal pathway in basal ganglia. Front Neuroanat 5:31
5. González S et al (2012) Dopamine D4 receptor, but not the ADHD-associated D4.7 variant, forms functional heteromers with the dopamine D2S receptor in the brain. Mol Psychiatry 17(6):650–662
6. Skieterska K et al (2013) Detection of G protein-coupled receptor (GPCR) dimerization by coimmunoprecipitation. Methods Cell Biol 117:323–340
7. Gahbauer S, Böckmann RA (2016) Membrane-mediated oligomerization of G protein coupled receptors and its implications for GPCR function. Front Physiol 7:494. https://doi.org/10.3389/fphys.2016.00494
8. Petersen J et al (2017) Agonist-induced dimer dissociation as a macromolecular step in G protein-coupled receptor signaling. Nat Commun 8:226. https://doi.org/10.1038/s41467-017-00253-9
9. Fuxe K et al (2008) Receptor-receptor interactions within receptor mosaics. Impact on neuropsychopharmacology. Brain Res Rev 58:415–452. https://doi.org/10.1016/j.brainresrev.2007.11.007
10. Bonifazi A et al (2017) Novel bivalent ligands based on the sumanirole pharmacophore reveal dopamine D2 receptor (D2R) biased agonism. J Med Chem 60:2890–2907. https://doi.org/10.1021/acs.jmedchem.6b01875
11. Rondard P et al (2008) Functioning of the dimeric GABAB receptor extracellular domain revealed by glycan wedge scanning. EMBO J 27:1321–1332. https://doi.org/10.1038/emboj.2008.64
12. Hiller C, Kühhorn J, Gmeiner P (2013) Class A G-protein-coupled receptor (GPCR) dimers and bivalent ligands. J Med Chem 56:6542–6559
13. Lecat-Guillet N et al (2017) FRET-based sensors unravel activation and allosteric modulation of the GABAB receptor. Cell Chem Biol 24(3):360–370
14. Bouvier M (2008) BRET analysis of GPCR oligomerization: newer does not mean better. Nat Methods 4:3–4. https://doi.org/10.1038/nmeth0107-3
15. Kaczor AA et al (2014) Application of BRET for studying G protein-coupled receptors. Mini Rev Med Chem 14(5):411–425
16. Paulmurugan R, Gambhir S (2007) Combinatorial library screening for developing an improved split-firefly luciferase fragment-assisted complementation system for studying protein–protein interactions. Anal Chem 79:2346–2353
17. Nakagawa C et al (2011) Improvement of a Venus-based bimolecular fluorescence complementation assay to visualize bFos-bJun interaction in living cells. Biosci Biotechnol Biochem 75(7):1399–1401
18. Kodama Y, Hu CD (2010) An improved bimolecular fluorescence complementation assay with a high signal-to-noise ratio. BioTechniques 49:793–805. https://doi.org/10.2144/000113519
19. Shyu YJ et al (2006) Identification of new fluorescent protein fragments for bimolecular fluorescence complementation analysis under physiological conditions. BioTechniques 40(1):61–66
20. Naqvi T et al (2004) β galactosidase enzyme fragment complementation as a high-throughput screening protease technology. J Biomol Screen 9(5):398–408
21. Kerppola TK, Hu CD (2003) Simultaneous visualization of multiple protein interactions in living cells using multicolor fluorescence complementation analysis. Nat Biotechnol 21(5):539–545
22. Hall MP et al (2012) Engineered luciferase reporter from a deep sea shrimp utilizing a novel imidazopyrazinone substrate. ACS Chem Biol 7:1848–1857
23. Dixon AS et al (2016) NanoLuc complementation reporter optimized for accurate measurement of protein interactions in cells. ACS Chem Biol 11:400–408. https://doi.org/10.1021/acschembio.5b00753
24. Cannaert A et al (2016) Detection and activity profiling of synthetic cannabinoids and their metabolites with a newly developed bioassay. Anal Chem 88(23):11476–11148

25. Guo W et al (2008) Dopamine D2 receptors form higher order oligomers at physiological expression levels. EMBO J 27:2293–2304. https://doi.org/10.1038/emboj.2008.153

26. Saini DK, Gautam N (2010) Live cell imaging for studying g protein-coupled receptor activation in single cells. In: Szallasi A (ed) Analgesia. Methods in molecular biology (methods and protocols), vol 617. Humana, Totowa, NJ. https://doi.org/10.1007/978-1-60327-323-7_16

Chapter 16

Proximity Biotinylation for Studying G Protein-Coupled Receptor Dimerization

Maxwell S. DeNies, Luciana K. Rosselli-Murai, Victoria L. Murray, Elisabeth M. Steel, and Allen P. Liu

Abstract

The importance of G protein-coupled receptor (GPCR) receptor–receptor interactions in regulating cell signaling is well documented. However, traditional methods to study these interactions are difficult to multiplex and often require extensive technical expertise and equipment. In this chapter, we will describe the major biochemical and fluorescence microscopy methods to study GPCR–GPCR interactions and introduce an alternative technique that utilizes proximity biotinylation to study these interactions. The goal of this chapter is to provide researchers with a comprehensive protocol to implement this new approach to study GPCR dimerization. Finally, we will compare and contrast the advantages and limitations of the proximity biotinylation assay with established methods to help researchers identify the best approach for their research application.

Key words GPCR, Homodimerization, Heterodimerization, Proximity biotinylation

1 Introduction

The ability of cells to sense and respond to the environment through receptors is essential for life. G protein-coupled receptors (GPCRs) are a large family of cell membrane receptors that play a critical role in cell physiology and disease. GPCRs are expressed in all known tissues and, among other functions in the brain, are essential for central nervous system (CNS) development and maintenance [1].

Evidence of receptor–receptor interactions and their importance in cell physiology was initially described in the 1980s [2]. Since then, identifying the molecular mechanisms of intra-membrane GPCR–GPCR interactions as well as the biological significance of these interactions has become an actively growing field for both basic science research [3–7] and drug development [8–10]. Receptor dimerization is the most elementary receptor–receptor interaction and is traditionally divided into two major categories:

Kjell Fuxe and Dasiel O. Borroto-Escuela (eds.), *Receptor-Receptor Interactions in the Central Nervous System*, Neuromethods, vol. 140, https://doi.org/10.1007/978-1-4939-8576-0_16, © Springer Science+Business Media, LLC, part of Springer Nature 2018

homodimerization defined by interaction between two of the same receptor and heterodimerization where two different receptors interact. GPCRs commonly form functional homo- and heterodimers in the brain. Brain GPCR homodimerization was initially observed with mGluR5 [5] and gonadotropin [11] receptors. GPCR heterodimerization is also very common in the CNS. Functional GPCR heterodimers were initially observed between GABA B1 and B2 receptors [12] as well as adenosine A1 and dopamine D1 receptors [13]. As observed with dopamine receptors, GPCRs commonly form both homo- [14] and heterodimers [13] depending on physiological condition and/or cell type. Recent technological advancements have spurred the discovery of increasingly complex and functional heteromeric receptor species commonly known as receptor mosaics that consist of more than two different GPCRs [15]. While early research primarily focused on identifying and studying receptor dimers individually, more recently the field has expanded to study the dynamics of multiple homo- and heterodimer species simultaneously (i.e., multiplexing) as well as more complex heteromeric receptor mosaics [16–18]. Throughout the rest of this chapter, we will use our work on chemokine receptor CXCR4 as a model to showcase a new proximity-based biotinylation method to study GPCR dimerization [16].

CXCR4 is an essential GPCR that is best known for its role in HIV viral infection, immune cell chemotaxis, and cancer metastasis [19]. Interestingly, it is also expressed in all major CNS cell types including astroglia [20], microglia [21], oligodendrocytes [22], and neurons [23, 24]. Upon CXCL12 stimulus, CXCR4 activates a myriad of downstream signaling pathways—pAKT, pERK, JAK/STAT, and NFkB signaling—and is essential for CNS development [1]. Interestingly, CXCR4 has been shown to form both homo- and heterodimers, each of which result in distinct signaling outcomes [1, 16, 25].

Traditional methods to study GPCR dimerization can be divided into two major categories: biochemical and imaging-based assays. GPCR dimerization was initially discovered and studied using co-immunoprecipitation (co-IP) assays [3, 5]. While this method implies direct physical interaction between receptors, the use of harsh detergents to solubilize membrane proteins creates a high probability for protein aggregation and perceived dimerization, as GPCRs are highly hydrophobic and readily aggregate upon cell lysis [17]. Consequently, rapid improvements in imaging technologies have limited the scope of this technology to primarily discovery science applications where it is coupled with mass spectrometry to identify unknown interacting proteins/receptors with a bait receptor.

The two leading imaging-based assays to study GPCR dimerization are Förster resonance energy transfer (FRET) and bioluminescence resonance energy transfer (BRET). Both techniques are based on the same underlying physical principle that upon donor excitation, energy is transferred from the donor to acceptor fluorophore resulting in acceptor fluorescence when the donor and acceptor are within Förster distance (approximately 10 nm) [26]. The major difference between FRET and BRET is the source of donor excitation [27]. FRET is reliant on external illumination (i.e., laser or mercury/xenon lamps) and consequently commonly has increased background and leads to fluorophore photobleaching. BRET overcomes this limitation by using bioluminescent luciferase, instead of fluorescent proteins or dyes, as the donor/energy source that is activated upon addition of the luciferase substrate [27]. While this removes the necessity of donor illumination, luciferase-fluorophore pairs are limited and BRET acceptor signals are traditionally weaker compared to FRET. Within the last decade, researchers have expanded these methods, notably sequential BRET-FRET (SRET) [28] and BRET-biomolecular fluorescence complementation (BiFC) [29], to explore more complex receptor interactions and stoichiometry.

Unlike the aforementioned biochemical and imaging-based techniques, proximity biotinylation can detect receptor dimerization by both fluorescence microscopy and Western blotting [16]. This approach leverages the ability of a bacterial biotin ligase (BirA) to selectively tag membrane receptors containing a genetically encoded acceptor peptide (AP) sequence with biotin. Together, the BirA enzyme and AP sequence are analogous to a FRET donor and acceptor pair. However, unlike FRET or BRET where energy transfer is required for dimerization quantification, in this assay BirA covalently biotinylates the AP sequence and therefore interactions (including *past* transient interactions) are registered and can be traced/quantified by microscopy, flow cytometry, or Western blot. These features allow the approach to be uniquely flexible as well as easily multiplexed for the bulk study of multiple GPCR homo- and heterodimerization species simultaneously.

To implement this technique, BirA-GPCR (target GPCR) and AP-GPCR (testing GPCR) constructs need to be generated and co-expressed in cells (Fig. 1a). Upon incubating cells with biotin, when the target and testing GPCRs are in close proximity (i.e., interacting), the AP sequence of the testing GPCR will be permanently biotinylated (Fig. 1b). Fluorescent streptavidin and subsequent imaging (microscopy or flow cytometry) can then be used to identify biotinylated receptors at the cell membrane and infer receptor dimerization. Receptor dimerization can additionally be deduced by probing cell lysates for biotinylated protein and comparing banding patterns. By tagging additional GPCRs with AP sequence, this approach can be easily multiplexed to simultaneously study bulk homo- and multiple heterodimers by Western blot.

Proximity biotinylation has several advantages compared to other methods. Firstly, due to the plethora of commercially

A.

Target GPCR

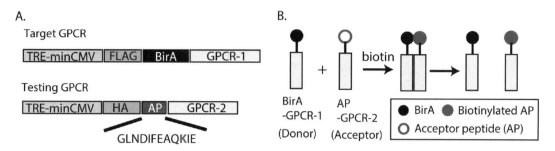

Fig. 1 Proximity biotinylation constructs for detecting GPCR dimerization. (**a**) Schematic of adenovirus constructs of target GPCR fused to Flag and BirA sequences and testing GPCR fused to HA and AP sequences. (**b**) Schematic of biotinylation of AP-GPCR-2 by BirA-GPCR-1. BirA-GPCR-1 is akin of donor and AP-GPCR-2 is akin of acceptor as in FRET assays

available fluorescent streptavidin conjugate options, specialized microscope filter sets are unnecessary and concerns of donor and acceptor fluorophore bleed-through abolished. Together these features make this approach less technically intensive and easier to implement on basic microscope systems. Furthermore, the ability to conduct fix and live cell microscopy experiments as well as probing receptor dimerization patterns by Western blot using the same research strategy makes this approach fundamentally more flexible than solely imaging-based methods. Additionally, unlike more complicated FRET-based approaches, this technique can be easily multiplexed by simply adding the AP sequence to additional testing GPCRs. The major limitation of this approach is that it cannot be used to study more complex receptor mosaics in its current state, as the microscopy readout is dichotomous and the presence of receptor mosaics cannot be distinguished from multiple receptor heterodimers by Western blotting.

2 Materials

2.1 Equipment

Benchtop centrifuge for cell culture.
Incubator (5% CO_2, 37 °C).
SDS-PAGE gel electrophoresis apparatus.
Immunoblotting.
Digital imaging system of fluorescent blots.
Flow-cytometer.
Fluorescence microscope.

2.2 Mammalian Cell Culture Supplies

Tissue culture plates and flasks.
Retinal pigment epithelial wild type cell (RPE WT, ATCC CRL-2302).
RPE cell growth media (F-12/DMEM media supplemented with 10% fetal serum (Sigma), 100 units/ml penicillin and streptomycin, and 20 mM HEPES).

0.25% Trypsin-EDTA.

Seeding media (F-12/DMEM media supplemented with 10% charcoal/dextran-treated fetal bovine serum (Thermo Scientific), 100 units/ml penicillin and streptomycin, and 20 mM HEPES).

Biotinylation media (F-12/DMEM, 100 mM biotin, 1 mM ATP, and 5 mM $MgCl_2$).

2.3 Reagents

GPCR and controls adenovirus vector DNA constructs.

Tetracycline transactivator (tTA) adenovirus.

Adenovirus containing a tetracycline-regulatable promoter (Tet-off) and encoding GPCR of interest.

Phosphate-buffered saline (PBS).

Bovine serum albumin (BSA).

Citric saline buffer (135 mM potassium chloride, 15 mM sodium chloride).

Lysis buffer (50 mM Tris-HCl pH 7.4, 2 mM EGTA, 150 mM NaCl, 1% NP-40, 10 mM NaF, 2 mM sodium vanadate, 0.25% sodium deoxycholate, 1× protease inhibitor cocktail (Roche)).

DTT (dithiothreitol).

TBS-T (137 mM NaCl, 2.7 mM KCl, 19 mM Tris pH 7.4, 0.1% Tween 20).

PFA (paraformaldehyde).

4× Laemmli sample loading buffer (Bio-RAD).

3 Methods

3.1 GPCR Adenovirus Constructs for Proximity Biotinylation Assay

The first step for developing a proximity biotinylation assay to probe GPCR homo- and heterodimerization is to create plasmid constructs of target GPCR and testing GPCRs. As shown in Fig. 1a, the target-GPCR construct is fused to BirA, a biotin ligase from *E. coli* that recognizes and biotinylates the lysine residue on a specific acceptor peptide (AP) sequence (GLNDIFEAQKIE). The testing GPCRs and a negative control (a receptor that is not expected to dimerize with the target GPCR, i.e., transferrin receptor (TfnR)) are fused to the AP sequence. In both constructs, the BirA and AP sequences are fused to the extracellular domains of the receptors with short flexible linkers (GSGSTSGSGK). These linkers are inserted to ensure high probability of enzyme–substrate interaction. The AP-GPCR constructs also have an HA tag preceding it while the BirA-GPCR construct has a Flag tag preceding the BirA to allow for easy quantification of each GPCR via immune-labeling.

Our group has used recombinant adenovirus system to co-express constructs and regulate their expression by using a transactivator (tTA)-regulated adenovirus vector [16, 30]. However, the aforementioned constructs can be cloned in any plasmid of interest for mammalian transient co-expression. The use of adenovirus vectors offers the advantage that infection efficiency is high and multiple constructs can be infected at the same time. In this protocol, all receptors fusions were created in an adenovirus vector and the generation of recombinant adenovirus was conducted in accordance with the adenovirus generation manual from Clontech. The detailed protocol can be found in Clontech Laboratories, Inc. website (Adeno-X Expression System User Manual).

3.2 Proximity-Based Biotinylation for Detecting Receptor Dimerization by Western Blot and Fluorescence Microscopy

The specificity of the BirA/AP system prompted us to investigate the possibility of using this for detecting protein–protein interaction, and specifically for examining homo- and heterodimerization of GPCRs through relatively simple experiments such as Western blot detection and fluorescence microscopy. Here we will describe a protocol to detect GPCRs interactions using cells infected with pairs of BirA-GPCR and AP-GPCR.

3.2.1 Western Blot Analysis of GPCR Dimerization

1. Seed RPE cells* at approximately 70% confluency in a 6-well plate.

2. Next day, add high titration of adenovirus BirA-GPCR and AP-GPCR to 1 ml of cell media.**

3. After 18 h of virus infection, treat cells with the biotinylation media*** for 15 min at 37 °C.

4. Place cell plate on ice and wash cells with 1 ml of cold PBS.

5. Add 150 µl of lysis buffer to each well and incubate on ice for 10 min.

6. Scrape cells with a cell scraper and transfer the cell lysate to a microcentrifuge tube.

7. Centrifuge at maximum speed for 15 min at 4 °C.

8. Mix supernatant with 4× Laemmli sample loading buffer containing 1 mM DTT for 10 min****.

9. Load samples onto a 10% SDS-PAGE gel and proceed with electrophoresis for 2 h at 90 V.

10. Transfer gel protein bands to a nitrocellulose membrane.

11. Block membrane with 5% dry milk diluted in TBS-T for 1 h at room temperature.

12. Incubate with primary antibodies diluted in 5% milk/TBS-T solution for 2–3 h at room temperature. For AP-GPCR detection, we used primary mouse anti-HA antibody (1:5000, Roche) and rabbit anti-actin (1:3000, Pierce) as a loading control.

13. Wash nitrocellulose membranes three times with TBS-T for 5 min for each wash on a rocker.

14. Incubate secondary antibodies diluted in 5% milk/TBS-T solution. We used goat anti-mouse IRDye-800CW antibody and anti-rabbit IRDye-680RD (Li-Cor) for imaging on a Li-Cor Odyssey Sa Infrared Imager. The membranes were scanned at 200 mm resolution, 3 mm offset, and intensity 7 for channels 700 and 800 (Fig. 2a). To probe for biotinylation, we used streptavidin-Alexa Fluor 750 (1:10,000, Life technologies).

3.2.2 Immunofluorescence Assay for Visualization of GPCR Dimerization

1. Seed RPE cells on coverslips, infect with adenoviruses for 18 h, and incubate in biotinylation media for 15 min at 37 °C.

2. Wash coverslip with PBS and fix in 4% PFA/PBS solution for 20 min at room temperature.

3. Wash three times with PBS.

4. Block with 1% BSA/PBS solution for 1 h at room temperature.

5. Incubate cells with 5 μg/ml streptavidin (SA) conjugated Alexa Fluor (AF) 568 (Invitrogen) for 1 h at room temperature.

6. Fixed cells can be imaged on a fluorescence microscope equipped with the proper light source to excite and detect SA-AF568 labeled cells (Fig. 2b).

Notes:

*To measure receptor dimerization, it is important to use cells that express low endogenous levels of receptors of interest.

**During infection protocol, the cell culture media should not contain any antibiotic.

***The fetal bovine serum (FBS) that is used for routine tissue culture contains greater than 20 nM of biotin, according to the manufacturer's specification. Although this is a low level of biotin, the presence of 10% FBS during the virus infection steps to express the receptor constructs is sufficient to biotinylate the receptors without exogenously added biotin. To circumvent this problem, charcoal/dextran-treated FBS that contains less than 2 nM of biotin should be used. When tested, this yielded no detectable biotinylation without addition of exogenous biotin. Charcoal/dextran-treated FBS was used in all experiments unless otherwise noted.

****For better detection of GPCRs and SDS-PAGE gel separation, the samples should not be heat denatured. Heat denaturation will cause GPCR aggregation and it will not enter the SDS-PAGE gel.

One of the most notable advantages of this assay is that it is suitable for bulk multiplex detection of receptor dimerization. As stated previously, BirA/AP is analogous to FRET donor/acceptor pairs. While multiplex detection is possible using FRET, fluorophore properties limit its feasibility, as it is often difficult to resolve

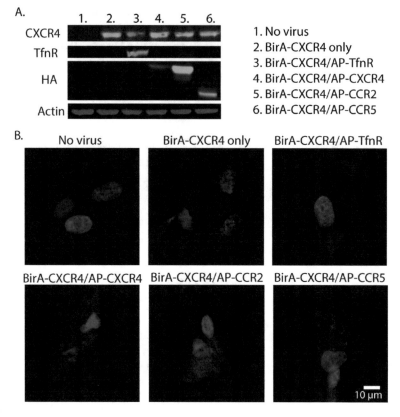

Fig. 2 Proximity biotinylation assays for detecting receptor dimerization. (**a**) Western blot of CXCR4 from RPE cells infected with different adenoviruses used in this study. BirA-CXCR4 was probed with anti-CXCR4 antibodies, and the AP-GPCRs were probed with anti-HA antibodies. Transferrin receptor (TfnR), used as a negative control for interaction, was probed with anti-TfnR. (**b**) Fluorescence images of SA-AF568 in RPE cells expressing BirA-CXCR4 and AP-GPCR and treated with biotin for 15 min. Samples with no virus, BirA-CXCR4, BirA-CXCR4/AP-TfnR all have background staining while samples with BirA-CXCR4 and either of AP-CXCR4, AP-CCR2, or AP-CCR5 all exhibited significant fluorescent labeling [16]

FRET signals using a single fluorescence donor and multiple acceptors. As shown in Fig. 3a, the presented proximity biotinylation assay overcomes this issue as a single BirA-GPCR can biotinylate multiple AP-GPCR. Additionally, in contrast to other approaches that focus on transient protein–protein interactions, the BirA/AP system registers interactions over time, as once biotinylated the AP-GPCR will remain biotinylated even after dissociation from the BirA-GPCR. This cumulative registration feature followed by Western blot detection (as described above) and subsequent quantification allows for this assay to be easily multiplexed and reveal bulk differential homo- and heterodimerization species of the target GPCR in the same experiment. In our experimental conditions, we co-expressed our target GPCR (BirA-CXCR4) with three testing GPCRs (AP-CXCR4, AP-CCR2, and AP-CCR5) and deduced the correspondent AP-GPCR band by systematically removing one AP-GPCR at a time (Fig. 3b). The biotinylated protein-banding pattern can then be quantified and normalized to the expression of each AP-GPCR tested to measure homo- and heterodimerization within the same cellular context [16].

3.3 Quantitative Analysis of Surface Biotinylated Receptor by Flow Cytometry

Different parameters that affect the receptor biotinylation state can be systematically explored using the proximity biotinylation assay followed by flow cytometry analysis. For instance, the amount of biotinylation varies with biotinylation time of the expressed AP-GPCR and once biotinylated, receptors remain biotinylated even in the case that the biotinylated receptor is no longer dimerized. Using flow cytometry to detect site-specific surface biotinylation with streptavidin-phycoerythrin (SA-PE), it is possible to detect the degree of labeling over time or any other experimental condition(s) (Fig. 3).

1. Seed RPE cells in 6-well dishes at about 70% confluency.

2. Next day, infect cells with adenovirus constructs as described previously in Sect. 3.2.

3. After 18 h infection, incubate cells with biotinylation media for a designated time (i.e., 0, 10, or 15 min) at 37 °C.

4. Rinse cells with PBS and add 500 μl of pre-warmed citric saline buffer to detach cells from the well. Citric saline solution should be used in this case because trypsin will cleave the extracellular portion of cell receptors. Incubate at 37 °C for approximately 5 min.

5. Transfer cells to microcentrifuge tubes on ice. Wash plates with chilled PBS and transfer to the same tube. The remaining steps are also performed on ice.

6. Centrifuge cell suspension at 250 g for 3 min at 4 °C.

7. For blocking, resuspend cell pellet in 500 μl of 1% BSA/PBS solution and incubate with continuous agitation for 15 min at 4 °C.

8. Centrifuge cell suspension at 250 g for 3 min at 4 °C.

9. Resuspend cells in 100 μl of labeling solution (SA-PE, Life Technologies). Antibody was prepared in accordance with Life Technologies manual.

10. Wash cells three times with PBS solution.

11. Resuspend the cell pellet in 1 volume PBS (i.e., 100 μl) and 1 volume of 4% PFA/PBS (i.e., 100 μl). At this point, the samples should be subject to flow cytometry analysis within the next 12 h and kept on ice.

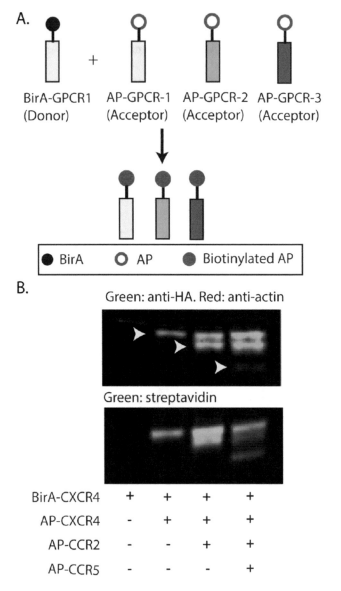

Fig. 3 Multiplex biotinylation by BirA-GPCR for measuring homo- and heterodimerization within the same cellular context. (**a**) Schematic of the multiplex assay of BirA-GPCR-1 and AP-GPCR-1, AP-GPCR-2, AP-GPCR-3. (**b**) Fluorescence Western blot of different combination of AP-GPCRs helps to delineate the specific band for each chemokine receptor. The proportion of biotinylation of each AP-GPCR can be achieved by quantifying the Western blot bands as described previously [16]

12. For flow cytometry analysis of biotinylated receptors, the 488 nm laser was used to excite PE and the emission was collected using the 575/24 filter. Forward scatter, side scatter, and emission should be collected for a minimum of 10,000 cells for each sample. Gating on the forward versus side scatter

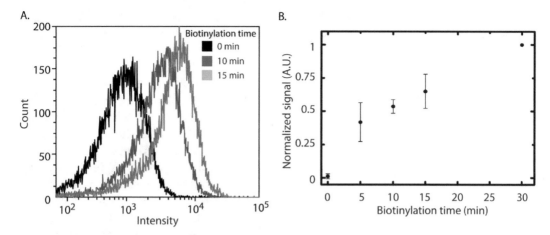

Fig. 4 Characterization of proximity biotinylation of CXCR4 surface homodimerization by flow cytometry. (**a**) Intensity histogram of CXCR4 biotinylation with increasing biotinylation time for a representative experiment. (**b**) Fluorescence signal increases with the amount of time cells are incubated with biotin before fixation and labeling. Intensities are computed as % positive x median intensity of % positive and normalized by value at 30 min ($n = 5$, mean $\pm$ S.E.) [16]

plot eliminated debris and doublet cells. Positive fluorescence is determined by gating compared to a negative control (Fig. 4a, b). To evaluate the degree of GPCR homodimerization and heterodimerization with other GPCRs, dual channel flow cytometry measurements were performed for cells expressing pairs of BirA-GPCR and a single AP-GPCR. In this case, SA-PE signals are normalized by anti-HA signals to properly account varying AP-GPCR expression between different conditions as previously described by our group [16].

4 Conclusion

In this chapter, we have presented some of the popular methodologies to study GPCR dimerization and highlighted the major advantages and limitations of each approach. We additionally presented an alternative proximity biotinylation assay that overcomes some of the key technological limitations of traditional biochemical and imaging assays. One of the limitations of the proximity biotinylation assay is that it is restricted to studying GPCR dimerization and in its current state is not well suited to definitively identify higher order GPCR heteromers. However, the recent development of a split BirA*, a related promiscuous BirA ligase commonly used for BioID proteomics experiments [31], provides the opportunity for this approach to be expanded to study trimeric GPCR interactions. Similarly to split-GFP, the split BirA* fragments are not functional and "dimerization" of BirA* fragments is required for biotinylation activity. This approach could be adopted to study heteromerization of three receptors by fusing each BirA* fragment as well as the AP sequence to different GPCRs. One caveat of this approach is that the original split BirA* construct is promiscuous and therefore does not selectively label the AP sequence. However, to overcome this issue researchers could mutate the promiscuous BirA back to the wild type sequence, which would in theory require the AP sequence for labeling. Although the concept of GPCR dimerization has been well accepted and appreciated, the development of techniques that facilitate the discovery and verification of receptor dimerization will still play a major role in unraveling the complexity of GPCR regulation. Importantly, understanding how GPCR dimerization, in the brain or other tissue systems, can give rise to unexpected pharmacological and signaling diversity will continue to be an important frontier in GPCR research.

References

1. Nash B, Meucci O (2014) Functions of the chemokine receptor CXCR4 in the central nervous system and its regulation by μ-opioid receptors. Int Rev Neurobiol 118:105–128

2. Fuxe K et al (1983) Evidence for the existence of receptor–receptor interactions in the central nervous system. Studies on the regulation of monoamine receptors by neuropeptides. J Neural Transm Suppl 18:165–179

3. Hebert TE et al (1996) A peptide derived from a beta2-adrenergic receptor transmembrane domain inhibits both receptor dimerization and activation. J Biol Chem 271(27):16384–16392

4. Lin H, Liu AP, Smith TH, Trejo J (2013) Cofactoring and dimerization of proteinase-activated receptors. Pharmacol Rev 65(4):1198–1213

5. Romano C, Yang WL, O'Malley KL (1996) Metabotropic glutamate receptor 5 is a disulfide-linked dimer. J Biol Chem 271(45):28612–28616

6. Smith TH, Li JG, Dores MR, Trejo J (2017) Protease-activated receptor-4 and P2Y12 dimerize, co-internalize and activate Akt signaling via endosomal recruitment of β-arrestin. J Biol Chem. https://doi.org/10.1074/jbc.M117.782359

7. Terrillon S, Bouvier M (2004) Roles of G-protein-coupled receptor dimerization. EMBO Rep 5(1):30–34

8. Portoghese PS (2001) From models to molecules: opioid receptor dimers, bivalent ligands, and selective opioid receptor probes. J Med Chem 44(14):2259–2269

9. Fuxe K, Borroto-Escuela DO (2016) Heteroreceptor complexes and their allosteric receptor–receptor interactions as a novel biological principle for integration of communication in the CNS: targets for drug development. Neuropsychopharmacology 41(1):380–382

10. Soriano A et al (2009) Adenosine A2A receptor-antagonist/dopamine D2 receptor-agonist bivalent ligands as pharmacological tools to detect A2A-D2 receptor heteromers. J Med Chem 52(18):5590–5602

11. Cornea A, Janovick JA, Maya-Núñez G, Conn PM (2001) Gonadotropin-releasing hormone receptor microaggregation rate monitored by fluorescence resonance energy transfer. J Biol Chem 276(3):2153–2158

12. White JH et al (1998) Heterodimerization is required for the formation of a functional GABA(B) receptor. Nature 396(6712):679–682

13. Ginés S et al (2000) Dopamine D1 and adenosine A1 receptors form functionally interacting heteromeric complexes. Proc Natl Acad Sci U S A 97(15):8606–8611

14. Lee SP, O'Dowd BF, Rajaram RD, Nguyen T, George SR (2003) D2 dopamine receptor homodimerization is mediated by multiple sites of interaction, including an intermolec-

ular interaction involving transmembrane domain 4. Biochemistry (Mosc) 42(37): 11023–11031

15. Fuxe K, Guidolin D, Agnati LF, Borroto-Escuela DO (2015) Dopamine heteroreceptor complexes as therapeutic targets in Parkinson's disease. Expert Opin Ther Targets 19(3): 377–398

16. Steel E, Murray VL, Liu AP (2014) Multiplex detection of homo- and heterodimerization of g protein-coupled receptors by proximity biotinylation. PLoS One 9(4):e93646

17. Fuxe K, Marcellino D, Guidolin D, Woods AS, Agnati LF (2008) Heterodimers and receptor mosaics of different types of G-protein-coupled receptors. Physiology (Bethesda MD) 23: 322–332

18. Fuxe K, Marcellino D, Guidolin D, Woods AS, Agnati L (2009) Brain receptor mosaics and their intramembrane receptor-receptor interactions: molecular integration in transmission and novel targets for drug development. J Acupunct Meridian Stud 2(1):1–25

19. Busillo JM, Benovic JL (2007) Regulation of CXCR4 signaling. Biochim Biophys Acta 1768(4):952–963

20. Bajetto A et al (1999) Glial and neuronal cells express functional chemokine receptor CXCR4 and its natural ligand stromal cell-derived factor 1. J Neurochem 73(6):2348–2357

21. Lipfert J, Ödemis V, Wagner D-C, Boltze J, Engele J (2013) CXCR4 and CXCR7 form a functional receptor unit for SDF-1/CXCL12 in primary rodent microglia. Neuropathol Appl Neurobiol 39(6):667–680

22. Maysami S et al (2006) Modulation of rat oligodendrocyte precursor cells by the chemokine CXCL12. Neuroreport 17(11): 1187–1190

23. Meucci O, Fatatis A, Simen AA, Bushell TJ, Gray PW, Miller RJ (1998) Chemokines regulate hippocampal neuronal signaling and gp120 neurotoxicity. Proc Natl Acad Sci U S A 95(24):14500–14505

24. van der Meer P, Ulrich AM, González-Scarano F, Lavi E (2000) Immunohistochemical analysis of CCR2, CCR3, CCR5, and CXCR4 in the human brain: potential mechanisms for HIV dementia. Exp Mol Pathol 69(3):192–201

25. Pello OM et al (2008) Ligand stabilization of CXCR4/delta-opioid receptor heterodimers reveals a mechanism for immune response regulation. Eur J Immunol 38(2):537–549

26. Sun Y, Rombola C, Jyothikumar V, Periasamy A (2013) Förster resonance energy transfer microscopy and spectroscopy for localizing protein-protein interactions in living cells. Cytometry A 83(9):780–793

27. Xu Y, Piston DW, Johnson CH (1999) A bioluminescence resonance energy transfer (BRET) system: Application to interacting circadian clock proteins. Proc Natl Acad Sci U S A 96(1):151–156

28. Carriba P et al (2008) Detection of heteromerization of more than two proteins by sequential BRET-FRET. Nat Methods 5(8):727–733

29. Gandia J et al (2008) Detection of higher-order G protein-coupled receptor oligomers by a combined BRET-BiFC technique. FEBS Lett 582(20):2979–2984

30. Liu AP, Aguet F, Danuser G, Schmid SL (2010) Local clustering of transferrin receptors promotes clathrin-coated pit initiation. J Cell Biol 191(7):1381–1393

31. De Munter S et al (2017) Split-BioID: a proximity biotinylation assay for dimerization-dependent protein interactions. FEBS Lett 591(2):415–424

Chapter 17

Conformational Profiling of the 5-HT$_{2A}$ Receptor Using FlAsH BRET

Pavel Powlowski, Kyla Bourque, Jace Jones-Tabah, Rory Sleno, Dominic Devost, and Terence E. Hébert

Abstract

Functional selectivity or biased agonism describes ligand-specific activation of particular downstream signaling pathways following drug treatment. This phenomenon may be exploited for the development of drugs with increased target selectivity and consequently better safety profiles. Analyzing bias at the level of signal transduction pathways is challenging due to cell type-dependent differences in the expression or activity of downstream effector proteins. Bioluminescence resonance energy transfer (BRET) has previously been used to characterize biased agonism at the level of the receptor using fluorescent biarsenical hairpin (FlAsH) binders as energy acceptors. By walking the FlAsH binding tetracysteine tag into different positions within the intracellular loops and carboxyl terminus of the 5-HT$_{2A}$ receptor along with a C-terminally fused *Renilla* luciferase, we generated a panel of seven conformation-sensitive biosensors that were able to capture conformational information in response to agonist. These FlAsH BRET-based biosensors were expressed in HEK 293 cells but also in the more relevant N2A cells, a neuronal-like cell line where 5-HT$_{2A}$ receptor biology can be captured. These biosensors were first validated for cell surface expression and normal signaling phenotypes, followed by conformational analysis in response to the full agonist 5-HT. These results demonstrate a potentially simple tool for drug discovery once conformational profiles are correlated with their downstream signaling pathways.

Key words G protein-coupled receptors, Resonance energy transfer, Conformational profiling, Drug discovery, Fluorescent biarsenical hairpin binders

1 Introduction

G protein-coupled receptors (GPCRs) are widely distributed in the central nervous system and participate in a broad range of neurophysiological processes including arousal, mood, and sensory perception. As such, GPCRs are also targeted in the pharmacological management of many neuropsychiatric diseases including depression and schizophrenia. Despite their broad physiological and clinical significance, the signaling properties of GPCRs in their native cellular context remain incompletely understood. Individual neurons express numerous GPCRs which can couple to diverse

Kjell Fuxe and Dasiel O. Borroto-Escuela (eds.), *Receptor-Receptor Interactions in the Central Nervous System*, Neuromethods, vol. 140, https://doi.org/10.1007/978-1-4939-8576-0_17, © Springer Science+Business Media, LLC, part of Springer Nature 2018

signaling pathways with convergent effects on downstream processes such as membrane excitability, synaptic strength, and gene expression [1, 2]. Moreover, downstream signaling by individual GPCRs has been shown to be both ligand- and cell-type dependent [3]. The ability to measure GPCR activity directly at the level of the receptor allows us to sidestep this complexity [4–6]. GPCRs are heptahelical transmembrane proteins that can adopt a wide range of conformational states in response to interactions with extracellular agonists or intracellular protein partners [7]. Specific conformational states are associated with the activation of different downstream signaling pathways. These conformational states can be detected in real time in live cells using the FlAsH BRET method described below.

FlAsH BRET conformational biosensors have now been built into several GPCRs (4–7) and can be used to characterize specific conformational changes that occur in response to ligand or protein interactions. Briefly, this technique relies on bioluminescence resonance energy transfer (BRET) between a bioluminescent enzyme, *Renilla* luciferase II (RlucII) and a small molecule FlAsH dye. The efficiency of energy transfer from the donor (RlucII) to the acceptor (FlAsH) is a function of distance and orientation of the donor and acceptor moieties and is thus sensitive to conformational rearrangements within the receptor. FlAsH only fluoresces when bound selectively to a tetracysteine tag which can be introduced at various locations within the receptor. By fusing RlucII to the receptor C-tail and inserting FlAsH binding motifs throughout different intracellular loops using insertional mutagenesis, a set of biosensors can be generated to capture different conformational "perspectives" of a receptor of interest.

Biosensors designed within the prostaglandin F (FP), angiotensin II type 1 (AT1R), and β_2-adrenergic receptors in our lab have shown this approach to be robust and reproducible [4, 5, 7]. These biosensors showed the ability to differentiate between balanced or biased ligands with different profiles demonstrating that these assays can be used to guide drug discovery efforts [5]. This technique can also be used to study the allosteric crosstalk occurring between receptor dimers and with intracellular protein partners [6]. The FlAsH BRET approach offers a simple and robust perspective into conformational dynamics of GPCRs in a live cell context and can be applied to understand the complex mechanisms underlying receptor conformation and function.

Here, we demonstrate the generation and validation of a panel of conformational biosensors in the serotonin 5-HT$_{2A}$ receptor. The FlAsH binding tetracysteine sequence, CCPGCC, was inserted at various locations in the intracellular loops of the 5-HT$_{2A}$ receptor to create a panel of biosensors which could capture conformational information from different vantage points. In this chapter, we will outline how to generate FlAsH BRET biosensor constructs, a basic methodology to validate the trafficking and signaling properties of the modified receptors, and how to perform and interpret BRET experiments.

2 Materials

2.1 Cell Culture

Materials and Reagents

1. T25 flasks (Thermo Scientific, 156367).
2. T75 flasks (Thermo Scientific, 156499).
3. Dulbecco's Modified Eagle Medium, DMEM (Wisent, 319-015-CL).
4. Fetal bovine serum, FBS (Wisent, QC).
5. 0.25% Trypsin-EDTA (Wisent, QC).
6. Hygromycin B (Wisent, QC, 450-141-XL).
7. 10 μM retinoic acid (Sigma, MI, R2625).
8. Dimethyl sulfoxide (DMSO) (Sigma-Aldrich).

HEK 293 cells were cultured in DMEM with 5% fetal bovine serum (vol./vol.) at 37 °C in a humidified atmosphere with 95% air and 5% CO_2. Stock cells were kept in T75 flasks and grown to ~80% confluency before being plated for transfection.

Neuro-2A (N2A) cells were cultured in DMEM with 10% fetal bovine serum (vol./vol.) at 37 °C in a humidified atmosphere with 95% air and 5% CO_2. The generation of N2A cells stably expressing the 5-HT_{2A} FlAsH BRET biosensors was accomplished by antibiotic selection with 200 μg/mL hygromycin B. The 5-HT_{2A} biosensors were sub-cloned into the pIRESH expression vector containing a hygromycin resistance cassette. Stock cells were kept in T25 flasks and grown to ~80% confluency before plating for experiments.

2.2 Transfection Reagents

Lipofectamine 2000 (Invitrogen, CA).

2.3 Ligands

1. Full agonist, 5-HT (Sigma-Aldrich): dissolve in ascorbic acid to make stock concentration at 10 mM. For stability purposes, aliquot and store at −20 °C.
2. Partial agonist, Aripiprazole (Sigma-Aldrich): dissolve in DMSO to make stock concentration at 10 mM. For stability purposes, aliquot and store at −20 °C.
3. Partial Agonist, TCB-2 (Tocris Bioscience, UK): dissolve in distilled H_2O to make stock concentration at 10 mM. For stability purposes, aliquot and store at −20 °C.

4. Antagonist, M100907 (Sigma-Aldrich): dissolve in DMSO to make stock concentration at 10 mM. For stability purposes, aliquot and store at −20 °C (*see* **Note 1**).

2.4 BRET Substrate

Coelenterazine h (NanoLight Technologies) is the ideal substrate for the bioluminescent enzyme *Renilla* Luciferase (RlucII) when using FlAsH as the acceptor in the BRET pair. *Renilla* luciferase oxidizes the substrate producing light with a peak emission of 485 nm. Coelenterazine h is reconstituted in 100% ethanol at 1 mM concentration and flushed with argon gas for storage. For long-term storage, aliquot and keep at −80 °C in an environment protected from light. Use at a working concentration of 2 μM (*see* **Note 2**).

2.5 Plate Reader

A plate reader with filter-based detection that can detect light sequentially at two wavelengths (F485, F530) with a relatively high read speed is optimal. For ideal performance, the plate reader should be equipped with injector units as they are essential for capturing kinetic readings pre- and post-injection of a ligand. A temperature control unit is also necessary to maintain physiological temperature for the duration of the assay.

3 Methods

3.1 Design of Tetracysteine-Tagged Conformation-Sensitive Biosensors

The positioning of the tetracysteine tags within the receptor coding sequence is of critical importance. Positions must be chosen in regions that are least likely to perturb receptor function. For instance, most GPCRs are known to have a D[E]RY motif within the vicinity of the second intracellular loop, it is thus important to avoid this site when inserting the tag (*see* Table 1 for a list of conserved motifs within GPCRs). In cases where crystal structures are not available, homology modeling may be used. Free software such as Phyre[2] [8] and I-TASSER [9] may be used to do this. For example, a receptor homology structure of the 5-HT$_{2A}$ receptor was modeled using Phyre[2] based on the human 5-HT$_{2A}$ isoform 1 amino acid sequence (NCBI reference sequence: NP_000612.1). The highest scoring model was based on the crystal structure of rhodopsin bound to arrestin (PDB: 4ZWJ), with a confidence of 100 and a percent ID of 21. The model coordinates were downloaded and viewed as a cartoon depiction in PyMol, after which intracellular loops were identified and colored. A prediction of transmembrane helix positions was made with TMHMM server v.2.0 [10] which was used to further verify the position of the intracellular loops within the receptor sequence.

Table 1

Highly conserved microswitches and motifs within some GPCRs that must be considered when designing FlAsH-based biosensors

Motif	Location
D[E]RY motif	Helix III
NPxxY motif	Helix VII
CWxP motif	Helix VI
Phosphorylation sites	Multiple
G protein binding	Receptor dependent
β-arrestin binding	Carboxyl terminus
Palmitoylation site	Receptor dependent

3.2 Generation of the Tetracysteine-Tagged Constructs

Materials and Reagents

1. PCR thermocycler.
2. 37 °C water bath.
3. NheI and BamHI restriction enzymes (New England Biolabs, MA).
4. Phusion HF-DNA polymerase (New England Biolabs, MA).

Seven FlAsH BRET-based biosensors were engineered in the 5-HT$_{2A}$ receptor, two within the second intracellular loop, three within the third, and two within the carboxyl terminus (*see* **Note 3**). Using PCR, a NheI restriction site, a Kozak consensus sequence (GCCACC) (*see* **Note 4**), and signal peptide were introduced to the 5′ end of the c-Myc tagged 5-HT$_{2A}$ coding sequence. At the 3′ end of the receptor, the stop codon was removed, and a BamHI restriction site was introduced (primers were synthesized by Integrated DNA Technologies Coralville, IA; *see* Table 2). A previously published pIRESH-hAT1R FlAsH BRET biosensor from our lab [5] was digested with the enzymes NheI and BamHI to remove hAT1R and provide compatible restriction sites for the 5-HT$_{2A}$ insertion while leaving RlucII in the vector backbone attached to a flexible linker in frame with the C-terminus of the receptor. The c-Myc tag was maintained for the purposes of detecting the receptor at the cell surface by immunofluorescence.

To insert the tetracysteine tag (CCPGCC) into the receptor coding sequence, overlapping extension PCR was used. A total of seven different constructs (SP-c-Myc-5-HT$_{2A}$-RlucII with one CCPGCC tag) were generated. The first round of PCR involved

Table 2

Primers used for the insertion of the tetracysteine tag (CCPGCC) with the 5-HT$_{2A}$ receptor coding sequence

FlAsH tags within ICL2	
ICL2 p1 forward	5'-TGC TGC CCC GGC TGC TGC AAT CCC ATC CAC-3'
ICL2 p1 reverse	5'-GCA GCA GCC GGG GCA GCA CTG GAT GGC GAC-3'
ICL2 p2 forward	5'-TGC TGC CCC GGC TGC TGC TTC AAC TCC AGA-3'
ICL2 p2 reverse	5'-GCA GCA GCC GGG GCA GCA GCG GCT GTG GTG-3'
FlAsH tags within ICL3	
ICL3 p1 forward	5'-TGC TGC CCC GGC TGC TGC TTA GCT TCT TTC-3'
ICL3 p1 reverse	5'-GCA GCA GCC GGG GCA GCA TTT GGC CCG TGT-3'
ICL3 p2 forward	5'-TGC TGC CCC GGC TGC TGC CAG CGG TCG ATC-3'
ICL3 p2 reverse	5'-GCA GCA GCC GGG GCA GCA GAA GAG CTT TTC-3'
ICL3 p3 forward	5'-TGC TGC CCC GGC TGC TGC AGG ACT ATG CAG-3'
ICL3 p3 reverse	5'-GCA GCA GCC GGG GCA GCA CCT GCC TGT GTA-3'
FlAsH tags within C-tail	
C-tail p1 forward	5'-TGC TGC CCC GGC TGC TGC ATT TTA GTG AAC-3'
C-tail p1 reverse	5'-GCA GCA GCC GGG GCA GCA TAA CTG CAA TGG-3'
C-tail p2 forward	5'-TGC TGC CCC GGC TGC TGC GGA AAG CAG CAT-3'
C-tail p2 reverse	5'-GCA GCA GCC GGG GCA GCA TAG AGC AAC CAT-3'
Universal end primers	
NheI-Kozak-SP-Partial-c-Myc fwd:	5'-GGA CGC TAG CGC CAC CAT GAA CAC GAT CAT CGC CCT GAG CTA CAT CTT CTG CCT GGT GTT CGC CGA ACA AAA ACT TAT TTC TGA A-3'

(continued)

Table 2
(continued)

Partial 5-HT_{2A} BamHI *no-stop rvs:*	5′-GGC CGG ATC CCA CAC AGC TCA CCT T-3′
Sequencing primers	
CMV forward	5′-CAC CAA AAT CAA CGG GAC TT-3′
For middle of 5-HT_{2A}	5′-CTC AAC TAC GAA CTC CCT AAT GCA A-3′
Rluc seq. reverse	5′-AGC ACG TTC ATC TGC TTG-3′

two separate reactions to introduce the tag sequence to the 3′ end of the coding strand or to the 5′ end of the template strand in between two amino acid codons, using both universal primers and the forward and reverse FlAsH tag insertion primers. Two fragments were obtained which were mixed together in a second PCR reaction with the universal end primers to generate the full receptor coding sequence. All constructs were verified by bidirectional sequencing (Génome Quebec) using CMV fwd, Rluc seq rvs, and a sequencing primer designed specifically to capture the central portion of the $5\text{-}HT_{2A}$ insert (*see* Table 2).

3.3 Validation of the FlAsH-Based Constructs

The tetracysteine-tagged receptors must be validated to ensure that the insertion of the six amino acid motif does not impair receptor function including cell surface localization to ensure proper folding and export of the tetracysteine-tagged receptor to the membrane. Agonist-induced receptor signaling must also be confirmed. For example, for $G\alpha q$-coupled receptors, calcium signaling can be examined whereas for $G\alpha s$-coupled receptors, cAMP production can be studied. More distal downstream signaling events including mitogen-activated protein kinase (MAPK) activation should also be examined.

3.3.1 Cell Surface Localization of Recombinant Receptors

Materials and Reagents

1. 6-well plate (Thermo Scientific, 140675).
2. Black clear bottom 96-well plate (Thermo Scientific, 165305).
3. Dulbecco's Modified Eagle Medium, DMEM (Wisent, 319-015-CL).
4. Fetal bovine serum, FBS (Wisent, QC).
5. 0.25% Trypsin-EDTA (Wisent, QC).
6. Poly-L-ornithine (Sigma-Aldrich).
7. Phosphate buffered saline, PBS.

8. 2% PFA solution. Adjust pH to 7.3. Store at −20 °C. Keep for approximately 2–3 months (*see* **Note 5**).

9. 1% bovine serum albumin (BSA) fraction V (Fisher Scientific).

10. Mouse anti-c-Myc antibody (BioLegend, CA, 9e10).

11. Anti-mouse Alexa fluor488 conjugated secondary antibody (Life Technologies, CA, A-10029).

12. Hoechst nuclear stain (Sigma, B2261).

Protocol

1. Plate 200,000 HEK 293 cells on a 6-well plate 24 h prior to transfection.

2. The next day, transiently transfect cells with 1.5 μg pcDNA3.1(−), 1 μg 5-HT$_{2A}$ wild-type or 5-HT$_{2A}$TC-tagged construct, completed to 1.5 μg total DNA with pcDNA3.1(−), using Lipofectamine 2000 according to the manufacturer's instructions.

3. The following day, detach cells with 0.25% Trypsin-EDTA and re-plate 10,000 cells onto a poly-L-ornithine treated black clear bottom 96-well plate.

4. The following day, fix cells with 100 μL of 2% paraformaldehyde (PFA) for 20 min on ice (*see* **Note 6**).

5. Block cells for 1 h in 100 μL of 1% bovine serum albumin (BSA) fraction V in 1× PBS to reduce nonspecific binding of the antibody.

6. Incubate cells with a primary mouse anti-c-Myc antibody (1:500) for 1 h at room temperature.

7. Wash cells three times with 1× PBS solution.

8. Incubate cells with an anti-mouse Alexa fluor 488 conjugated secondary antibody (1:1000) and Hoechst DNA stain (1:10000) for 1 h at room temperature.

9. Wash cells three times with 1× PBS solution.

10. Cell surface expression of the recombinant tetracysteine-tagged receptors can be confirmed using a variety of fluorescent microscopy platforms equipped with the appropriate excitation channels and emission filters. In the present example, images were captured using the Opera Phenix High Content Screening System (Perkin Elmer, MA) with 20× WD objective, 488 nm laser line with confocal scanning capability and compatible filter sets for Alexa fluor 488 detection.

3.3.2 Assessing Functionality of Recombinant Receptors

Materials and Reagents

1. 6-well plates (Thermo Scientific, 140675).

2. 12-well clear plates (Falcon, 353043).

3. Dulbecco's Modified Eagle Medium, DMEM (Wisent, 319-015-CL).

4. Fetal bovine serum, FBS (Wisent, QC).

5. 0.25% Trypsin-EDTA (Wisent, QC).

6. Heat block.

7. Laemmli buffer: 2% SDS, 10% glycerol, 60 mM Tris pH 6.8, 0.02% bromophenol blue, 5% β-mercaptoethanol. Store at −20 °C.

8. Guanidine hydrochloride stripping solution: 6 M guanidine hydrochloride, 20 mM Tris-HCl pH 7.5, 0.26% Nonidet P-40 (NP-40). Store at room temperature.

9. Anti-p44/42 ERK antibody (Cell Signaling Technologies, MA, 9101).

10. Total ERK1 antibody (Santa Cruz Biotech, CA, K-23).

11. A secondary anti-rabbit antibody conjugated to horseradish peroxidase (1:20000, Sigma, MI, A0545).

12. Western-Lighting ECL (PerkinElmer, MA).

13. Enhanced ECL (GE Healthcare, UK, RPN2235).

Protocol

1. Plate 200,000 HEK 293 cells in 6-well plates 24 h prior to transfection.

2. The next day, transiently transfect cells with 1 μg wild-type 5-HT$_{2A}$ or TC-tagged 5-HT$_{2A}$, completed to 1.5 μg total DNA with pcDNA3.1(−), using Lipofectamine 2000 according to the manufacturer's instructions.

3. Twenty-four hours later detach cells with 0.25% Trypsin-EDTA and re-plate 400 μL of the cell suspension into 12-well clear plates.

4. The following day, starve cells in DMEM lacking supplementation for 5 h.

5. Treat cells with vehicle (1 μM ascorbic acid) or 1 μM 5-HT diluted in DMEM for 5 min (*see* **Note 7**).

6. Put cells on ice and wash them once with ice cold 1× PBS.

7. Lyse cells in 200 μL Laemmli buffer (*see* **Note 8**).

8. Sonicate cell lysates for 10 s at 1 W power.

9. Heat sample to 95 °C for 5 min.

10. Run samples on a 10% polyacrylamide gel using western blot technique.

11. Use anti-p44/42 ERK antibody to blot for phospho-ERK (1:1000).

12. Use secondary anti-rabbit antibody conjugated to horseradish peroxidase (1:20,000) for visualization of the bands via chemiluminescence.

13. Visualize the bands using enhanced ECL reagent.

14. Strip the PVDF membrane with guanidine hydrochloride stripping solution.

15. Use total ERK1 antibody to control for equal loading (1:2000).

16. Use secondary anti-rabbit antibody conjugated to horseradish peroxidase (1:20,000) for visualization of the bands via chemiluminescence.

17. Visualize the bands using Western-Lighting (ECL) reagent.

Here we show the example of cell surface localization of ICL3p3 tetracysteine-tagged receptor by immunofluorescence (Fig. 1a, also see Table 2 for location) and validate the ability of all the tetracysteine-tagged receptors to activate ERK1/2 in response to treatment with 5-HT (Fig. 1b). The tetracysteine-tagged receptor in ICL2p1 was not functional and would not be used for further profiling.

3.4 FlAsH Labeling Protocol

Materials and Reagents

1. Dulbecco's Modified Eagle Medium, DMEM (Wisent, 319-015-CL).

2. Fetal bovine serum, FBS (Wisent, QC).

3. 0.25% Trypsin-EDTA (Wisent, QC).

4. Poly-L-ornithine (Sigma-Aldrich).

5. White 96-well plates (ThermoFisher, 236105).

6. Chemical fume hood.

7. 37 °C incubator.

8. 1,2-Ethanedithiol (EDT) (Sigma-Aldrich).

9. Dimethylsulfoxide, DMSO (Sigma-Aldrich).

10. FlAsH reagent (Life Technologies). Store at −20 °C protected from light.

11. Hank's balanced salt solution 1× (HBSS) without phenyl red, with sodium bicarbonate, calcium, and magnesium. Store at 4 °C and keep sterile (Wisent, 311-513-CL).

12. 25 mM 2, 3-dimercapto-1-propanol (BAL wash buffer, Fluka). Store at 4 °C.

13. Krebs buffer: 146 mM NaCl, 4.2 mM KCl, 0.5 mM MgCl2, 1 mM CaCl$_2$, 10 mM HEPES pH 7.4, 0.1% glucose. Store at room temperature, in an environment protected from light.

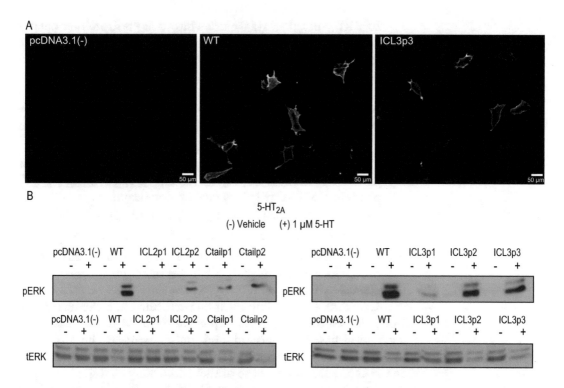

Fig. 1 Validation of the FlAsH BRET-based 5-HT$_{2A}$ conformation-sensitive biosensors. (a) Tetracysteine-tagged 5-HT$_{2A}$ receptor constructs demonstrate the ability to traffic to the cell surface. Immunofluorescence images of transiently transfected HEK 293 cells showing cell surface localization of SP-c-Myc-5-HT$_{2A}$-WT-RlucII compared to the mutant construct where the CCPGCC tag was inserted in SP-c-Myc-5-HT$_{2A}$-ICL3-p3-RlucII. Non-permeabilized cells were labeled with an anti-c-Myc primary antibody (1:1000) and an Alexa fluor 488-conjugated secondary antibody (1:1000). Cells were imaged using the Opera Phenix High Content Screening System (Perkin Elmer). Scale bar, 50 μm. (b) Tetracysteine-tagged 5-HT$_{2A}$ receptor constructs demonstrate their ability to function similar to the wild-type receptor. HEK 293 cells transiently transfected with wild-type SP-c-Myc-5-HT$_{2A}$-WT-RlucII or CCPGCC-tagged 5-HT$_{2A}$-RlucII constructs were treated with vehicle (1 μM ascorbic acid) or 1 μM 5-HT for 5 min followed by cell lysis. SDS-PAGE was performed on the cell lysates, followed by western blotting for phospho-ERK1/2 (pERK) and total ERK1/2 (tERK) as a loading control

HEK 293 cells

1. Plate 200,000 HEK 293 cells in 6-well plates 24 h prior to transfection.

2. The next day, transiently transfect cells with 1 μg wild-type 5-HT$_{2A}$ or TC-tagged 5-HT$_{2A}$, completed to 1.5 μg total DNA with pcDNA3.1(−), using Lipofectamine 2000 according to the manufacturer's instructions.

3. Twenty-four hours later, detach cells with 0.25% Trypsin-EDTA and re-plate 40,000 cells per well of a 96-well plate.

In BRET-based assays we use white, solid bottom microplates (*see* Materials) to maximize the amount of light collected by the detector (white plates reflect the light to enrich the signal as opposed to black plates which absorb light).

N2A cells

N2A are a murine neuroblastoma-derived cell line that can be differentiated into neuron-like cell types and are a commonly used cell line for the study of neuronal processes.

1. Plate 40,000 cells stably expressing the FlAsH-based biosensor position of interest in poly-L-ornithine treated white 96-well plates in DMEM with 10% fetal bovine serum.

2. 24 h later, the media was changed to DMEM 2.5% FBS supplemented with 10 µM retinoic acid to induce differentiation.

3. Culture the cells for an additional 48 h to allow for differentiation before performing BRET measurements.

Protocol

Caution: The labeling must be performed in a chemical fume hood due to the strong and unpleasant odor of EDT. FlAsH is a biarsenical-based compound and must therefore be discarded appropriately. It is thus important to collect all solutions containing FlAsH-EDT and dispose of them according to disposal regulations set by waste management facilities at your institution. The entire FlAsH labeling procedure takes roughly 1.5–2 h [11]. TC-FlAsH™ in-cell tetracysteine tag detection kit can be purchased by Life Technologies (T34561).

1. Pre-warm the HBSS in a water bath to reach a working temperature of 37 °C.

2. Prepare a 1 M solution of 1,2-ethanedithiol (EDT) by diluting it in dimethyl sulfoxide. Vortex.

3. Dilute the 1 M EDT solution with DMSO to make a 25 mM solution of EDT. Vortex.

4. Add one volume of FlAsH reagent (2 mM) to two volumes of 25 mM EDT to make a 667 µM FlAsH-EDT$_2$ solution (*see* **Notes 9** and **10**).

5. Incubate the FlAsH-EDT$_2$ solution for 10 min at room temperature.

6. Following the incubation, add 100 µL of HBSS to the 667 µM FlAsH solution. Continue the incubation for 5 min at room temperature.

7. Complete the volume with HBSS to make a solution with final concentration of 750 nM FlAsH-EDT$_2$. Vortex.

8. Wash cells with 150 µL of HBSS prior to the FlAsH labeling.

9. Label cells with 60 µL of the 750 nM FlAsH-EDT$_2$ solution and incubate for 1 h at 37 °C, protected from any source of direct light (*see* **Note 11**).

10. Following the incubation, wash cells once with 100 μL of a 100 μM BAL wash buffer diluted in HBSS buffer and then incubated for 10 min at 37 °C.

11. Perform another BAL wash without incubation.

12. Wash cells once with 150 μL of Krebs assay buffer.

13. Keep the cells in 80 μL of Krebs for 2 h at room temperature, in an environment protected from light, prior to the BRET assay (*see* **Notes 12** and **13**).

3.5 Assay Protocol: Setting Up and Collecting the Data

1. Prepare the ligands. Dilute the concentrated ligand stock to make a saturating concentration of the drug (*see* **Note 14**).

2. Set the temperature of the plate reader between 25 and 28 °C.

3. Wash the injectors once with Krebs assay buffer.

4. Prime the injectors with the vehicle and agonist.

5. Set up the kinetic protocol as detailed below in Table 3.

6. Prepare a 20 μM solution of coelenterazine h diluted in Krebs assay buffer.

7. Add 10 μL of the 20 μM coelenterazine h solution on 6 wells and incubate 5 min at room temperature (*see* **Notes 16** and **17**).

8. Measure the BRET signal.

9. Repeat steps 6–9 until finished.

Table 3

Kinetic protocol for capturing agonist-induced conformational changes within the 5-HT$_{2A}$ receptor

Kinetics 1	⏩	Dispense	⏩	Kinetics 2	
Total time (s)	41	Injector 1	Vehicle	Total time(s)	82
Counting time (s)	0.20	Volume (μL)	10	Counting time (s)	0.20
Repeats	50	Speed	3	Repeats	100
Delay (s)	0	Meas. Operation	by Well	Delay (s)	0
Emission filter	F485	Repeated operation	No	Emission filter	F485
Second measurement	Yes	Injector 2	Agonist	Second measurement	Yes
Emission filter 2	F530	Volume (μL)	10	Emission filter2	F530
		Speed	3	*See* **Note 15**	
		Meas. operation	By well		
		Repeated operation	No		

3.6   Data Analysis

3.6.1   Calculating the BRET Ratio

The BRET ratio is computed by dividing the fluorescence (acceptor channel, F530) by the luminescence (donor channel, F485) as follows: $\dfrac{\text{F530Acceptor}}{\text{F485Donor}}$.

3.6.2   Calculating Basal BRET

Basal BRET is the degree of BRET exhibited by the intramolecular biosensors under ligand-free conditions. It is a rough indication of the distance between the donor and acceptor at baseline in a given environment. It can also infer the relative expression levels of the biosensors and their location within cells. The higher the basal BRET, the closer the two probes are at baseline, whereas the BRET is smaller when the two probes are farther apart. In order to compute this measurement, we need to measure the BRET ratio of the SP-c-myc-5-HT$_{2A}$-WT-RlucII construct without the FlAsH labeling step. The corresponding ratio will give us the filter bleed-through, the light originating from RlucII being captured by the acceptor F530 filter. We then use this ratio and subtract it from the pre-injection BRET measurement of each tetracysteine-tagged biosensor position (*see* Fig. 2a).

3.6.3   Calculating ΔBRET

The delta BRET (ΔBRET), as referred to in this protocol, refers to the change in BRET in response to the addition of an agonist. In order to compute this value, we need to calculate the pre-injection and post-injection BRET, the BRET before and after the addition of vehicle or agonist, respectively. Since kinetic readings are recorded, according to our assay protocol, our pre-injection BRET is the average of the first 50 repeats, whereas the post-injection BRET is represented by the last 100 repeats. Then, to obtain the change in BRET, we subtract the pre-injection BRET from the post-injection BRET, as follows: $\Delta\text{BRET} = \text{avg}(\text{BRET}_{\text{post-injection}}) - \text{avg}(\text{BRET}_{\text{pre-injection}})$ (*see* **Note 18**) (*see* Fig. 2b–d). We also localize ICL3p3 in N2A cells and show a similar response to 5-HT as when the sensor was expressed in HEK 293 cells (Fig. 3).

4   Notes

1. Ligands were dissolved in ascorbic acid to prevent oxidation which maintains their stability for longer periods of time. Avoid excessive freeze/thaw cycles; small aliquots of 10–25 µL are the best method to store the ligands.

2. An alternative to RlucII as a donor is NanoLuc luciferase. NanoLuc produces tenfold more light compared to RlucII and is also smaller in size (RlucII 26 kDa, NanoLuc 19.1 kDa). The optimal coelenterazine derivative for NanoLuc is furimazine.

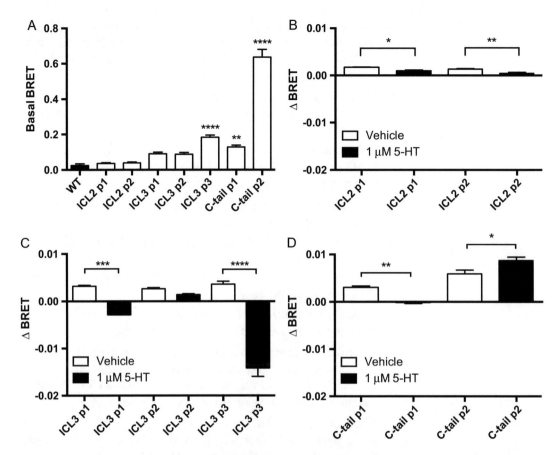

Fig. 2 Conformational profiling of the tetracysteine-tagged 5-HT$_{2A}$ constructs transiently transfected in HEK 293 cells. (**a**) Basal BRET demonstrates the ligand-independent BRET signal of the conformational biosensor constructs. Basal BRET values were obtained by subtracting the BRET ratio of unlabeled wild-type SP-c-Myc-5-HT$_{2A}$-WT-RlucII from the pre-injection BRET ratio of the CCPGCC-tagged constructs. Changes in BRET as a response to the full agonist, 1 μM 5-HT, as reported by the different sensor positions (**b**) ICL2, (**c**) ICL3, (**d**) C-tail. All readings were taken using the Tristar multimode plate reader (Berthold Technologies). Data represents the mean of three independent experiments. Values shown are mean ± s.e.m. Graphs were generated using GraphPad Prism 6 software. One-way ANOVA was performed followed by Sidak's multiple comparisons test. Asterisks represent *$p \leq 0.05$, **$p \leq 0.01$, ***$p \leq 0.001$, ****$p \leq 0.0001$

3. It is always preferable to generate multiple distinct insertion sites providing that the mutants may not all pass the validation stage since the introduction of the six amino acid tag may perturb receptor function.

4. For efficient transgene expression, the addition of a Kozak consensus sequence (GCCGCCACC) is recommended as it aids the ribosomal machinery to initiate translation.

5. PFA is a fixative, wear appropriate safety glasses and N-95 mask when handling it in powder form. Prepare a double water bath and heat water to 45 °C. Dissolve powder in 80–90 mL 1× PBS in fume hood. Add 1–3 drops of 1 M

NaOH until PFA begins to dissolve. Let mix for about 15 min. Let solution cool, then pH to 7.3 and complete volume to 100 mL. After use, discard according to regulations set by the waste management facilities at your institution.

6. Incubating the cells on ice reduces permeabilization of the cells, thus reducing the background noise when imaging the cells.

7. Thirty minutes prior to the introduction of the agonist, the incubator should be blocked off in order to reduce temperature and CO_2 fluctuations.

8. For ease, cell scraping in a 12-well plate is best accomplished by using the back end of a p200 microtip.

9. All work with EDT should be performed in a chemical fume hood, due to its unpleasant odor and toxicity. As FlAsH is light-sensitive, all work was done without artificial lighting in the chemical fume hood.

10. Since the volumes added are very small, it is best to keep them in a drop on the side of a 15 mL conical tube.

11. To avoid unwanted accidents or odor, wrap the microplate with parafilm prior to incubation. Incubation can take place in a basic 37 °C incubator, not necessarily with a CO_2 chamber. Do not incubate in your regular cell culture incubator.

12. The final volume that the cells are left in depends on the assay to be performed. Typically, in our assays, 80 μL is standard since we add the BRET substrate (10 μL) and then the ligand (10 μL) for a final volume of 100 μL. If pre-treatments are required, the volume can be adjusted.

13. Incubate assay microplate in same room as the assay measurements will be made to allow the cells to equilibrate to ambient room temperature. The ideal location to incubate the microplate is in the plate reader. This would allow the cells to get accustomed and equilibrate to the same environment as where the assay will take place. If not possible, incubate the microplate in the same room as the plate reader.

14. To determine the saturating concentration of the ligand in question, a review of the literature is suggested. Bear in mind that some of the ligand might "stick" to the interior tubing of the injectors, therefore, if the suggested concentration is 1 μM, try the recommended concentration; however, if no change in BRET is observed consider using a tenfold higher concentration such as 10 μM.

15. The kinetic protocol can be altered if need be. If the light output is too low, the counting time at each filter can be increased. If more or less kinetic readings are required, the total time can be increased or decreased accordingly.

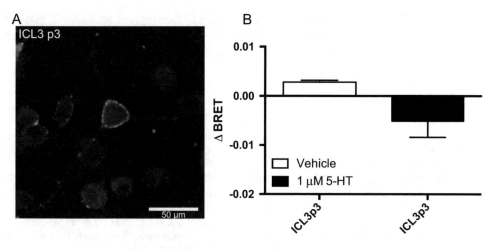

Fig. 3 Validation and conformational profiling of the tetracysteine-tagged 5-HT$_{2A}$ constructs stably expressed in Neuro-2A cells. (**a**) SP-c-Myc-5-HT$_{2A}$-ICL3-p3-RlucII construct stably expressed in N2A cells exhibits cell surface localization. Non-permeabilized cells were labeled with an anti-c-Myc primary antibody (1:1000) and an Alexa fluor 488 conjugated secondary antibody (1:1000). Cells were imaged using the Opera Phenix High Content Screening System (Perkin Elmer). Scale bar, 50 μm. (**b**) Third loop biosensor position P3 in the 5-HT$_{2A}$ reports on conformational changes in response to the administration of 1 μM 5-HT similar to that seen in HEK 293 cells. All readings were taken using the Tristar multimode plate reader (Berthold Technologies). Data represents the mean of three independent experiments. Values shown are mean ± s.e.m. Graphs were generated using GraphPad Prism 6 software

16. When adding the coelenterazine h, ensure that it is added to the bottom of the well and gently shake the microplate, so that the substrate diffuses evenly throughout the well. Likewise, reading 6 wells at a time ensures that the luminescence output remains stable since the luminescence signal degrades over time. For this reason, it is important to prepare the 20 μM coelenterazine h solution 1 min before its addition onto the cells.

17. Within the first 5 min of addition of the substrate, the luminescence signal is increasing and therefore it is best to wait 5 min until the luminescence output is relatively constant.

18. The equation provided as to how to compute the ΔBRET is merely a suggestion. The change in BRET can also be calculated by taking the average of the last 50 measurements and subtracting them by the first 50 measurements. Plotting the BRET ratio as a function of time is also a way to gain a better comprehension of how the BRET ratio is changing over time.

Acknowledgements

c-Myc-5-HT$_{2A}$ was a gift from Javier Gonzalez-Maeso (Addgene plasmid # 67944). P.P. and K.B. were supported by GEPROM trainee fellowship awards and J.J.T. was supported by NSERC. R.S.

was funded by traineeships from the McGill CIHR Drug Development Training Grant. The work was funded by a grant to T.E.H. from the Canadian Institutes for Health Research (MOP-130309).

References

1. Borroto-Escuela DO et al (2013) G protein-coupled receptor heterodimerization in the brain. Methods Enzymol 521:281–294

2. Qi Z, Miller GW, Voit EO (2010) The internal state of medium spiny neurons varies in response to different input signals. BMC Syst Biol 4:26

3. Devost D et al (2016) Cellular and subcellular context determine outputs from signaling biosensors. Methods Cell Biol 132:319–337

4. Bourque K et al (2017) Distinct conformational dynamics of three G protein-coupled receptors measured using FlAsH-BRET biosensors. Front Endocrinol (Lausanne) 8:61

5. Devost D et al (2017) Conformational profiling of the AT1 angiotensin II receptor reflects biased agonism, G protein coupling, and cellular context. J Biol Chem 292(13):5443–5456

6. Sleno R et al (2017) Conformational biosensors reveal allosteric interactions between heterodimeric AT1 angiotensin and prostaglandin F2alpha receptors. J Biol Chem 292(29):12139–12152

7. Sleno R et al (2016) Designing BRET-based conformational biosensors for G protein-coupled receptors. Methods 92:11–18

8. Kelley LA et al (2015) The Phyre2 web portal for protein modeling, prediction and analysis. Nat Protoc 10(6):845–858

9. Yang J et al (2015) The I-TASSER Suite: protein structure and function prediction. Nat Methods 12(1):7–8

10. Krogh A et al (2001) Predicting transmembrane protein topology with a hidden Markov model: application to complete genomes. J Mol Biol 305(3):567–580

11. Hoffmann C et al (2010) Fluorescent labeling of tetracysteine-tagged proteins in intact cells. Nat Protoc 5(10):1666–1677

Chapter 18

Searching the GPCR Heterodimer Network (GPCR-hetnet)
Database for Information to Deduce the Receptor–
Receptor Interface and Its Role in the Integration
of Receptor Heterodimer Functions

Ismel Brito, Manuel Narvaez, David Savelli, Kirill Shumilov, Michael Di
Palma, Stefano Sartini, Kamila Skieterska, Kathleen Van Craenenbroeck,
Ismael Valladolid-Acebes, Rauner Zaldivar-Oro, Malgorzata Filip,
Riccardo Cuppini, Alicia Rivera, Fang Liu, Patrizia Ambrogini,
Miguel Pérez de la Mora, Kjell Fuxe, and Dasiel O. Borroto-Escuela

Abstract

The G protein-coupled receptor heterocomplex network database (GPCR-hetnet) is a database designed
to store information on GPCR heteroreceptor complexes and their allosteric receptor–receptor interactions. It is an expert-authored and peer-reviewed, curated collection of well-documented GPCR–GPCR
interactions that span the gamut from classical GPCR–GPCR interactions to more complex receptor–
receptor interactions (GPCR-Receptor Tyrosine Kinase and GPCR-ionotropic receptor/ligand gated ion
channel). Although GPCR-hetnet contains interactions among GPCR from several different species, the
curators have initially focused on receptor–receptor interactions in humans. Currently (August 2017)
GPCR-hetnet contains information on 250 receptors (192 GPCR, 52 RTK, and 6 ionotropic receptors)
and >1023 interactions. The GPCR-hetnet provides four searchable datasets: the hetnet, the non-hetnet,
the rtknet, and the ionnet. Other supporting datasets include information about receptors that are present
in GPCR-hetnet such as literature citations. This chapter describes in a basic protocol how to use, navigate,
and browse through the GPCR-hetnet database to identify the clusters in which a receptor protomer of
interest is involved, while further applicability are also described and introduced.

Key words G protein-coupled receptors, Ionotropic receptor/ligand gated ion channel, Receptor
tyrosine kinase, Network, Heterodimerization, Heteromers, Dimerization, Oligomerization, Hubs,
Receptor–receptor interactions, Clusters, Architecture

Kjell Fuxe and Dasiel O. Borroto-Escuela (eds.), *Receptor-Receptor Interactions in the Central Nervous System*, Neuromethods, vol. 140,
https://doi.org/10.1007/978-1-4939-8576-0_18, © Springer Science+Business Media, LLC, part of Springer Nature 2018

1   Introduction

The study and understanding of GPCR receptor–receptor interactions in the last 30 years has led to an explosion of information about GPCR oligomerization phenomena. A series of important contributions have confirmed the relevance of homo- and heterodimerization processes within the GPCR superfamily. Special mention deserves the pioneering work of Fuxe [1–10], Franco [11–13], Bouvier [14–16], Javitch [17, 18], Hebert [16, 19], Reynolds [20–22], Devi [23], Kenakin [24], George [25, 26], Wess [27], Blumer [28], Bockaert [29], Portoghese [30, 31], van Rijn [32, 33], Schellekens [34, 35] groups, as some relevant examples. This information explosion has been accelerated by the invention of cutting-edge and high-throughput techniques that assess receptor–receptor interactions such as co-immunoprecipitation, fluorescence and bioluminescence energy transfer methods, in situ proximity ligation assay, and the fluorescence cross-correlation spectroscopy (FCCS) [19, 36–44]. These methods now allow researchers to generate huge experimental data sets that cannot easily be interpreted by simple inspection. And it is here, during the analysis and interpretation of such data sets, where the GPCR-hetnet databases come to play a key role.

The GPCR-hetnet represents results from extensive research carried out to identify and to understand the underlying principles of GPCR–GPCR interactions based on the postgenomic emerging concept that the cell must be viewed to be built of complex networks of interacting macromolecules instead of individual cellular components with their own, specific functions [45–48]. The GPCR-hetnet has been proven to be a useful mathematical artifact for studying complex systems of GPCR oligomerization. It captures what is already known about the interaction between the protomers of the homo- and heteroreceptor complexes using a data model that is accessible to computation. It expands our understanding of the protomer interface interactions by allowing us multiple comparisons considering the physicochemical, structural, and functional properties of the protomers involved. The topological analysis of the GPCR-hetnet has contributed to the functional prediction of biological and pharmacological propensities of several GPCRs [49]. However, several functions of GPCR heteroreceptor complexes have yet to be fully elucidated or even predicted.

The GPCR-hetnet project, covered in this chapter, represents a curated collection of well-documented GPCR–GPCR interactions gathered by experts in the field, peer reviewed and edited by professional staff members prior to being published in the database. In addition to the first released GPCR-hetnet edition in 2014 which only contained the GPCR–GPCR interactions, at least three other public GPCR interaction databases are currently accessible on the Web (www.gpcr-hetnet.com). The GPCR–GPCR of noninteracting protomers (non-hetnet), the GPCR–RTK interaction network database (rtknet), and the GPCR-ionotropic receptor interaction network database (ionnet) represent a somewhat larger scope and store information about GPCR receptor–receptor interactions and illustrate better the molecular mechanisms of signaling integration at the membrane level.

Protocol describes how to navigate and browse through the GPCR-hetnet over the Internet, using the GPCR-hetnet viewer. It also shows how to identify the interacting receptor partner in which a receptor of interest is involved using either the common name or the accession number, respectively. Furthermore, the protocol describes how a suite of data analysis tools can assist with the interpretation of user-supplied experimental data sets to obtain information on the putative interface interaction, function, and specificity of a GPCR protomer.

2 Methods Considerations: How was built the GPCR-hetnet?

2.1 GPCR Receptor–Receptor Interaction Dataset

Interaction data for each of the individual human GPCR protomers were obtained manually from the Search Tool for the Retrieval of Interacting Genes (STRING; http://string-db.org/) database and literature (SCOPUS database). Only protomers that have been validated by one or more independent publications (experimentally verified interactions) were used to create the graph [49]. Experimental methods consider as indicators of physical receptor–receptor interactions the methods of co-immunoprecipitation, BRET/FRET/SRET/TR-FRET/BiFC, in situ PLA, and FCCS among others.

2.2 Network Construction and Analysis

The GPCR-HetNet can be considered as a network of interactions among GPCR protomers. Each protomer in a receptor complex is considered as a node in the network and the connections between the nodes are the edges. The interaction data were used to build and analyze the network using Cytoscape (Version 3.6.0; http://www.cytoscape.org/), a network visualization and analysis platform that supports a wide variety of plug-ins relative to network analysis and manipulation. Duplicated edges and self-loops resulting from reciprocal interaction detection (e.g., homodimerization) were removed prior to the analysis. The network was treated as undirected throughout the study, meaning that there were no distinctions implied between the nodes. In order to overcome this misinterpretation introduced by the use of different graphs, we determined the topological features of networks instead of analyzing their graphical representations [50–52]. A topological feature of a network is an invariant property whose value is the same no matter the network graph chosen. Cytoscape was used to calculate the basic network metrics such as the number of nodes and edges, density, diameter, degree distribution, path length, and clustering coefficient.

2.3 Topological Features

The most elementary topological feature of a network is the node *degree*, *k*, which measures the number of connections or links the node maintains with the other nodes. The node *degree distribution* of a network, $P(k)$, is the fraction of nodes that have exactly *k* connections to other nodes, i.e., $P(k) = degree_k / N$, where $degree_k$ is the number of nodes with degree equal to *k*. A *path* between a pair of nodes is a set of adjacent edges and nodes that we need to visit in order to travel from one node to the other. The *path distance* is the number of edges the path contains. The *shortest path* between a pair of nodes is the path that has the smallest path distance. *Clustering coefficient* assesses the trend of the nodes of the network to form clusters. The local clustering coefficient of node *n* is given by the following formulae: $C(n) = 2n_l / k(k-1)$, where n_l is the number of connections among the nodes that *n* is connected to. The clustering coefficient of a network is the average of the local clustering coefficients of all nodes in the network. The *clustering coefficient distribution*, $C(k)$, is defined by the average of the clustering coefficient of the nodes with degree equal to *k*. The so-called *network density D* assesses how connected the network is, $D = average_k / N - 1$, where $average_k$ is the average degree of the network. A *connected component* is a subgraph in which every pair of nodes is connected to each other by at least one path. Another feature of the topological connectivity of a network measures the relative size of the largest component of a network. This is computed by dividing the number of nodes in the largest component by the number of total nodes in the network. This measure is referred to as the *relative connectivity*, *f* (for further details, see [50]).

2.4 Network Models

Knowing the model of our biological network is essential to further understand the complex system that is modeled. To this aim we need to examine two of the aforementioned topological features: node degree distribution, $P(k)$, and clustering coefficient, $C(k)$. Barabási et al. describe three models of biological networks labeled as: random, scale-free, and hierarchical [50]. In a random network, $P(k)$ follows a Poisson distribution and $C(k)$ is independent of the node degree *k*. In other words, the majority of nodes have roughly the same number of connections, and their tendency to form clusters is the same no matter the node degree. In contrast, node degrees show a power-law distribution, $P(k) \sim k^\gamma$, in scale-free and hierarchical models. In these models, networks have many nodes with small degrees and allow nodes with high degrees. The most notable characteristic of these two models is the so-called preferential attachment property, which implies that a newly added node is more likely to interact with nodes of higher degrees. Scale-free and hierarchical models, however, differ from each other in the way $C(k)$ is expressed. Similar to random networks, $C(k)$ is independent of *k* in scale-free networks. In contrast to scale-free networks, $C(k)$ in hierarchical networks can be expressed as a function of the degree in the following way: k^{-1}. Hierarchical networks can be seen as a special type of scale-free networks with a large clustering coefficient.

3 Materials or Necessary Resources

1. Hardware: Computer capable of supporting a Web browser and an Internet connection.

2. Software: Any modern Web browser such as Firefox, Safari, Chrome, and Internet Explorer will work to display the GPCR-hetnet Web pages.

3. Files: No local files required.

4 Protocol: Browsing a GPCR Heteroreceptor Complex in the GPCR-hetnet

This protocol will introduce the basic navigational techniques needed to browse the GPCR-hetnet database.

1. Go to the GPCR-hetnet home page at www.gpcr-hetnet.com. The home page (Fig. 1) has several elements. The **navigation bar**, at the very top left of the page, provides access to the top-level sections of the global Hetnet, Non-HetNet, RTK-HetNet and ION-HetNet tools, and resources of the GPCR-hetnet site. **"Instructions"** is a description of the project as a whole and a short user guides information about the web site and the different database and their accessibilities; **"About us"** provides access to assistance with a GPCR-HetNet question or troubleshooting a problem and also the email address for submission of new data (see step 9). The **"For further info"** section at the right bottom site of the Web provides information on how to cite GPCR-HetNet contents in journal articles and access training and tutorial resources. The **"Search protomer,"** located on the left side of the home page, provides access to a simple search tool which allows to introduce queries using flexible keyword or UniProt accession numbers. All the databases included in the GPCR-HetNET are explored through a graphic interface. This tool was developed and implemented in Java and is freely available at www.gpcr-hetnet-com. Although this tool continues to evolve, the following functions are implemented at the time of writing.

2. To begin browsing a protomer, check on the top left bars to see all GPCR-hetnet datasets, and select, by clicking, the one you are interested in (the global Hetnet, Non-HetNet, RTK-HetNet, and ION-HetNet). The top left bars which contain all the GPCR-hetnet datasets is shown in Figs. 1 and 3. The selection

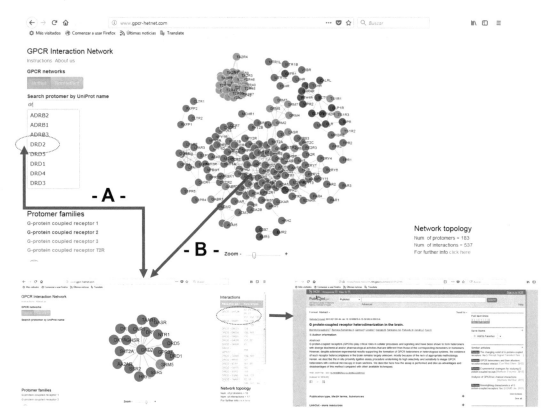

Fig. 1 The GPCR-HetNet home page (www.gpcr-hetnet.com) features a navigation bar and side panel providing access to protomers analysis tools and interaction data. A and B represent different ways to localize the protomer of interest

of the bar depends on the protomer of interest. For instance, if we are looking for specific GPCR–GPCR receptor–receptor interactions the information will appear in the whole HetNet. However, if our main interest is to localize the receptor tyrosine kinase protomer which interacts with a GPCR protomer partner, we should first select the RTK-HetNet bar.

3. Once selected the database of interest, for example, select "HetNet" (the default option), click on the "Search protomer by uniprot name" on the home page and enter the UniProt name (e.g., for D2R protomer: DRD2). You are allowed also to just write the initial letters of the gene or protein name of the protomer of interest (e.g., for the D2R protomer is enough to write "D"). Scroll down the protomer hierarchy panel, and click on the "protomer of interest" link and enter.

4. You will see the search results page for your protomer dataset. The results page for the example query is shown in Fig. 1. Click on the right top panels adjacent to the central connection network to open and check the reference list and the

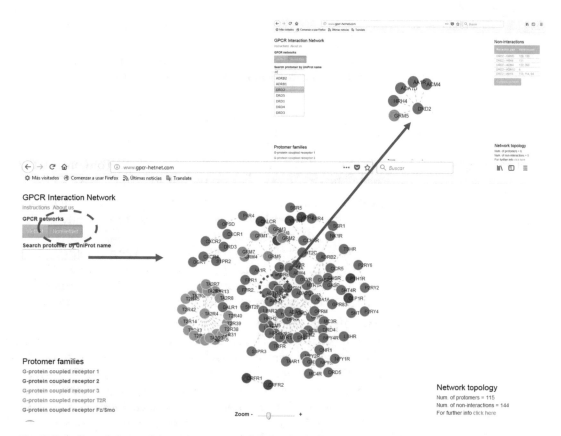

Fig. 2 Selection of the noninteracting partners for the dopamine D2 receptor. Notice that the noninteracting connections are highlighted by a dash lines between the nodes (receptors protomers)

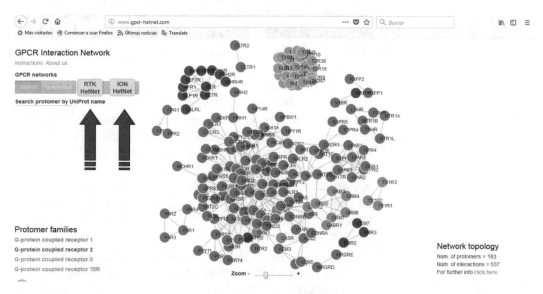

Fig. 3 The GPCR-HetNet home page (www.gpcr-hetnet.com) also providing access to two more GPCR dataset: the RTK-HetNet, which describes the receptor–receptor interacting panorama between GPCR and RTK receptor superfamilies and the ION-HetNet, which contains a collection of all ionotropic receptor protomers interacting with GPCR protomers. To access to the RTK-HetNet or the ION-HetNet, click on the bars indicated by red arrows on the web site

experimental observations which validated the existence of a selected protomer interacting partner (Fig. 1).

5. Notice that it is possible to drill down to a paired receptor–receptor interaction by just clicking on one of the interacting partners in the loaded page, which contains our query in relationship with their interacting partners. The receptor–receptor diagram will re-center on the selected object. To return to the main "hetnet" page, click on the left top bar (HetNET).

6. A very exciting field on GPCR receptor–receptor interactions is to consider the level of specificity and promiscuity of the interaction of our protomer of interests. By clicking on the left top Non-HetNet bar and searching our protomer of interest as described in steps 3 and 4, it is possible to visualize all the potential "receptors" belonging to the GPCR superfamily which do not interact with our protomer of interest (Fig. 2).

7. Two new receptor–receptor interaction networks have been introduced this year (Fig. 3). The RTK-HetNet, which describes the receptor–receptor interactions panorama between GPCR and RTK receptor superfamilies and the ION-HetNet, which contains a collection of all ionotropic receptor protomers interacting with GPCR protomers. As described above in Step 2, to begin browsing an ionotropic receptor or RTK protomer, we must select first the database of interest on the top left bars (the RTK-HetNet or the ION-HetNet). Once selected the database of interest, as indicated in Step 3, click on the "Search protomer by uniprot name" button on the home page and follow the same general browsing principle described in Steps 3 and 4.

8. How to submit GPCR receptor–receptor interaction data? GPCR-HetNet entries are curated by a team of seven researchers enrolled as PhD or postdoctoral students at Karolinska Institutet, University of Urbino, Cuban Neuroscience Observatory and the University of Malaga. However, any scientist is encouraged to submit GPCR interaction information in his/her own field of interest. Prospective contributors are advised to read the online documentation before submitting new interactions. The following details must be provided by email (dasiel.borroto.escuela@ki.se, ismelbr@gmail.com, kjell.fuxe@ki.se, or dasmel@gmail.com) to submit new interaction data:

• The Uniprot code number of the interacting protomers. This can easily be retrieved by searching UniProt (www.expacy.ch).

• The PMID code number of the appropriate reference. This can easily be retrieved by searching NCBI Entrez (http://www.ncbi.nlm.nih.gov/entrez/query.fcgi?db=PubMed), using a couple of authors' names, or a fragment of the title.

- The biochemical or biophysical methods employed to demonstrate the existence of the submitted receptor–receptor interaction are obtained by writing names such as co-immunoprecipitation, BRET/FRET/SRET/TR-FRET/BiFC, in situ PLA, and FCCS but is not limited to these names.

5 Further Applications

The GPCR-HetNet not only provides information about a binary interaction between two GPCR protomers, but also provides information about larger ensembles of receptors that participate in a receptor complex. Also combining the datasets of each of these networks with other relevant repositories it is possible to get a better understanding of the common properties of the hub versus non-hub receptors, on the relevant clusters and motives, and on the analysis of receptor–receptor interface interactions.

5.1 Hubs and Non-hubs

Preferential attachment property in scale-free and hierarchical models leads to the origin of hubs, a relatively small set of highly connected nodes, which have biological significance in protein interaction networks [53–55]. However, despite its simple definition, there is no consensus on when a node is a hub [54, 56, 57]. Vallabhajosyula et al. [58] propose an objective characterization of hubs, which relies on the idea that hubs have lower connectivity among themselves than non-hub nodes. These hub parameters and procedures are the ones used by us to characterize the hub components in the GPCR-HetNet in 2014 (see [49]). First, we create a systematic list of the network nodes by the decreasing order of their degree of connectivity. Secondly, we generate successive subgraphs adding, each time, one node from the degree list. For instance, we first generate G_1, which consists of just one node: the one that appears at the front of the degree list. Then we add the second node from the list and generate subgraph G_2, and so on. For each subgraph we compute its relative connectivity f. This process continues until we obtain a subgraph G_k whose relative connectivity f_k is larger than f_{k-1}. Value k is interpreted as the natural boundary between hub and non-hub nodes, and nodes from G_{k-1} are the hubs of the network. In our paper, following all these four selection criteria, we found that most of the receptors within the network are linked to each other by a small number of edges. However, DRD2, OPRM, ADRB2, AA2AR, AA1R, OPRK, OPRD, and GHSR were identified as hubs.

5.2 Clusters and Motifs

GPCR function is likely to be carried out in a highly modular manner. From the point of view of the GPCR-HetNet, modularity refers to a group of physically or functionally linked protomers (nodes) that work together to achieve a distinct function. In a network representation, a module (cluster) appears as a highly interconnected group of nodes that can be determined by the clustering coefficient, the signature of a network's potential modularity. The clustering coefficient quantifies the number of connected pairs between a node and its neighbors and it is important because it can provide insight into the overall organization of the relationships within a network (network hierarchical character). It may also indicate the presence of physical/functional modules which, in the case of a receptor–receptor interaction network, can represent higher order heteroreceptor complexes or receptor mosaics.

Clusters were found by means of the cluster search algorithm Molecular Complex Detection (MCODE) (Version 1.2; http://baderlab.org/Software/MCODE) using the haircut option which identifies nodes that have limited connectivity at the cluster periphery. A value of 2.0 was used for the degree of cutoff, representing the minimum number of edges for a node to be scored. The K-Core value, which is used to filter out clusters lacking a maximally interconnected core, was specified for two edges [59]. The global clustering coefficient of the GPCR-HetNet (the average of the clustering coefficients for all nodes in the network) is 0.25.

5.3 Analysis of Structural Motifs and Scaffold or Interacting Proteins Associated with Our Complexes

Studies of the dimerization of transmembrane (TM) helices in GPCR homo- and heteroreceptor complexes have been ongoing for many years [60–69] and have provided clues to the fundamental principles behind GPCR–GPCR receptor–receptor interface interaction characteristics and conformation. Although the understanding of GPCR–GPCR TM interface dimerization has been dominated by the idea that sequence motifs, simple recognizable amino acid sequences that drive lateral interaction, can be used to explain and predict the lateral interactions between TM helices in GPCR protomers. Nowadays it is becoming clear that the sequence motif paradigm is incomplete and further structural studies are needed.

Despite this complexity, it is still very useful to approach at a first glance the study of the GPCR–GPCR receptor–receptor interface interactions from the analysis of sequence motifs, or simple recognizable amino acid sequences that can drive the helix–helix lateral interaction. Combining the information from the HetNet and Non-HetNet datasets and the analysis of the presence of specific helix–helix interaction motifs, previously reported in the literature [70–73], we can obtain useful information. For instance, it was found that the GPCR–GPCR heteroreceptor complexes share with higher or lower probability at their lateral TM interface the GxxxG, QxxS, VxxGxxxGxxLL, AxxxA, WxxW, YxxY motifs but not the LIxxGVxxGVxxT, LTxxISAxVGI, AVxxGLxxGAxxLL, LxxVMxxIxxxG motifs.

The analysis of the receptor–receptor interactions may be scaled up or down combining the GPCR-HetNet with other relevant interacting protein dataset (e.g., the Biomolecular Interaction

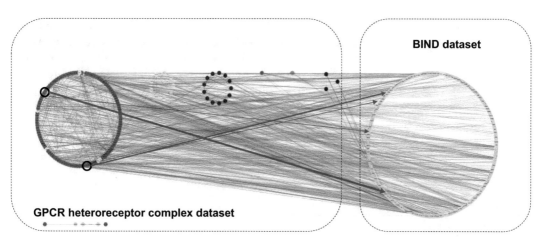

Fig. 4 Illustration of the integration of the GPCR-HetNet with the Biomolecular Interaction Network Database (BIND: http://bind.ca repository). The combination of the GPCR-HetNet with other relevant experimental dataset may allow us to identify a series of potential scaffolds or interacting proteins that are shared within a heteroreceptor complex

Network Database (BIND: http://bind.ca repository) (Fig. 4). BIND stores information about interactions, molecular complexes, and pathways. If we integrate it with the GPCR-HetNet, it may allow us to identify a series of potential scaffolds or interacting proteins that are shared within a heteroreceptor complex and the ones which are only specific of one of the protomers within that complex. With this information in our hands, several hypotheses will be opened.

6   Comments and Background Information

GPCR-HetNET aids scientific discovery by collecting, interpreting, and organizing information on GPCR receptor–receptor interaction phenomena so that it is easy to access and use. It saves researchers countless hours of work in monitoring and collecting this information themselves. It represents a powerful searching and filtering features to help users find the exact data they are interested in. This is further enhanced by flexible and effective groups of datasets, which allow users to define their search in terms on receptor–receptor interaction superfamilies and build their queries.

The HetNET is the central hub for the collection of functional and structural information and other rich annotations on GPCR–GPCR interactions. It is further divided into three more manually annotated dataset sections (Non-HetNet, RTK-HetNet, and ION-HetNet). The Non-HetNET is a non-redundant archive containing all the publicly available information on GPCR receptor noninteracting protomers in the world of GPCR interactions. The RTK-HetNet provides information on GPCR-RTK receptor–receptor interactions and the ION-HetNET dataset provides information on GPCR-ionotropic receptors complexes and their receptor–receptor interactions sets. Supporting datasets are a collection of meta-information about protomers entries such as literature citations and cross-referenced databases. The GPCR-HetNet provides training material through the Karolinska Institutet online portal, including short video tutorials embedded in the web site and also available on the YouTube channel.

Acknowledgments

The work was supported by the Swedish Medical Research Council (62X-00715-50-3) to K.F., by ParkinsonFonden 2015, AFA Försäkring (130328) to K.F., and by Hjärnfonden (FO2016-0302) and Karolinska Institutet Forskningsstiftelser (2016–2017) to D.O.B-E. D.O.B-E. belongs to the "Academia de Biólogos Cubanos" group.

References

1. Fuxe K, Ferre S, Zoli M, Agnati LF (1998) Integrated events in central dopamine transmission as analyzed at multiple levels. Evidence for intramembrane adenosine A2A/dopamine D2 and adenosine A1/dopamine D1 receptor interactions in the basal ganglia. Brain Res Brain Res Rev 26(2-3):258–273

2. Fuxe K, Borroto-Escuela DO (2016) Heteroreceptor complexes and their allosteric receptor-receptor interactions as a novel biological principle for integration of communication in the CNS: targets for drug development. Neuropsychopharmacology 41(1):380–382. https://doi.org/10.1038/npp.2015.244

3. Fuxe K, Agnati LF, Borroto-Escuela DO (2014) The impact of receptor-receptor interactions in heteroreceptor complexes on brain plasticity. Expert Rev Neurother 14(7):719–721. https://doi.org/10.1586/14737175.2014.922878

4. Fuxe K, Borroto-Escuela D, Fisone G, Agnati LF, Tanganelli S (2014) Understanding the role of heteroreceptor complexes in the central nervous system. Curr Protein Pept Sci 15(7):647

5. Fuxe K, Borroto-Escuela DO, Ciruela F, Guidolin D, Agnati LF (2014) Receptor-receptor interactions in heteroreceptor complexes: a new principle in biology. Focus on their

role in learning and memory. Neurosci Discov 2(1):6. https://doi.org/10.7243/2052-6946-2-6

6. Borroto-Escuela DO, Li X, Tarakanov AO, Savelli D, Narvaez M, Shumilov K, Andrade-Talavera Y, Jimenez-Beristain A, Pomierny B, Diaz-Cabiale Z, Cuppini R, Ambrogini P, Lindskog M, Fuxe K (2017) Existence of brain 5-HT1A-5-HT2A isoreceptor complexes with antagonistic allosteric receptor-receptor interactions regulating 5-HT1A receptor recognition. ACS Omega 2(8):4779–4789. https://doi.org/10.1021/acsomega.7b00629

7. Borroto-Escuela DO, DuPont CM, Li X, Savelli D, Lattanzi D, Srivastava I, Narvaez M, Di Palma M, Barbieri E, Andrade-Talavera Y, Cuppini R, Odagaki Y, Palkovits M, Ambrogini P, Lindskog M, Fuxe K (2017) Disturbances in the FGFR1-5-HT1A heteroreceptor complexes in the Raphe-hippocampal 5-HT system develop in a genetic rat model of depression. Front Cell Neurosci 11:309. https://doi.org/10.3389/fncel.2017.00309

8. Borroto-Escuela DO, Wydra K, Pintsuk J, Narvaez M, Corrales F, Zaniewska M, Agnati LF, Franco R, Tanganelli S, Ferraro L, Filip M, Fuxe K (2016) Understanding the functional plasticity in neural networks of the basal ganglia in cocaine use disorder: a role for allosteric

receptor-receptor interactions in A2A-D2 heteroreceptor complexes. Neural Plast 2016:4827268. https://doi.org/10.1155/2016/4827268

9. Borroto-Escuela DO, Romero-Fernandez W, Mudo G, Perez-Alea M, Ciruela F, Tarakanov AO, Narvaez M, Di Liberto V, Agnati LF, Belluardo N, Fuxe K (2012) Fibroblast growth factor receptor 1- 5-hydroxytryptamine 1A heteroreceptor complexes and their enhancement of hippocampal plasticity. Biol Psychiatry 71(1):84–91. https://doi.org/10.1016/j.biopsych.2011.09.012

10. Borroto-Escuela DO, Tarakanov AO, Guidolin D, Ciruela F, Agnati LF, Fuxe K (2011) Moonlighting characteristics of G protein-coupled receptors: focus on receptor heteromers and relevance for neurodegeneration. IUBMB Life 63(7):463–472. https://doi.org/10.1002/iub.473

11. Franco R, Ferre S, Agnati L, Torvinen M, Gines S, Hillion J, Casado V, Lledo P, Zoli M, Lluis C, Fuxe K (2000) Evidence for adenosine/dopamine receptor interactions: indications for heteromerization. Neuropsychopharmacology 23(4 Suppl):S50–S59. https://doi.org/10.1016/S0893-133X(00)00144-5

12. Gines S, Hillion J, Torvinen M, Le Crom S, Casado V, Canela EI, Rondin S, Lew JY, Watson S, Zoli M, Agnati LF, Verniera P, Lluis C, Ferre S, Fuxe K, Franco R (2000) Dopamine D1 and adenosine A1 receptors form functionally interacting heteromeric complexes. Proc Natl Acad Sci U S A 97(15):8606–8611. https://doi.org/10.1073/pnas.150241097

13. Navarro G, Borroto-Escuela D, Angelats E, Etayo I, Reyes-Resina I, Pulido-Salgado M, Rodriguez-Perez AI, Canela EI, Saura J, Lanciego JL, Labandeira-Garcia JL, Saura CA, Fuxe K, Franco R (2018) Receptor-heteromer mediated regulation of endocannabinoid signaling in activated microglia. Role of CB1 and CB2 receptors and relevance for Alzheimer's disease and levodopa-induced dyskinesia. Brain Behav Immun 67:139–151. https://doi.org/10.1016/j.bbi.2017.08.015

14. Angers S, Salahpour A, Bouvier M (2001) Biochemical and biophysical demonstration of GPCR oligomerization in mammalian cells. Life Sci 68(19-20):2243–2250

15. Angers S, Salahpour A, Joly E, Hilairet S, Chelsky D, Dennis M, Bouvier M (2000) Detection of beta 2-adrenergic receptor dimerization in living cells using bioluminescence resonance energy transfer (BRET). Proc Natl Acad Sci U S A 97(7):3684–3689. https://doi.org/10.1073/pnas.060590697

16. Hebert TE, Loisel TP, Adam L, Ethier N, Onge SS, Bouvier M (1998) Functional rescue of a constitutively desensitized beta2AR through receptor dimerization. Biochem J 330(Pt 1):287–293

17. Han Y, Moreira IS, Urizar E, Weinstein H, Javitch JA (2009) Allosteric communication between protomers of dopamine class A GPCR dimers modulates activation. Nat Chem Biol 5(9):688–695. https://doi.org/10.1038/nchembio.199

18. Guo W, Urizar E, Kralikova M, Mobarec JC, Shi L, Filizola M, Javitch JA (2008) Dopamine D2 receptors form higher order oligomers at physiological expression levels. EMBO J 27(17):2293–2304. https://doi.org/10.1038/emboj.2008.153

19. Goupil E, Laporte SA, Hebert TE (2013) A simple method to detect allostery in GPCR dimers. Methods Cell Biol 117:165–179. https://doi.org/10.1016/B978-0-12-408143-7.00009-8

20. Dean MK, Higgs C, Smith RE, Bywater RP, Snell CR, Scott PD, Upton GJ, Howe TJ, Reynolds CA (2001) Dimerization of G-protein-coupled receptors. J Med Chem 44(26):4595–4614

21. Gouldson PR, Higgs C, Smith RE, Dean MK, Gkoutos GV, Reynolds CA (2000) Dimerization and domain swapping in G-protein-coupled receptors: a computational study. Neuropsychopharmacology 23(4 Suppl):S60–S77. https://doi.org/10.1016/S0893-133X(00)00153-6

22. Gouldson PR, Snell CR, Bywater RP, Higgs C, Reynolds CA (1998) Domain swapping in G-protein coupled receptor dimers. Protein Eng 11(12):1181–1193

23. Devi LA (2001) Heterodimerization of G-protein-coupled receptors: pharmacology, signaling and trafficking. Trends Pharmacol Sci 22(10):532–537

24. Kenakin T (2002) Drug efficacy at G protein-coupled receptors. Annu Rev Pharmacol Toxicol 42:349–379. https://doi.org/10.1146/annurev.pharmtox.42.091401.113012

25. Lee SP, Xie Z, Varghese G, Nguyen T, O'Dowd BF, George SR (2000) Oligomerization of dopamine and serotonin receptors. Neuropsychopharmacology 23(4 Suppl):S32–S40. https://doi.org/10.1016/S0893-133X(00)00155-X

26. Xie Z, Lee SP, O'Dowd BF, George SR (1999) Serotonin 5-HT1B and 5-HT1D receptors form homodimers when expressed alone and heterodimers when co-expressed. FEBS Lett 456(1):63–67

27. Zeng F, Wess J (2000) Molecular aspects of muscarinic receptor dimerization. Neuropsychopharmacology 23(4 Suppl):S19–S31. https://doi.org/10.1016/S0893-133X(00)00146-9

28. Overton MC, Blumer KJ (2000) G-protein-coupled receptors function as oligomers in vivo. Curr Biol 10(6):341–344

29. Bockaert J, Pin JP (1999) Molecular tinkering of G protein-coupled receptors: an evolutionary success. EMBO J 18(7):1723–1729. https://doi.org/10.1093/emboj/18.7.1723

30. Portoghese PS (2001) From models to molecules: opioid receptor dimers, bivalent ligands, and selective opioid receptor probes. J Med Chem 44(14):2259–2269

31. Waldhoer M, Fong J, Jones RM, Lunzer MM, Sharma SK, Kostenis E, Portoghese PS, Whistler JL (2005) A heterodimer-selective agonist shows in vivo relevance of G protein-coupled receptor dimers. Proc Natl Acad Sci U S A 102(25):9050–9055. https://doi.org/10.1073/pnas.0501112102

32. van Rijn RM, Whistler JL, Waldhoer M (2010) Opioid-receptor-heteromer-specific trafficking and pharmacology. Curr Opin Pharmacol 10(1):73–79. https://doi.org/10.1016/j.coph.2009.09.007

33. van Rijn RM, Chazot PL, Shenton FC, Sansuk K, Bakker RA, Leurs R (2006) Oligomerization of recombinant and endogenously expressed human histamine H(4) receptors. Mol Pharmacol 70(2):604–615. https://doi.org/10.1124/mol.105.020818

34. Schellekens H, De Francesco PN, Kandil D, Theeuwes WF, McCarthy T, van Oeffelen WE, Perello M, Giblin L, Dinan TG, Cryan JF (2015) Ghrelin's orexigenic effect is modulated via a serotonin 2C receptor interaction. ACS Chem Neurosci 6(7):1186–1197. https://doi.org/10.1021/cn500318q

35. Schellekens H, Dinan TG, Cryan JF (2013) Taking two to tango: a role for ghrelin receptor heterodimerization in stress and reward. Front Neurosci 7:148. https://doi.org/10.3389/fnins.2013.00148

36. Borroto-Escuela DO, Romero-Fernandez W, Garriga P, Ciruela F, Narvaez M, Tarakanov AO, Palkovits M, Agnati LF, Fuxe K (2013) G protein-coupled receptor heterodimerization in the brain. Methods Enzymol 521:281–294. https://doi.org/10.1016/B978-0-12-391862-8.00015-6

37. Borroto-Escuela DO, Flajolet M, Agnati LF, Greengard P, Fuxe K (2013) Bioluminescence resonance energy transfer methods to study G protein-coupled receptor-receptor tyrosine kinase heteroreceptor complexes. Methods Cell Biol 117:141–164. https://doi.org/10.1016/B978-0-12-408143-7.00008-6

38. Fernandez-Duenas V, Llorente J, Gandia J, Borroto-Escuela DO, Agnati LF, Tasca CI, Fuxe K, Ciruela F (2012) Fluorescence resonance energy transfer-based technologies in the study of protein-protein interactions at the cell surface. Methods 57(4):467–472. https://doi.org/10.1016/j.ymeth.2012.05.007

39. Skieterska K, Duchou J, Lintermans B, Van Craenenbroeck K (2013) Detection of G protein-coupled receptor (GPCR) dimerization by coimmunoprecipitation. Methods Cell Biol 117:323–340. https://doi.org/10.1016/B978-0-12-408143-7.00017-7

40. Achour L, Kamal M, Jockers R, Marullo S (2011) Using quantitative BRET to assess G protein-coupled receptor homo- and heterodimerization. Methods Mol Biol 756:183–200. https://doi.org/10.1007/978-1-61779-160-4_9

41. Lohse MJ, Nuber S, Hoffmann C (2012) Fluorescence/bioluminescence resonance energy transfer techniques to study G-protein-coupled receptor activation and signaling. Pharmacol Rev 64(2):299–336. https://doi.org/10.1124/pr.110.004309

42. Hink MA, Postma M (2013) Monitoring receptor oligomerization by line-scan fluorescence cross-correlation spectroscopy. Methods Cell Biol 117:197–212. https://doi.org/10.1016/B978-0-12-408143-7.00011-6

43. Herrick-Davis K, Grinde E, Cowan A, Mazurkiewicz JE (2013) Fluorescence correlation spectroscopy analysis of serotonin, adrenergic, muscarinic, and dopamine receptor dimerization: the oligomer number puzzle. Mol Pharmacol 84(4):630–642. https://doi.org/10.1124/mol.113.087072

44. Kuhn C, Bufe B, Batram C, Meyerhof W (2010) Oligomerization of TAS2R bitter taste receptors. Chem Senses 35(5):395–406. https://doi.org/10.1093/chemse/bjq027

45. Xia Y, Yu H, Jansen R, Seringhaus M, Baxter S, Greenbaum D, Zhao H, Gerstein M (2004) Analyzing cellular biochemistry in terms of molecular networks. Annu Rev Biochem 73:1051–1087. https://doi.org/10.1146/annurev.biochem.73.011303.073950

46. Borroto-Escuela DO, Agnati LF, Fuxe K, Ciruela F (2012) Muscarinic acetylcholine receptor-interacting proteins (mAChRIPs): targeting the receptorsome. Curr Drug Targets 13(1):53–71

47. Borroto-Escuela DO, Correia PA, Romero-Fernandez W, Narvaez M, Fuxe K, Ciruela F,

Garriga P (2011) Muscarinic receptor family interacting proteins: role in receptor function. J Neurosci Methods 195(2):161–169. https://doi.org/10.1016/j.jneumeth.2010.11.025

48. Choura M, Rebai A (2010) Application of computational approaches to study signalling networks of nuclear and Tyrosine kinase receptors. Biol Direct 5:58. https://doi.org/10.1186/1745-6150-5-58

49. Borroto-Escuela DO, Brito I, Romero-Fernandez W, Di Palma M, Oflijan J, Skieterska K, Duchou J, Van Craenenbroeck K, Suarez-Boomgaard D, Rivera A, Guidolin D, Agnati LF, Fuxe K (2014) The G protein-coupled receptor heterodimer network (GPCR-HetNet) and its hub components. Int J Mol Sci 15(5):8570–8590. https://doi.org/10.3390/ijms15058570

50. Barabasi AL, Oltvai ZN (2004) Network biology: understanding the cell's functional organization. Nat Rev Genet 5(2):101–113. https://doi.org/10.1038/nrg1272

51. Yook SH, Oltvai ZN, Barabasi AL (2004) Functional and topological characterization of protein interaction networks. Proteomics 4(4):928–942. https://doi.org/10.1002/pmic.200300636

52. Zhu X, Gerstein M, Snyder M (2007) Getting connected: analysis and principles of biological networks. Genes Dev 21(9):1010–1024. https://doi.org/10.1101/gad.1528707

53. Albert R, Jeong H, Barabasi AL (2000) Error and attack tolerance of complex networks. Nature 406(6794):378–382. https://doi.org/10.1038/35019019

54. Han JD, Bertin N, Hao T, Goldberg DS, Berriz GF, Zhang LV, Dupuy D, Walhout AJ, Cusick ME, Roth FP, Vidal M (2004) Evidence for dynamically organized modularity in the yeast protein-protein interaction network. Nature 430(6995):88–93. https://doi.org/10.1038/nature02555

55. Wuchty S, Almaas E (2005) Peeling the yeast protein network. Proteomics 5(2):444–449. https://doi.org/10.1002/pmic.200400962

56. Batada NN, Reguly T, Breitkreutz A, Boucher L, Breitkreutz BJ, Hurst LD, Tyers M (2006) Stratus not altocumulus: a new view of the yeast protein interaction network. PLoS Biol 4(10):e317. https://doi.org/10.1371/journal.pbio.0040317

57. Ekman D, Light S, Bjorklund AK, Elofsson A (2006) What properties characterize the hub proteins of the protein-protein interaction network of Saccharomyces cerevisiae? Genome Biol 7(6):R45. https://doi.org/10.1186/gb-2006-7-6-r45

58. Vallabhajosyula RR, Chakravarti D, Lutfeali S, Ray A, Raval A (2009) Identifying hubs in protein interaction networks. PLoS One 4(4):e5344. https://doi.org/10.1371/journal.pone.0005344

59. Delprato A (2012) Topological and functional properties of the small GTPases protein interaction network. PLoS One 7(9):e44882. https://doi.org/10.1371/journal.pone.0044882

60. Borroto-Escuela DO, Romero-Fernandez W, Tarakanov AO, Gomez-Soler M, Corrales F, Marcellino D, Narvaez M, Frankowska M, Flajolet M, Heintz N, Agnati LF, Ciruela F, Fuxe K (2010) Characterization of the A2AR-D2R interface: focus on the role of the C-terminal tail and the transmembrane helices. Biochem Biophys Res Commun 402(4):801–807. https://doi.org/10.1016/j.bbrc.2010.10.122

61. Maggio R, Barbier P, Colelli A, Salvadori F, Demontis G, Corsini GU (1999) G protein-linked receptors: pharmacological evidence for the formation of heterodimers. J Pharmacol Exp Ther 291(1):251–257

62. Harikumar KG, Wootten D, Pinon DI, Koole C, Ball AM, Furness SG, Graham B, Dong M, Christopoulos A, Miller LJ, Sexton PM (2012) Glucagon-like peptide-1 receptor dimerization differentially regulates agonist signaling but does not affect small molecule allostery. Proc Natl Acad Sci U S A 109(45):18607–18612. https://doi.org/10.1073/pnas.1205227109

63. Yanagawa M, Yamashita T, Shichida Y (2011) Comparative fluorescence resonance energy transfer analysis of metabotropic glutamate receptors: implications about the dimeric arrangement and rearrangement upon ligand bindings. J Biol Chem 286(26):22971–22981. https://doi.org/10.1074/jbc.M110.206870

64. Elsner A, Tarnow P, Schaefer M, Ambrugger P, Krude H, Gruters A, Biebermann H (2006) MC4R oligomerizes independently of extracellular cysteine residues. Peptides 27(2):372–379. https://doi.org/10.1016/j.peptides.2005.02.027

65. Arachiche A, Mumaw MM, de la Fuente M, Nieman MT (2013) Protease-activated receptor 1 (PAR1) and PAR4 heterodimers are required for PAR1-enhanced cleavage of PAR4 by alpha-thrombin. J Biol Chem 288(45):32553–32562. https://doi.org/10.1074/jbc.M113.472373

66. Leger AJ, Jacques SL, Badar J, Kaneider NC, Derian CK, Andrade-Gordon P, Covic L, Kuliopulos A (2006) Blocking the protease-activated receptor 1-4 heterodimer in platelet-

mediated thrombosis. Circulation 113(9): 1244–1254. https://doi.org/10.1161/ CIRCULATIONAHA.105.587758

67. Laroche G, Lepine MC, Theriault C, Giguere P, Giguere V, Gallant MA, de Brum-Fernandes A, Parent JL (2005) Oligomerization of the alpha and beta isoforms of the thromboxane A2 receptor: relevance to receptor signaling and endocytosis. Cell Signal 17(11):1373–1383. https:// doi.org/10.1016/j.cellsig.2005.02.008

68. Gao F, Harikumar KG, Dong M, Lam PC, Sexton PM, Christopoulos A, Bordner A, Abagyan R, Miller LJ (2009) Functional importance of a structurally distinct homodimeric complex of the family B G protein-coupled secretin receptor. Mol Pharmacol 76(2):264–274. https://doi.org/10.1124/ mol.109.055756

69. Schulz A, Grosse R, Schultz G, Gudermann T, Schoneberg T (2000) Structural implication for receptor oligomerization from functional reconstitution studies of mutant V2 vasopressin receptors. J Biol Chem 275(4):2381–2389

70. Li E, Wimley WC, Hristova K (2012) Transmembrane helix dimerization: beyond the search for sequence motifs. Biochim Biophys Acta 1818(2):183–193. https://doi. org/10.1016/j.bbamem.2011.08.031

71. Lemmon MA, Treutlein HR, Adams PD, Brunger AT, Engelman DM (1994) A dimerization motif for transmembrane alpha-helices. Nat Struct Biol 1(3):157–163

72. Kay BK, Williamson MP, Sudol M (2000) The importance of being proline: the interaction of proline-rich motifs in signaling proteins with their cognate domains. FASEB J 14(2): 231–241

73. Sal-Man N, Gerber D, Bloch I, Shai Y (2007) Specificity in transmembrane helix-helix interactions mediated by aromatic residues. J Biol Chem 282(27):19753–19761. https://doi. org/10.1074/jbc.M610368200

Chapter 19

Detection, Analysis, and Quantification of GPCR Homo- and Heteroreceptor Complexes in Specific Neuronal Cell Populations Using the In Situ Proximity Ligation Assay

Dasiel O. Borroto-Escuela, Manuel Narvaez, Ismael Valladolid-Acebes, Kirill Shumilov, Michael Di Palma, Karolina Wydra, Thorsten Schaefer, Irene Reyes-Resina, Gemma Navarro, Giuseppa Mudó, Malgorzata Filip, Stefano Sartini, Kristina Friedland, Harriët Schellekens, Sarah Beggiato, Luca Ferraro, Sergio Tanganelli, Rafael Franco, Natale Belluardo, Patrizia Ambrogini, Miguel Pérez de la Mora, and Kjell Fuxe

Abstract

GPCR's receptosome operates via coordinated changes between the receptor expression, their modifications and interactions between each other. Perturbation in specific heteroreceptor complexes and/or their balance/equilibrium with other heteroreceptor complexes and corresponding homoreceptor complexes is considered to have a role in pathogenic mechanisms. Such mechanisms lead to mental and neurological diseases, including drug addiction, depression, Parkinson's disease, and schizophrenia. To understand the associations of GPCRs and to unravel the global picture of their receptor–receptor interactions in the brain, different experimental detection techniques for receptor–receptor interactions have been established (e.g., co-immunoprecipation based approach). However, they have been criticized for not reflecting the cellular situation or the dynamic nature of receptor–receptor interactions. Therefore, the detection and visualization of GPCR homo- and heteroreceptor complexes in the brain remained largely unknown until recent years, when a well-characterized in situ proximity ligation assay (in situ PLA) was adapted to validate the receptor complexes in their native environment. The in situ PLA protocol presented here can be used to visualize GPCR receptor–receptor interactions in cells and tissues in a highly sensitive and specific manner. We have developed a combined method using immunohistochemistry and PLA, particularly aimed to monitor interactions between GPCRs in specific neuronal cell populations. This allows the analysis of homo- and heteroreceptor complexes at a cellular and subcellular level. The method has the advantage that it can be used in clinical specimens, providing localized, quantifiable homo- and heteroreceptor complexes detected in single cells. We compare the advantages and limitations of the methods, underlining recent progress and the growing importance of these techniques in basic research. We discuss also their potential as tools for drug development and diagnostics.

Key words G protein-coupled receptors, Immunohistochemistry, In situ proximity ligation assay, Heteroreceptor complexes, Dimerization, Receptor–receptor interaction, Stoichiometry

Kjell Fuxe and Dasiel O. Borroto-Escuela (eds.), *Receptor-Receptor Interactions in the Central Nervous System*, Neuromethods, vol. 140, https://doi.org/10.1007/978-1-4939-8576-0_19, © Springer Science+Business Media, LLC, part of Springer Nature 2018

1   Introduction

Mental and neurological diseases including drug addiction, depression, Parkinson's disease, and schizophrenia are highly complex in their etiology [1–7]. It is not surprising that the underlying multiple molecular mechanisms are poorly understood and treatment possibilities are inadequate. One emerging hypothesis is that direct physical interactions of different receptors named homo-/heteroreceptor complexes may be involved with disease onset and progression [8–16]. Thus, these homo- and heteroreceptor complexes could serve as a biomarker and/or drug target of the disease [11, 17–23]. Recent experimental evidence has contributed to the development of the concept of GPCR heteroreceptor complexes, in which GPCRs physically interact with each other or produce an integrated regulation with receptor tyrosine kinases [11, 24, 25] and ionotropic receptors [26] leading to an integrated activation of intracellular signaling cascades generating inter alia changes in gene expression [20, 27, 28].

The majority of identified GPCR homo- and heteroreceptor complexes have been found using the yeast two-hybrid screen [25], co-immunoprecipitation [24, 29, 30], fluorescence (Förster) or bioluminescence resonance energy transfer (FRET), and BRET [31–37]. Each approach used has provided precise and valuable information which was considered with caution in view of their varying number of false-positive results and technical limitations. Some controversy regarding some approaches also emerged [37, 38]. However, when these methods are properly assessed it is possible to safely demonstrate the direct interactions between membrane receptors [39, 40]. Improvement of the previous methods and introduction of novel techniques have also been developed, such as real-time FRET experiments in living cells [41] and dual-color fluorescence cross-correlation spectroscopy [42]. However, despite the extensive experimental results obtained with these biophysical/biochemical techniques, supporting the formation of GPCR homo- and heteroreceptor complexes in overexpressing systems (cell lines), the existence of GPCR complexes at endogenous expression levels (their native environment) was not demonstrated until 2010–2013 [11, 20–22, 24, 30, 43–47].

In this chapter, we present an approach that combines the well-established immunohistochemical (IHC) technique for immunofluorescence labeling of cells and the recently introduced in situ PLA. Combining IHC with PLA analysis allows a simultaneous visualization of a homo- and heteroreceptor complexes of interest in specific neuronal and non-neuronal cell populations (cellular level) or cellular components like for instance the terminal, the soma or the dendritic spines (subcellular level). The introduction of a neuronal or glial cell marker during the in situ

PLA technique simultaneously allows an additional readout of GPCR homo- and heteroreceptor complexes by combining the advantages of the two methods. At the end of the chapter, we will discuss the advantages and disadvantages of this method compared to other available techniques.

2 The In Situ PLA: Principle of the Assay

In situ proximity ligation assay (in situ PLA) was first described by Fredriksson and colleagues in 2002 [48], further developed for protein–protein interactions by Gullberg et al. and Soderberg et al. [49–54]. It was optimized for GPCR receptor–receptor interaction detection by the Fuxe laboratory [30, 43, 46, 47, 55] and the Javitch group [44]. It became commercialized by Olink Bioscience (http://www.olink.com/) and nowadays its reagents are sold by Sigma-Aldrich, Biomol, among other companies becoming an established and generally applicable immunohistochemical tool for advanced and precise protein–protein interactions and protein modification analysis.

The in situ PLA technique combines the dual recognition of a probe-targeted assay with a split-reporter approach, creating a selective and sensitive method for specific detection of two receptors in close proximity forming a homo- or heteroreceptor complex [24, 30, 46, 47] (Fig. 1a). This innovative method utilizes one pair of oligonucleotide-labeled antibodies binding in close proximity (20–30 nm apart) to different epitopes of the same receptor or two receptors in a homo- or heteroreceptor complex, respectively. The assay is used for localized detection, visualization, and quantification of a single receptor (useful to detect receptor posttranslational modification, e.g., phosphorylation, palmitoylation, etc.), receptor-scaffold or chaperon interactions or receptor–receptor interactions in adherent cell lines, cytospin preparations or tissues, including frozen or paraffin-embedded patient samples. The cells or tissue need to be fixed with a fixative appropriate for antibodies used in the in situ PLA protocol and if necessary antigen retrieval and antibody-specific blocking must be performed.

There are two different approaches [51, 53, 54]; one method uses direct primary antibody conjugation and the other uses detection with secondary conjugates. In the detection of PLA with secondary conjugates, the unmodified primary antibodies are detected with two secondary antibodies that are equipped or conjugated to short DNA oligonucleotides. Two additional DNA strands, called connector oligonucleotides, are then introduced. The two DNA strands on the secondary antibodies are ligated to the two additionally introduced connector oligonucleotides, leading to the formation of a circular, single stranded DNA molecule. In the created DNA circle, one of the secondary antibody conjugated DNA

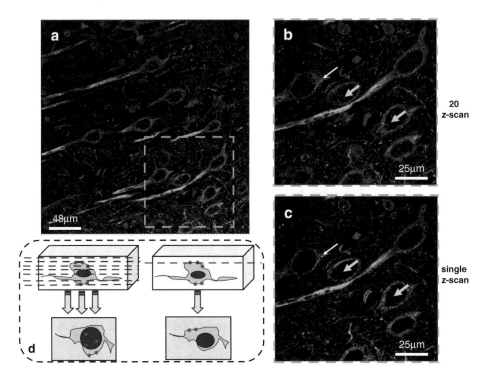

Fig. 1 Examples of combined in situ PLA assay and neuronal labeling, using the Milli-Mark™ Pan Neuronal marker antibody AlexaFluor488 conjugate, in the cerebral cortex. A high density of PLA positives profiles (red) is shown in the internal pyramidal nerve cell layer V (green) of the prefrontal cortex, representing the adenosine A2AR-dopamine D4R (A2AR-D4R) heteroreceptor complex. (**a**) Microphotograph from transverse sections of the rat anterior cingulate cortex (Bregma level: 1.2 mm) showing the distribution of the A2AR-D4R heteroreceptor complexes using the in situ proximity ligation assay (in situ PLA) technique [18, 24, 43] combined with the Neuro-Chrom™ Pan neuronal marker antibody-Alexa488 conjugate immunostaining. The combined use of in situ PLA and neuronal labeling indicates the expression of the A2AR-D4R mainly at the somatic membrane level. White arrows point to red PLA clusters. The nuclei are shown in blue by DAPI. (**b** and **c**) High magnifications of this microphotograph are shown in the right panels and the pictures are taken and visualized using multiple z-scan (20 z-scan) and single z-scan projection, respectively. (**d**) Schematic representation of the image taking from multiple z-scan or a single z-scan. Note that if you use a multiple z-scan projections, it will result in a combined image where some positive PLA blobs/clusters appear to be inside the nuclei blue DAPI staining, although they are actually located on the cytoplasmic membrane (see yellow arrows in panel **b** and **c**). The scale bars are shown in the lower-right panels

probes serves as a primer for the rolling circle amplification, the so-called positive probe (+). As a result, when adding a DNA polymerase, a long DNA product is formed and it remains covalently attached to the PLA probes (+). After finalizing the rolling circle amplification, the concatemeric repetitions of the same sequence enable hybridization of multiple detection fluorescent-labeled oligonucleotides that can be visualized under a fluorescent microscope and quantified. The principle is the same for the detection of PLA with direct primary antibody conjugation; however, in this method the secondary antibodies that are equipped with DNA strands are not required.

3 Materials and Buffers

1. *Hoffman solution* (cryoprotection for free-floating brain sections): 250 ml 0.4 M PBS, Ethylene Glycol 300 ml, 300 g Sucrose, 10 g Polyvinylpyrrolidone, 9 g NaCl. Add high purity water to 1000 ml. Keep the resulting solution in a freezer (Cold storage: −20 °C.).

2. *Brain tissue samples and their preparation*. For the study of GPCR homo- and heteroreceptor complexes in the rat brain, we highly recommend the use of formaldehyde-fixed frozen free-floating sections. First, animals should be deeply anesthetized by an intraperitoneal (i.p.) injection of a high dose of pentobarbital (60 mg/ml, [0.1 ml/100 g]) and then perfused intracardially with 30–50 ml ice-cold 4% paraformaldehyde (PFA) in 0.1 M phosphate-buffered saline (PBS, pH 7.4) solution. After perfusion, brains are collected and transferred into well-labeled glass vials filled with 4% PFA fixative solution for 6–12 h. Then, the brains are placed in 10% and 30% sucrose (0.1 M PBS, pH 7.4) and incubated for 1 day (10% sucrose) and a number of days (30% sucrose) at 4 °C with several sucrose buffer changes, until freezing the brain (in a bowl with dry ice: put inside a beaker with isopentane; when the isopentane reaches −45 °C, enter the mold with tissue; once frozen store at −80 °C). Proceed to generate the tissue sections (10–30 µm-thick) using a cryostat (Stored tissue at −20 °C on the day before cutting). After cutting, store them in Hoffman solution (e.g., in a 24-well plate). Alternatively, to the use of fixed free-floating sections, you can use fixed frozen sections attached to microscopy slides. The mounted sections on slides must be kept at −20 °C until use.

3. *Phosphate-buffered saline* (PBS). PBS is prepared by mixing: 0.23 g NaH_2PO_4 (anhydrous; 1.90 mM), 1.15 g Na_2HPO_4 (anhydrous; 8.10 mM), 9.00 g NaCl (154 mM). Then, add H_2O to 900 ml and if needed, adjust to desired pH (7.2–7.4) with 1 M NaOH or 1 M HCl. Finally, add H_2O to 1 l, filter, sterilize, and store indefinitely at 4 °C. PBS could also be prepared at a 10× concentration (commercially available at Sigma-Aldrich (Cat. No: P5493-1L)), and stored until dilution into the working solution.

4. *Glycine Buffer* (10 mM): dissolve 0.75 g glycine in 100 ml PBS. Store at 4 °C.

5. *Citric acid Buffer* (1 mM) in water, pH 7.4. Store at 4 °C.

6. *Permeabilization buffer*. 0.1% Triton X-100 in PBS (e.g., 0.1 ml Triton X-100 in 100 ml PBS). Store at 4 °C.

7. *Blocking solution*. Prepare the blocking solution by preparing 0.2% BSA in PBS (e.g., 0.2 g in 100 ml PBS, stored at 4 °C). Adjust the amount of reagents accordingly, so that the total

volume is kept at 400 µl × well (12-well plate). Prepare this solution fresh. This blocking solution can be replaced by e.g., the Duolink Blocking solution (Sigma-Aldrich) or the Odyssey Blocking buffer (Licor Biosciences). However, choose the blocking agent best suited for the antibodies used. If animal serum is used to replace the BSA, like for example the use of 5% sterile-filtered goat serum, make sure that it is sterile filtered, as unfiltered serum may increase the amount of background signals.

8. *Primary antibodies and proximity probes (oligonucleotide-labeled secondary antibodies) incubation buffer.* We strongly recommend to dilute the antibodies in the blocking solution to be used (see above 5).

- Proximity probes are created through the attachment of oligonucleotides to antibodies. The antibodies are functionalized by either direct covalent coupling of an oligonucleotide [54] or non-covalent coupling by incubating biotinylated antibodies with a streptavidin-modified oligonucleotide [50]. The oligonucleotide component of the proximity probes can be functionalized to either primary antibodies or secondary antibodies. The latter approach avoids the need to conjugate the oligonucleotide components to each primary antibody protomer pair, saving time and costs. The conjugation reactions can severely affect the antibodies specificities and functionality. For this reason, we strongly recommend to use validated proximity probes from specialized companies like Sigma-Aldrich, which can be bought directly.

- *Primary antibodies validation* [56]. GPCR antibodies are among the most frequently used tools in basic neurochemistry research and in clinical assays. However, the quality and consistency of data generated through the use of GPCR antibodies vary. This poses an impediment to the rigor and reproducibility that are the cornerstones of the advancement of GPCR research. Therefore, a validation of each antibody used in our in situ PLA experiments for its specificity and reproducibility is strongly recommended. *How did we determine the primary antibody specificity?* As a general principle a highly specific primary antibody recognizes its target with minimal crossreactivity (off-target binding) within a given application and experimental context. Therefore, the evaluation of potential crossreactivity under the conditions tested is needed. (1) It can be assessed by measuring the relevant signal in control cells or tissues in which the target receptor has been knocked out or knocked down using techniques such as CRISPR–Cas9 or RNA interference (RNAi). In this way, the expression of the target receptor is either eliminated or reduced; any signal observed after substantial reduction of receptor levels indicates crossreactivity. This approach is powerful and particularly useful for examining antibody specificity for GPCR receptors which originate from a related multigene family. However, this strategy cannot be used for some applications and certain types of samples like the human tissue samples and body fluids. (2) Also two (or more) independent antibodies (they must be able to bind to different epitopes/regions of the receptor) that recognize the same receptor target can be used to assess antibody specificity in a range of assays. This approach requires that the expression pattern generated by the two antibodies correlates within a given application environment. (3) Primary antibodies may be validated by expressing a receptor containing an affinity tag (such as 3XHA, myc, flag, etc.) or a fluorescent protein (such as GFP2, YFP, sRed). This will allow for parallel detection via additional well-validated immunoreagents or direct observation through the fluorescent protein. The detection pattern of the antibody being validated must correlate with the pattern demonstrated by the anti-affinity tag antibody or the fluorescent signal. If substantial discord between these patterns comes out, it will suggest crossreactivity.

9. *Ligation buffer*: 10 mM Tris–acetate, 10 mM magnesium acetate, 50 mM potassium acetate, pH 7.5 adjusted with HCl. Stored at −20 °C [54].

10. *Hybridization-ligation solution*: BSA (250 g/ml), T4 DNA ligase (final concentration of 0.05 U/µl), Tween-20 (0.05%), NaCl 250 mM, ATP 1 mM, and the circularization or connector oligonucleotides (125–250 nM). Circularization or connector oligonucleotides can be designed and synthesized as described previously [54]. Before usage vortex briefly to mix the ligase with the solution. Alternatively, the ligation buffer and the hybridization-ligation solution can be ordered from Olink Bioscience or Sigma-Aldrich (Cat No. DUO92008).

11. *Washing Buffer A*: 8.8 g NaCl, 1.2 g Tris Base, 0.5 ml Tween 20. Dissolve in 800 ml high purity water and adjust the pH to 7.4 using HCl. Adjust with high purity water to 1000 ml and filter through a 0.22 µm filter. Store at 4 °C.

12. *Amplification solution*: Instead of preparing the solutions described below (11–13), one amplification solution can be purchased (Olink Bioscience or Sigma-Aldrich (Cat No. DUO92008)) and used.

13. *Rolling circle amplification (RCA) buffer*: 50 mM Tris–HCl, 10 mM $MgCl_2$, 10 mM $(NH_4)_2SO_4$, pH 7.5 adjusted with HCl. Stored at −20 °C.

14. *RCA solution* (final concentration: phi-29 polymerase 0.125–0.200 U/μl, BSA (250 g/ml), 1× RCA buffer, Tween-20 (0.05%) and dNTP (250 M for each)).

15. *Detection solution* (final concentration: BSA (250 g/ml), 2 μl Sodium citrate 20× (A 20× stock solution consists of 3 M sodium chloride and 300 mM trisodium citrate (adjusted to pH 7.0 with HCl)), 4 μl Tween-20 (0.5%), and the fluorescence detection by fluorophores (e.g., Texas Red or Alexa 555)-oligonucleotide strand (5 μM) see [54]).

16. *Washing Buffer B*: 5.84 g NaCl, 4.24 g Tris Base, 26.0 g Tris–HCl. Dissolve in 500 ml high purity water and adjust the pH to 7.5 using HCl. Then add again high purity water to 1000 ml. Filter through a 0.22 μm filter. Store at 4 °C.

17. *Mounting medium* (e.g., VectaShield, Vector Labs).

4   Assay Protocol

1. Wash the fixed free-floating sections (storage at −20 °C in Hoffman solution) four times with PBS.

2. Enhance the exposition of the receptor epitope by keeping your brain slices in citric acid buffer, for 45–60 min at 65 °C.

3. Wash twice, for 5 min each, with PBS at room temperature.

4. Quench your brain slices with 10 mM glycine buffer, for 20 min at room temperature.

5. Wash twice, for 5 min each, with PBS at room temperature.

6. Incubate with the permeabilization buffer (10% fetal bovine serum (FBS) and 0.5% Triton X-100 or Tween 20 in Tris buffer saline (TBS), pH 7.4) for 30 min at room temperature.

7. Wash twice, for 5 min each, with PBS at room temperature.

8. Incubate with the blocking buffer (0.2% BSA in PBS) for 30 min at room temperature.

9. Turn on the incubator and pre-heat the humidity chamber until usage.

10. Incubate the brain sections with a mixture of the primary antibodies diluted in a suitable concentration in the blocking solution (or at 1:100 in the Duolink II Antibody Diluent (1×)) for 1–2 h at 37 °C or at 4 °C overnight. Prepare primary antibodies alone at the same concentration to be used for the test group. In one well add only Antibody Diluent as an additional negative control.

11. Tap off the primary antibody solution from the slides. Wash twice, for 5 min each, with the blocking solution at room temperature under gentle agitation to remove the excess of primary antibodies.

12. In the mean time, dilute the secondary antibodies proximity probes to a suitable concentration in the blocking solution. Alternatively, dilute the two commercially available PLA probes (e.g., Duolink II anti-Mouse MINUS and Duolink II anti-Rabbit PLUS) 1:5 in Antibody Diluent. It is important to use the same buffer as those for the primary antibody to avoid background staining. Apply the proximity probe mixture to the sample and incubate for 1 h at 37 °C in a humidity chamber. It should be checked regularly that the reaction does not dry out since this will cause a high background signal.

13. To remove the unbound proximity probes, wash the slides twice, 5 min each time, with blocking solution at room temperature under gentle agitation.

14. Prepare the hybridization-ligation solution. To ensure optimal conditions for the enzymatic reactions, the sections should be soaked for 1 min in ligation buffer (10 mM Tris–acetate, 10 mM magnesium acetate, 50 mM potassium acetate, pH 7.5 [54]), prior to addition of the hybridization-ligation solution (final concentration: BSA (250 g/ml), T4 DNA ligase buffer, Tween-20 (0.05%), NaCl 250 mM, ATP 1 mM, and the circularization or connector oligonucleotides (125–250 nM). Remove the soaking solution (ligation buffer), and add T4 DNA ligase at a final concentration of 0.05 U/μl to the hybridization-ligation solution. Vortex briefly to mix the ligase with the solution. Apply the mixture immediately to the sections and incubate slides in a humidity chamber at 37 °C for 30 min. Alternatively, the ligation buffer and the hybridization-ligation solution can be ordered from Olink Bioscience or Sigma-Aldrich (Cat No. DUO92008). Vortex the Duolink II Ligation stock (5×) and dilute 1:5 in high purity water and mix. Take out the ligase from the freezer using a freezing block (−20 °C). Add the ligase to the ligation solution at a 1:40 dilution immediately before addition to brain samples, vortex and incubate the brain slides with this solution in a preheated humidity chamber for 30 min at 37 °C.

15. To remove the excess of connector oligonucleotides, wash twice, for 5 min each, with the washing buffer A at room temperature under gentle agitation. Tap off all excess washing solution.

16. Add to the brain slices the neuronal or cell process marker antibody conjugated with AlexaFluor (e.g., Neuro-Chrom™ Pan neuronal marker antibody-Alexa488 conjugate) and diluted in the rolling circle amplification mixture (see below). Directly conjugated antibodies already have the Alexa Fluor® dye so you can skip that secondary antibody staining step. A selection of direct conjugates against neural markers can be

found in: http://www.abcam.com/neuroscience/directly-conjugated-neural-markers or buy it to Millipore. The use of antibodies to neuronal proteins has become critical tools for identifying neurons and discerning morphological characteristics in culture and complex tissue. While the labeling from GFP constructs yields excellent cytoarchitectural detail, this approach is technically challenging and impractical for many neuroscience research needs. Neuron-specific antibodies are convenient precision tools useful in revealing cytoarchitecture. They can be used to labeling a single protein target within the neuron or to achieve a morphological staining across all parts of neurons (e.g., the Millipore Neuro-Chrom™ Pan neuronal marker antibody, a polyclonal antibody which reacts against key somatic, nuclear, dendritic, and axonal proteins distributed across the pan-neuronal architecture) that can then be detected by a single secondary antibody or used directly conjugate with AlexaFluor488 (see Figs. 1 and 2).

17. Prepare the rolling circle amplification mixture. Soak the sections in RCA buffer for 1 min. Remove the soaking solution, add the RCA solution (final concentration: phi-29 polymerase 0.125–0.200 U/µl, BSA (250 g/ml), RCA buffer, Tween-20 (0.05%) and dNTP (250 M for each)). Vortex briefly the RCA solution and incubate in a humidity chamber for 100 min at 37 °C. Prepare the detection solution and incubate the sections in a humidity chamber at 37 °C for 30 min. Keep the detection solution in the dark to prevent fluorophore bleaching. From now on, all reactions and washing steps should be performed in the dark. Alternatively to preparing these buffers and solutions by yourself, the amplification mixture and the detection solution can be ordered from Olink Bioscience or Sigma-Aldrich (Cat No. DUO92008). Dilute the Duolink II Amplification stock (5×) 1:5 in high purity water and mix. Take out the polymerase from the freezer using a freezing block (−20 °C). Add the polymerase to the amplification solution at a 1:80 dilution and vortex. Add the amplification-polymerase solution to each well. Incubate the brain slides in a pre-heated humidity chamber for 100 min at 37 °C.

18. Wash the sections twice in the dark, for 10 min each, with the washing buffer B at room temperature under gentle agitation. Light sensitive reagents. Keep the slides protected from light.

19. Dip the sections in a washing buffer B dilution of 1:10 and let sections dry at room temperature in the dark. Light sensitive reagents. Keep the slides protected from light.

20. The free-floating sections are put on a microscope slide and a drop of appropriate mounting medium (e.g., VectaShield or Dako) is applied. The cover slip is placed on the section and sealed with nail polish. The sections should be protected against light and can be stored for several days at 4 °C, or for

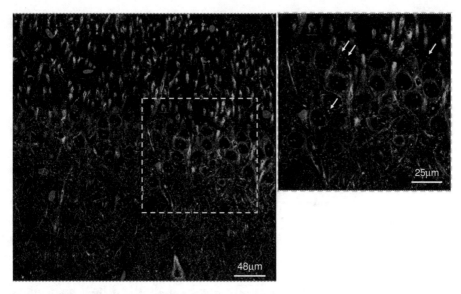

Fig. 2 Illustration of the A2AR-D4R heteroreceptor complexes in the dorsal hippocampus of the rat brain. (*Left panel*) Microphotograph from transverse sections of the rat dorsal hippocampus (Bregma level: −3.6 mm) showing the distribution of the 5 A2AR-D4R heteroreceptor complexes in CA1 using the in situ proximity ligation assay (in situ PLA) technique [18, 24, 43]. They are shown as red PLA blobs (clusters) found in high densities per cell in a large number of nerve cells in the pyramidal cell layer using confocal laser microscopy. The nuclei are shown in blue by DAPI. The neurons are shown in green by the use of the Neuro-Chrom™ Pan neuronal marker antibody-Alexa488 conjugate. To achieve as complete a morphological staining as possible across all parts of neurons, Millipore has developed this polyclonal antibody blend that reacts against key somatic, nuclear, dendritic, and axonal proteins distributed across the pan-neuronal architecture that can then be detected by a single secondary antibody or using as a polyclonal antibody conjugate with AlexaFluor488 as shown in this figure. This antibody cocktail has been validated in diverse methods, cell culture and immuno-histochemistry, giving researchers a convenient and specific qualitative and quantitative tool for studying neuronal morphology. (*Right panel*) In the right inset the PLA blobs in the pyramidal cell layer are shown in higher magnification. The square outlines the CA1 area from which the picture was taken

several months/years at −20 °C. Light sensitive reagents. Keep the slides protected from light.

21. Visualize the samples with a confocal or fluorescent microscope equipped with excitation/emission filters compatible with the fluorophores used. The in situ PLA signals have a very characteristic bright sub-micrometer sized punctate appearance that is easily recognized and distinguishable from potential background fluorescence (see Figs. 1a–c and 2). While moving the focus through the sample tissue, appearance and disappearance of PLA signals should be noticeable (Fig. 1d). Up to a certain density of PLA signals they appear as discrete dots (puncta, blobs) that can be easily counted/quantified using image analysis software. Obtain digital images.

22. Analyze the captured images using image techniques to quantify the number of dots (Fig. 1d). Many commercial image

analysis software packages can be used (e.g., Duolink ImageTool (Sigma-Aldrich)) but also free software packages are available (e.g., Blob Finder, Cell profiles). Usually four important parameters should be kept in mind for a proper analysis and result interpretation: (1) the number of DAPI nuclei in the sample field, (2) the number of positive PLA/dots per sample field (Fig. 1d), (3) the total number of positive PLA cells/ nuclei per sample field, and (4) the diameter sizes of the individual PLA blobs (the diameter may indicate if aggregates (higher order) of receptor complexes exist). Within these four values it will be possible to get an overall view of the expression/enrichment of GPCR heteroreceptor complexes in the different brain areas analyzed and extract relevant conclusions from the comparisons between brain areas (Figs. 1 and 2).

23. Compare the measurements and make a graph. We have proposed the molecular phenomenon of receptor–receptor interactions as a fruitful way to understand how brain function can increase through molecular integration of signals. An alteration in specific receptor–receptor interactions or their balance/equilibrium (with the corresponding monomers-homomers) is indeed considered to have a role in the pathogenic mechanisms that lead to various brain diseases. Therefore, targeting protomer–protomer interactions in heteroreceptor complexes or changing the balance with their corresponding homoreceptor complexes in discrete brain regions may become an important field for developing novel drugs, including hetero-bivalent drugs and optimal types of combined treatments. The analysis of animal or human brain material with in situ PLA can reveal if the relative abundance of specific homo-and heteroreceptor complexes in discrete brain regions is altered in brain diseases or under certain drug treatments, for instance, chronic L-dopa treatment in Parkinson's disease [57]. In this analysis, it is important also to determine the ratio between heteroreceptor complex populations versus total number of the two participating protomer populations, using in addition to Western blots, receptor autoradiography and biochemical binding methods. The two latter methods show the densities and affinities of the two functional receptor populations. The relationship between these parameters will help to normalize the heteroreceptor complexes values for comparison between groups in addition to evaluating the potential changes in the total number of the two protomer populations [43]. Certainly, we cannot compare or determine directly a balance between the homo- and heteroreceptor complexes populations in the same tissue using the in situ PLA approach, because of a technical limitation of the procedure itself. But the method could help us determine each population independently and compare their relative expression levels after an appropriate numerical analysis. Furthermore, of increasing

importance will be to determine the agonist/antagonist regulation of these homo-/heteroreceptor complexes in order to understand their potential roles as drug targets or as markers of brain disease progression.

5   Advantages and Disadvantages of the PLA Method

In situ PLA can offer advantages by:

1. Giving the opportunity to study the existence of any potential homo- and heteroreceptor complexes, for which a pair of suitable antibodies is available without the need of employing genetic constructs.

2. The method could be performed in both cells and tissue samples, including human specimens collected from biobanks.

3. The in situ PLA could be useful to monitor the effects of different compounds like agonists and antagonists or their combined treatment on the balance of homo- and heteroreceptor complexes in cells and tissue.

4. The information obtained by the in situ PLA is at the resolution of individual cells or even of subcellular compartments providing profound insights into cellular heterogeneity in tissues.

5. The method also provides an enhanced sensitivity and selectivity compared to many other methods since powerful rolling circle amplification and dual target recognition are used.

As with any method there are limitations, for instance:

1. The in situ PLA cannot be used in live cells, as fixation it is a pre-requisite for the cell/tissue material employed.

2. When studying receptor–receptor interactions, it is important to remember that the method, like many other methods for studying protein–protein interactions, can show that two proteins are in close proximity and therefore, likely directly interact. Proteins can also interact indirectly through an adapter protein. The maximal distance between two epitopes to give rise to a signal with in situ PLA is 10–30 nm with direct-conjugated proximity probes, and slightly longer when secondary proximity probes are used. This distance will be dependent on the size of the receptors/antibodies and the respective length of the attached oligonucleotides. By changing the length of the oligonucleotides, the maximal distance limits can be reduced or increased. However, in general, the functional distance is usually close to the one detected in a FRET assay [58].

3. Another critical parameter for achieving good results is the use of excellent antibodies. Importantly, the antibodies have to be used under optimal conditions taking into consideration parameters

such as antibody concentration, epitopes targeted by the antibodies, fixation, antigen retrieval, blocking conditions, etc.

4. A range of controls, both positive and negative ones should be used to guarantee the specificity of the PLA signal. Positive controls can include cells where the protein is known to be expressed, such as in certain cells or tissues or in cells transfected to express the protein. Negative controls include cells or tissues that do not express the protein or where the protein has been knocked out or downregulated by, e.g., siRNA.

Acknowledgements

This work has been supported by the Karolinska Institutets Forskningsstiftelser 2014/2015 to D.O.B-E., by the Swedish Medical Research Council (62X-00715-50-3) and AFA Försäkring (130328) to K.F. and D.O.B-E. D.O.B-E. belongs to Academia de Biólogos Cubanos.

References

1. Borroto-Escuela DO, Wydra K, Pintsuk J, Narvaez M, Corrales F, Zaniewska M, Agnati LF, Franco R, Tanganelli S, Ferraro L, Filip M, Fuxe K (2016) Understanding the functional plasticity in neural networks of the basal ganglia in cocaine use disorder: a role for allosteric receptor-receptor interactions in A2A-D2 heteroreceptor complexes. Neural Plast 2016:4827268. https://doi.org/10.1155/2016/4827268

2. Borroto-Escuela DO, Wydra K, Ferraro L, Rivera A, Filip M, Fuxe K (2015) Role of D2-like heteroreceptor compelxes in the effects of cocaine, morphine and hallucinogens. In: Preedy V (ed) Neurophatology of drug addictions and substance misuse, vol 1. Elsevier, London, pp 93–101. https://doi.org/10.15379/2409-3564.2015.02.01.5

3. Bjork K, Svenningsson P (2011) Modulation of monoamine receptors by adaptor proteins and lipid rafts: role in some effects of centrally acting drugs and therapeutic agents. Annu Rev Pharmacol Toxicol 51:211–242. https://doi.org/10.1146/annurev-pharmtox-010510-100520

4. Bockaert J, Perroy J, Becamel C, Marin P, Fagni L (2010) GPCR interacting proteins (GIPs) in the nervous system: roles in physiology and pathologies. Annu Rev Pharmacol Toxicol 50:89–109. https://doi.org/10.1146/annurev.pharmtox.010909.105705

5. Aarsland D, Pahlhagen S, Ballard CG, Ehrt U, Svenningsson P (2011) Depression in Parkinson disease—epidemiology, mechanisms and management. Nat Rev Neurol 8(1):35–47. https://doi.org/10.1038/nrneurol.2011.189

6. Artigas F (2015) Developments in the field of antidepressants, where do we go now? Eur Neuropsychopharmacol 25(5):657–670. https://doi.org/10.1016/j.euroneuro.2013.04.013

7. Blier P (2013) Neurotransmitter targeting in the treatment of depression. J Clin Psychiatry 74(Suppl 2):19–24. https://doi.org/10.4088/JCP.12084su1c.04

8. Borroto-Escuela DO, Fuxe K (2017) Diversity and bias through dopamine D2R heteroreceptor complexes. Curr Opin Pharmacol 32:16–22. https://doi.org/10.1016/j.coph.2016.10.004

9. Fuxe K, Borroto-Escuela DO (2016) Heteroreceptor complexes and their allosteric receptor-receptor interactions as a novel biological principle for integration of communication in the CNS: targets for drug development. Neuropsychopharmacology 41(1):380–382. https://doi.org/10.1038/npp.2015.244

10. Borroto-Escuela DO, Brito I, Di Palma M, Jiménez-Beristain A, Narvaez M, Corrales F, Pita-Rodríguez M, Sartini S, Ambrogini P, Lattanzi D, Cuppini R, Agnati LF, Fuxe K

(2015) On the role of the balance of GPCR homo/heteroreceptor complexes in the brain. J Adv Neurosci Res 2:36–44

11. Borroto-Escuela DO, Narvaez M, Perez-Alea M, Tarakanov AO, Jimenez-Beristain A, Mudo G, Agnati LF, Ciruela F, Belluardo N, Fuxe K (2015) Evidence for the existence of FGFR1-5-HT1A heteroreceptor complexes in the midbrain raphe 5-HT system. Biochem Biophys Res Commun 456(1):489–493. https://doi.org/10.1016/j.bbrc.2014.11.112

12. Fuxe K, Guidolin D, Agnati LF, Borroto-Escuela DO (2015) Dopamine heteroreceptor complexes as therapeutic targets in Parkinson's disease. Expert Opin Ther Targets 19(3):377–398. https://doi.org/10.1517/14728222.2014.981529

13. Fuxe K, Agnati LF, Borroto-Escuela DO (2014) The impact of receptor-receptor interactions in heteroreceptor complexes on brain plasticity. Expert Rev Neurother 14(7):719–721. https://doi.org/10.1586/14737175.2014.922878

14. Fuxe K, Borroto-Escuela D, Fisone G, Agnati LF, Tanganelli S (2014) Understanding the role of heteroreceptor complexes in the central nervous system. Curr Protein Pept Sci 15(7):647

15. Fuxe K, Borroto-Escuela DO, Ciruela F, Guidolin D, Agnati LF (2014) Receptor-receptor interactions in heteroreceptor complexes: a new principle in biology. Focus on their role in learning and memory. Neurosci Discov 2(1):6. https://doi.org/10.7243/2052-6946-2-6

16. Fuxe K, Tarakanov A, Romero Fernandez W, Ferraro L, Tanganelli S, Filip M, Agnati LF, Garriga P, Diaz-Cabiale Z, Borroto-Escuela DO (2014) Diversity and bias through receptor-receptor interactions in gpcr heteroreceptor complexes. Focus on examples from dopamine D2 receptor heteromerization. Front Endocrinol 5:71. https://doi.org/10.3389/fendo.2014.00071

17. Borroto-Escuela DO, Narvaez M, Wydra K, Pintsuk J, Pinton L, Jimenez-Beristain A, Di Palma M, Jastrzebska J, Filip M, Fuxe K (2017) Cocaine self-administration specifically increases A2AR-D2R and D2R-sigma1R heteroreceptor complexes in the rat nucleus accumbens shell. Relevance for cocaine use disorder. Pharmacol Biochem Behav 155:24–31. https://doi.org/10.1016/j.pbb.2017.03.003

18. Pintsuk J, Borroto-Escuela DO, Lai TK, Liu F, Fuxe K (2016) Alterations in ventral and dorsal striatal allosteric A2AR-D2R receptor-receptor interactions after amphetamine challenge: relevance for schizophrenia. Life Sci 167:92–97

19. Pintsuk J, Borroto-Escuela DO, Pomierny B, Wydra K, Zaniewska M, Filip M, Fuxe K (2016) Cocaine self-administration differentially affects allosteric A2A-D2 receptor-receptor interactions in the striatum. Relevance for cocaine use disorder. Pharmacol Biochem Behav 144:85–91. https://doi.org/10.1016/j.pbb.2016.03.004

20. Narvaez M, Millon C, Borroto-Escuela D, Flores-Burgess A, Santin L, Parrado C, Gago B, Puigcerver A, Fuxe K, Narvaez JA, Diaz-Cabiale Z (2014) Galanin receptor 2-neuropeptide Y Y1 receptor interactions in the amygdala lead to increased anxiolytic actions. Brain Struct Funct 220:2289–2301. https://doi.org/10.1007/s00429-014-0788-7

21. Borroto-Escuela DO, Romero-Fernandez W, Narvaez M, Oflijan J, Agnati LF, Fuxe K (2014) Hallucinogenic 5-HT2AR agonists LSD and DOI enhance dopamine D2R protomer recognition and signaling of D2-5-HT2A heteroreceptor complexes. Biochem Biophys Res Commun 443(1):278–284. https://doi.org/10.1016/j.bbrc.2013.11.104

22. Borroto-Escuela DO, Narvaez M, Di Palma M, Calvo F, Rodriguez D, Millon C, Carlsson J, Agnati LF, Garriga P, Diaz-Cabiale Z, Fuxe K (2014) Preferential activation by galanin 1-15 fragment of the GalR1 protomer of a GalR1-GalR2 heteroreceptor complex. Biochem Biophys Res Commun 452(3):347–353. https://doi.org/10.1016/j.bbrc.2014.08.061

23. Millon C, Flores-Burgess A, Narvaez M, Borroto-Escuela DO, Santin L, Parrado C, Narvaez JA, Fuxe K, Diaz-Cabiale Z (2014) A role for galanin N-terminal fragment (1-15) in anxiety- and depression-related behaviours in rats. Int J Neuropsychopharmacol 18:pii: pyu064. https://doi.org/10.1093/ijnp/pyu064

24. Borroto-Escuela DO, Romero-Fernandez W, Mudo G, Perez-Alea M, Ciruela F, Tarakanov AO, Narvaez M, Di Liberto V, Agnati LF, Belluardo N, Fuxe K (2012) Fibroblast growth factor receptor 1- 5-hydroxytryptamine 1A heteroreceptor complexes and their enhancement of hippocampal plasticity. Biol Psychiatry 71(1):84–91. https://doi.org/10.1016/j.biopsych.2011.09.012

25. Flajolet M, Wang Z, Futter M, Shen W, Nuangchamnong N, Bendor J, Wallach I, Nairn AC, Surmeier DJ, Greengard P (2008) FGF acts as a co-transmitter through adenosine A(2A) receptor to regulate synaptic plasticity. Nat Neurosci 11(12):1402–1409. https://doi.org/10.1038/nn.2216

26. Nai Q, Li S, Wang SH, Liu J, Lee FJ, Frankland PW, Liu F (2009) Uncoupling the D1-N-methyl-D-aspartate (NMDA) receptor complex promotes NMDA-dependent long-term potentiation and working memory. Biol Psychiatry 67(3):246–254. https://doi.org/10.1016/j.biopsych.2009.08.011

27. Borroto-Escuela DO, Romero-Fernandez W, Tarakanov AO, Ciruela F, Agnati LF, Fuxe K (2011) On the existence of a possible A2A-D2-beta-Arrestin2 complex: A2A agonist modulation of D2 agonist-induced beta-arrestin2 recruitment. J Mol Biol 406(5):687–699. https://doi.org/10.1016/j.jmb.2011.01.022

28. Laroche G, Lepine MC, Theriault C, Giguere P, Giguere V, Gallant MA, de Brum-Fernandes A, Parent JL (2005) Oligomerization of the alpha and beta isoforms of the thromboxane A2 receptor: relevance to receptor signaling and endocytosis. Cell Signal 17(11):1373–1383. https://doi.org/10.1016/j.cellsig.2005.02.008

29. Borroto-Escuela DO, Garcia-Negredo G, Garriga P, Fuxe K, Ciruela F (2010) The M(5) muscarinic acetylcholine receptor third intracellular loop regulates receptor function and oligomerization. Biochim Biophys Acta 1803(7):813–825. https://doi.org/10.1016/j.bbamcr.2010.04.002

30. Borroto-Escuela DO, Van Craenenbroeck K, Romero-Fernandez W, Guidolin D, Woods AS, Rivera A, Haegeman G, Agnati LF, Tarakanov AO, Fuxe K (2010) Dopamine D2 and D4 receptor heteromerization and its allosteric receptor-receptor interactions. Biochem Biophys Res Commun 404(4):928–934. https://doi.org/10.1016/j.bbrc.2010.12.083

31. Comps-Agrar L, Maurel D, Rondard P, Pin JP, Trinquet E, Prezeau L (2011) Cell-surface protein-protein interaction analysis with time-resolved FRET and snap-tag technologies: application to G protein-coupled receptor oligomerization. Methods Mol Biol 756:201–214. https://doi.org/10.1007/978-1-61779-160-4_10

32. Cottet M, Faklaris O, Falco A, Trinquet E, Pin JP, Mouillac B, Durroux T (2013) Fluorescent ligands to investigate GPCR binding properties and oligomerization. Biochem Soc Trans 41(1):148–153. https://doi.org/10.1042/BST20120237

33. Cottet M, Faklaris O, Maurel D, Scholler P, Doumazane E, Trinquet E, Pin JP, Durroux T (2012) BRET and time-resolved FRET strategy to study GPCR oligomerization: from cell lines toward native tissues. Front Endocrinol 3:92. https://doi.org/10.3389/fendo.2012.00092

34. Schellekens H, De Francesco PN, Kandil D, Theeuwes WF, McCarthy T, van Oeffelen WE, Perello M, Giblin L, Dinan TG, Cryan JF (2015) Ghrelin's orexigenic effect is modulated via a serotonin 2C receptor interaction. ACS Chem Neurosci 6(7):1186–1197. https://doi.org/10.1021/cn500318q

35. Borroto-Escuela DO, Flajolet M, Agnati LF, Greengard P, Fuxe K (2013) Bioluminescence resonance energy transfer methods to study G protein-coupled receptor-receptor tyrosine kinase heteroreceptor complexes. Methods Cell Biol 117:141–164. https://doi.org/10.1016/B978-0-12-408143-7.00008-6

36. Borroto-Escuela DO, Romero-Fernandez W, Tarakanov AO, Marcellino D, Ciruela F, Agnati LF, Fuxe K (2010) Dopamine D2 and 5-hydroxytryptamine 5-HT((2)A) receptors assemble into functionally interacting heteromers. Biochem Biophys Res Commun 401(4):605–610. https://doi.org/10.1016/j.bbrc.2010.09.110

37. Bouvier M, Heveker N, Jockers R, Marullo S, Milligan G (2007) BRET analysis of GPCR oligomerization: newer does not mean better. Nat Methods 4(1):3–4.; author reply 4. https://doi.org/10.1038/nmeth0107-3

38. James JR, Oliveira MI, Carmo AM, Iaboni A, Davis SJ (2006) A rigorous experimental framework for detecting protein oligomerization using bioluminescence resonance energy transfer. Nat Methods 3(12):1001–1006. https://doi.org/10.1038/nmeth978

39. Marullo S, Bouvier M (2007) Resonance energy transfer approaches in molecular pharmacology and beyond. Trends Pharmacol Sci 28(8):362–365. https://doi.org/10.1016/j.tips.2007.06.007

40. Audet M, Lagace M, Silversides DW, Bouvier M (2010) Protein-protein interactions monitored in cells from transgenic mice using bioluminescence resonance energy transfer. FASEB J 24(8):2829–2838. https://doi.org/10.1096/fj.09-144816

41. Fernandez-Duenas V, Gomez-Soler M, Jacobson KA, Santhosh Kumar T, Fuxe K, Borroto-Escuela DO, Ciruela F (2012) Molecular determinants of a(2a) r-d(2) r allosterism: role of the intracellular loop 3 of the d(2) r. J Neurochem 123:373–384. https://doi.org/10.1111/j.1471-4159.2012.07956.x

42. Herrick-Davis K, Grinde E, Cowan A, Mazurkiewicz JE (2013) Fluorescence correlation spectroscopy analysis of serotonin, adrenergic, muscarinic, and dopamine receptor dimerization: the oligomer number puzzle. Mol Pharmacol 84(4):630–642. https://doi.org/10.1124/mol.113.087072

43. Borroto-Escuela DO, Romero-Fernandez W, Garriga P, Ciruela F, Narvaez M, Tarakanov AO, Palkovits M, Agnati LF, Fuxe K (2013) G protein-coupled receptor heterodimerization in the brain. Methods Enzymol 521:281–294. https://doi.org/10.1016/B978-0-12-391862-8.00015-6

44. Trifilieff P, Rives ML, Urizar E, Piskorowski RA, Vishwasrao HD, Castrillon J, Schmauss C, Slattman M, Gullberg M, Javitch JA (2011) Detection of antigen interactions ex vivo by proximity ligation assay: endogenous dopamine D2-adenosine A2A receptor complexes in the striatum. BioTechniques 51(2):111–118. https://doi.org/10.2144/000113719

45. Romero-Fernandez W, Borroto-Escuela DO, Agnati LF, Fuxe K (2013) Evidence for the existence of dopamine D2-oxytocin receptor heteromers in the ventral and dorsal striatum with facilitatory receptor-receptor interactions. Mol Psychiatry 18(8):849–850. https://doi.org/10.1038/mp.2012.103

46. Borroto-Escuela DO, Li X, Tarakanov AO, Savelli D, Narvaez M, Shumilov K, Andrade-Talavera Y, Jimenez-Beristain A, Pomierny B, Diaz-Cabiale Z, Cuppini R, Ambrogini P, Lindskog M, Fuxe K (2017) Existence of brain 5-HT1A-5-HT2A isoreceptor complexes with antagonistic allosteric receptor-receptor interactions regulating 5-HT1A receptor recognition. ACS Omega 2(8):4779–4789. https://doi.org/10.1021/acsomega.7b00629

47. Borroto-Escuela DO, DuPont CM, Li X, Savelli D, Lattanzi D, Srivastava I, Narvaez M, Di Palma M, Barbieri E, Andrade-Talavera Y, Cuppini R, Odagaki Y, Palkovits M, Ambrogini P, Lindskog M, Fuxe K (2017) Disturbances in the FGFR1-5-HT1A heteroreceptor complexes in the Raphe-hippocampal 5-HT system develop in a genetic rat model of depression. Front Cell Neurosci 11:309. https://doi.org/10.3389/fncel.2017.00309

48. Fredriksson S, Gullberg M, Jarvius J, Olsson C, Pietras K, Gustafsdottir SM, Ostman A, Landegren U (2002) Protein detection using proximity-dependent DNA ligation assays. Nat Biotechnol 20(5):473–477. https://doi.org/10.1038/nbt0502-473

49. Gullberg M, Fredriksson S, Taussig M, Jarvius J, Gustafsdottir S, Landegren U (2003) A sense of closeness: protein detection by proximity ligation. Curr Opin Biotechnol 14(1):82–86

50. Gullberg M, Gustafsdottir SM, Schallmeiner E, Jarvius J, Bjarnegard M, Betsholtz C, Landegren U, Fredriksson S (2004) Cytokine detection by antibody-based proximity ligation. Proc Natl Acad Sci U S A 101(22):8420–8424. https://doi.org/10.1073/pnas.0400552101

51. Soderberg O, Gullberg M, Jarvius M, Ridderstrale K, Leuchowius KJ, Jarvius J, Wester K, Hydbring P, Bahram F, Larsson LG, Landegren U (2006) Direct observation of individual endogenous protein complexes in situ by proximity ligation. Nat Methods 3(12):995–1000. https://doi.org/10.1038/nmeth947

52. Darmanis S, Kahler A, Spangberg L, Kamali-Moghaddam M, Landegren U, Schallmeiner E (2007) Self-assembly of proximity probes for flexible and modular proximity ligation assays. BioTechniques 43(4):443–444. 446, 448 passim

53. Soderberg O, Leuchowius KJ, Kamali-Moghaddam M, Jarvius M, Gustafsdottir S, Schallmeiner E, Gullberg M, Jarvius J, Landegren U (2007) Proximity ligation: a specific and versatile tool for the proteomic era. Genet Eng (N Y) 28:85–93

54. Soderberg O, Leuchowius KJ, Gullberg M, Jarvius M, Weibrecht I, Larsson LG, Landegren U (2008) Characterizing proteins and their interactions in cells and tissues using the in situ proximity ligation assay. Methods 45(3):227–232. https://doi.org/10.1016/j.ymeth.2008.06.014

55. Borroto-Escuela DO, Hagman B, Woolfenden M, Pinton L, Jiménez-Beristain A, Oflijan J, Narvaez M, Di Palma M, Feltmann K, Sartini S, Ambrogini P, Ciruela F, Cuppini R, Fuxe K (2016) In situ proximity ligation assay to study and understand the distribution and balance of GPCR homo- and heteroreceptor complexes in the brain. In: Lujan R, Ciruela F (eds) receptor and ion channel detection in the brain. vol 10. Neuromethods. Springer, Berlin, pp 109–126. https://doi.org/10.1515/revneuro-2015-0024

56. Uhlen M, Bandrowski A, Carr S, Edwards A, Ellenberg J, Lundberg E, Rimm DL, Rodriguez H, Hiltke T, Snyder M, Yamamoto T (2016) A proposal for validation of antibodies. Nat Methods 13(10):823–827. https://doi.org/10.1038/nmeth.3995

57. Antonelli T, Fuxe K, Agnati L, Mazzoni E, Tanganelli S, Tomasini MC, Ferraro L (2006) Experimental studies and theoretical aspects on A2A/D2 receptor interactions in a model of Parkinson's disease. Relevance for L-dopa induced dyskinesias. J Neurol Sci 248(1-2):16–22. https://doi.org/10.1016/j.jns.2006.05.019

58. Weibrecht I, Leuchowius KJ, Clausson CM, Conze T, Jarvius M, Howell WM, Kamali-Moghaddam M, Soderberg O (2010) Proximity ligation assays: a recent addition to the proteomics toolbox. Expert Rev Proteomics 7(3):401–409. https://doi.org/10.1586/epr.10.10

Chapter 20

Unraveling the Functions of Endogenous Receptor Oligomers in the Brain Using Interfering Peptide: The Example of D1R/NMDAR Heteromers

Andry Andrianarivelo, Estefani Saint-Jour, Pierre Trifilieff, and Peter Vanhoutte

Abstract

Decoding signaling pathways in different brain structures is crucial to develop pharmacological strategies for neurological diseases. In this perspective, the targeting of receptors by selective ligands is one of the classical therapeutic strategies. Nonetheless, this approach often results in a decrease of efficiency over time and deleterious side effects because physiological functions can be affected. An emerging concept has been to target mechanisms that fine-tune receptor signaling, such as heteromerization, the process by which physical receptor–receptor interaction at the membrane allows the reciprocal modulation of receptors' signaling. Because of the central role of the synergistic transmission mediated by dopamine (DA) and glutamate (Glu) in brain physiology and pathophysiology, heteromerization between DA and Glu receptors has received a lot of attention. However, the study of endogenous heteromers has been challenging because of the lack of appropriate tools. Over the last years, progress has been made in the development of techniques to study their expression in the brain, regulation and function. In this chapter, we provide a methodological framework for the design and use of interfering peptides to study endogenous receptor oligomers through the example of the dopamine type 1 receptor (D1R) and the GluN1 subunit of NMDA receptor heteromers.

Key words Oligomers, Protein–protein interaction, Interfering peptide, Heteromerization, Dopamine receptor, NMDA receptor

1 Introduction

In mammals, many brain structures involved in fundamental physiological functions, ranging from motor coordination, cognition, learning and memory to goal-directed behavior, are the targets of imbricated and convergent dopamine (DA) and glutamate projections. As a witness of the crucial role played by dopamine-mediated modulation of excitatory glutamate transmission in these brain regions, an imbalance of dopamine and glutamate transmissions has been incriminated in a vast array of neurological or psychiatric disorders, such as Parkinson's and Huntington's diseases, schizo-

Kjell Fuxe and Dasiel O. Borroto-Escuela (eds.), *Receptor-Receptor Interactions in the Central Nervous System*, Neuromethods, vol. 140, https://doi.org/10.1007/978-1-4939-8576-0_20, © Springer Science+Business Media, LLC, part of Springer Nature 2018

phrenia, depression, obsessive compulsive disorders, or addiction [1–6]. Based on these observations, current treatments for the aforementioned pathological conditions primarily aim at targeting cognate neurotransmitter receptors but the loss of therapeutic efficacy over time as well as the appearance of severe deleterious side effects calls for alternative strategies. Intracellular signaling pathways recruited downstream from individual DA and Glu receptors stimulation are imperative for physiological cell responses and normal brain functions. Therefore, the targeting of molecular mechanisms specifically involved in dopamine–glutamate interactions, rather than the targeting of the receptors themselves, appears as an alternative strategy. This approach holds great potential to pave the way for new therapeutic avenues [7–9].

DA and glutamate receptors of the NMDA (N-methyl-D-aspartic acid) subtype mutually modulate their functions. This synergy takes place through mechanisms involving the recruitment of specific signaling cascades but their direct physical interactions (i.e., heteromerization or oligomerization) progressively appear as a complementary mechanism by which these receptors can reciprocally finely tune their functions and trafficking [10–18].

Receptor oligomers in general possess pharmacological and functional properties that are distinct from individual component receptors and allow a spatiotemporal modulation of the receptors' functions. For these reasons, heteromers have been the subject of intense research and, in particular, a number of studies showed the critical role of the physical interaction between DA and NMDA receptor in vitro for the modulation of their subcellular distribution and functions [7, 11, 19–21].

Until the past few years, the existence and roles of endogenous heteromers in the brain were still under debate, mostly due to a lack of mechanistic understanding of the heteromerization process and appropriate tools. The emergence of techniques such as the proximity ligation assay (see [22, 23]) now allows the detection of endogenous receptor heteromers in their native environment from fixed tissues and confirmed their presence in the brain [7, 24–26]. To study the functional roles of receptor heteromers, one option that emerged was to develop bivalent compounds that are capable of preferentially targeting a receptor when it forms an heteromer, rather than the receptor on its own, as validated for the heteromers formed by DA type 2 (D2R) and adenosine A2 (A2R) receptors [27]. As an alternative, the precise characterization of protein–protein interaction domains required for the formation of heteromers offers the possibility of using peptidic sequences designed to interfere with endogenous receptor heteromer formation though a competition mechanism. This is particularly applicable to DA and NMDA receptors, thanks to the seminal work from Dr. Fang Liu and others, which led to the characterization of short sequences in DAR and NMDAR subunits that are essential for the interaction of

D1R with either GluN1 or GluN2A or D2R with GluN2B subunits [10, 11, 14].

Based on the knowledge of the interaction domains involved in the binding of D1R to the obligatory subunit of NMDAR GluN1 [10, 13], we developed an interfering peptide-based strategy to alter D1R/GluN1 heteromerization, while preserving the functions of individual component receptors. Using this strategy, we demonstrated that D1R/GluN1 heteromers appear as a molecular bridge linking DA to the facilitation of NMDAR-mediated currents and NMDAR-dependent targeting of the Extracellular signal-regulated kinase (ERK) downstream from D1R and NMDAR. We also demonstrate that the recruitment of D1R to NMDAR was mandatory for long-term potentiation in the striatum as well as for some adaptations induced by exposure to cocaine [18].

Through the example of D1R/GluN1 heteromers, we propose in this chapter to provide a comprehensive and generally applicable methodological framework for the use of interfering peptides, and appropriate controls, to study the specific roles of endogenous receptor oligomers in the brain.

2   Methods

2.1   Design of Cell-Penetrating Interfering Peptide Sequence to Disrupt D1R/GluN1 Heteromers, While Preserving the Functions of Individual D1R and NMDAR

As a general statement, when the interaction domains between two proteins have been characterized in both partners, two peptides can potentially disrupt their interaction. A peptide corresponding to the region of the protein A that binds a protein B or the region of protein B interacting with protein A (Fig. 1a). A priori, these two strategies should yield a disruption of the binding of both partners; however, the functional consequences on downstream signaling may be different, and even opposite.

The interaction between D1R and the GluN1 subunit of NMDAR involves identified domains located within the intracellular c-terminus ends of both receptors: the t2 domain (L^{387}-L^{416}) of D1R and the C1 cassette (D^{864}-T^{900}) of GluN1 [10, 11] (Fig. 1b). Synthetic peptides corresponding to the t2 domain of D1R [15, 17] (Fig. 1c) or the C1 cassette of GluN1 [18] (Fig. 1d) have thus been used to study D1R/GluN1 functions in the brain.

These two strategies led to an efficient disruption of D1R/GluN1 complexes [10, 11, 18]. When using the peptide corresponding to the C1 cassette of GluN1, we observed a slight decrease of basal ERK activity and a complete blockade of ERK activation downstream from D1R and NMDAR (Fig. 2a). By contrast, we found that the interfering peptide corresponding to the t2 domain of D1R induced per se an increase of NMDAR-mediated activation of the ERK pathway and was not able to significantly diminish the increase of ERK activity induced by a co-stimulation of D1R and NMDAR of cultured neurons (Fig. 2b). These discrepancies

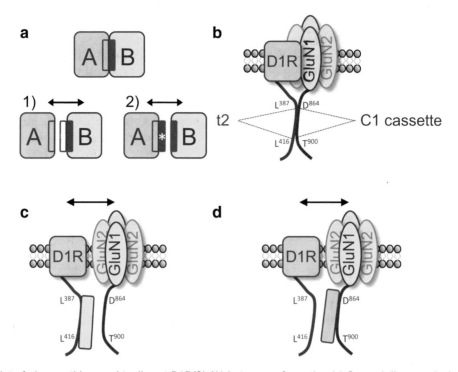

Fig. 1 Interfering peptides used to disrupt D1R/GluN1 heteromer formation (**a**) General diagram depicting the two strategies that can be used to disrupt the interaction between proteins A and B when the interaction domains (dark blue/red) are identified within both proteins. Panels (1) and (2) illustrate the mechanisms of action of both possible synthetic interfering peptides (*). Schematic representation of the mechanisms of action of the synthetic peptide corresponding to (**c**) the C1 cassette (GluN1C1*) that binds to D1R (**c**) [18] or (**d**) the t2 domain of D1R (t2*), which interacts with GluN1 (**c**) used in [15, 17]

between the two strategies could be attributed to the fact that the t2 peptide, and not the C1 mirror peptide, may compete with the binding of the protein Calcium Calmodulin, which constitutively inhibits NMDAR through interaction domains comprising a portion of the C1 cassette [28] (see **Note 1**). It is also possible that the binding of the t2 peptide at the receptor by itself mimics heteromerization with D1R, exerting an effect as an allosteric modulator of the NMDAR. This could explain why studies using these two strategies yielded opposing results regarding the role of D1R/GluN1 on (ERK-dependent) long-term neuronal plasticity [15, 17, 18]. Since we verified that the C1 cassette was able to disrupt heteromer formation without interfering with functions of individual D1R and NMDAR (see below), we pursued our work using the peptide GluN1C1 to selectively interfere with D1R/GluN1 formation and functions.

To achieve intracellular delivery of the GluN1C1 peptide we fused it to the trans-activating transcriptional activator sequence (TAT; sequence: GRKKRRQRRR) (see **Note 2**). A "spacer," which is composed of two proline residues, has been added

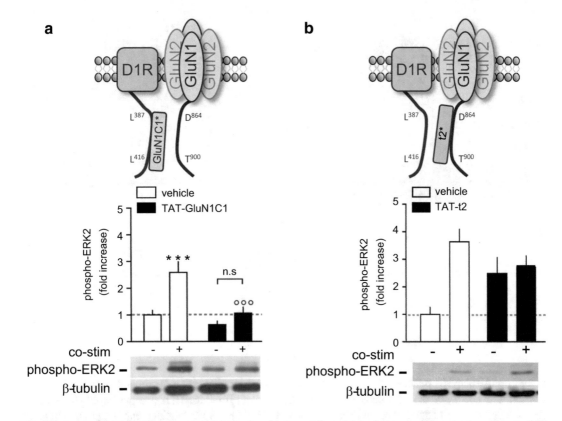

Fig. 2 Impact of D1R/GluN1 heteromer blockade with two different interfering peptides on downstream ERK signaling. Cultured striatal neurons were incubated or not with 5 μM of TAT-GluN1C1 (**a**) or TAT-t2 (**b**) prior to a co-stimulation (co-stim) of D1R and NMDAR with 3 μM of SKF38393 together with 0.3 μM of glutamate for 10 min (+) or a vehicle solution (−). The activity of ERK downstream from D1R and NMDAR was assessed by western blot using an antibody that specifically recognizes the phosphorylated (i.e., activated) form of ERK1/2. Quantifications show that TAT-GluN1C1 fully blocks ERK activation induced by the co-stimulation (**a**) whereas the TAT-t2 increases by itself basal ERK activity levels and does not inhibit ERK activation induced by the co-stimulation. Two-way ANOVA; Bonferroni post-hoc test; $^{**}p < 0.01$; $^{***}p < 0.0001$ (compared to control); $^{\circ\circ\circ}p < 0.0001$ (compared to co-stim); *n.s* not significant. Adapted from [18]

between the TAT and the GluN1C1 sequence to increase flexibility between these two parts of the peptide.

The design of a control peptide has been inspired by the work of Woods and coworkers who established that the C1 cassette of GluN1 was containing a stretch of 9 arginine-rich amino acids ($_{890}$SFKRRRSSK$_{898}$), which plays a critical role for the electrostatic interactions mediating the interactions between D1R with GluN1 [29]. A peptide corresponding to the C1 sequence deleted of these nine amino acids (TAT-GluN1C1Δ) has thus been used as a control and we were able to confirm its inability to alter D1R/GluN1 heteromer formation (Fig. 3a). However, alternative options should be considered to design control peptides (see **Note 3**).

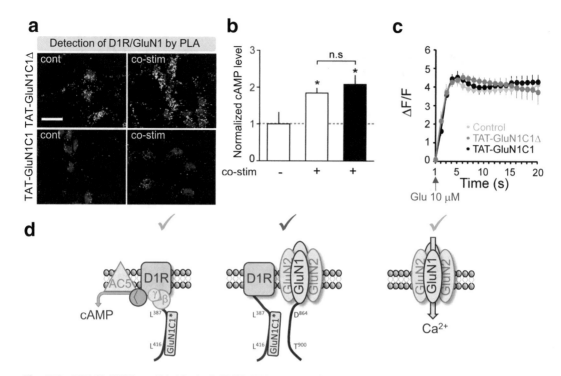

Fig. 3 The TAT-GluN1C1 peptide blocks D1R/GluN1 heteromer formation, while preserving individual D1R- and NMDAR-dependent signaling. Neurons were pre-incubated with TAT-GluN1C1Δ (top panels) or TAT-GluN1C1 (bottom panels) prior to and during an incubation with 3 μM of SKF38393 together with 0.3 μM of glutamate (co-stim) or a vehicle solution (cont) for 10 min. (**a**) Representative confocal images illustrating the punctate pattern of the PLA signal corresponding to D1R/GluN1 heteromers. Note that the co-stimulation favors D1R/GluN1 heteromer in the presence of the control peptide, whereas this phenomenon is fully blocked in neurons pre-incubated with the TAT-GluN1C1 peptide. (**b**) cAMP production (fold increase) in control condition and after 10 min of the co-stimulation paradigm. Note the lack of effect of TAT-GluN1C1 on D1R-mediated cAMP production. One-Way ANOVA, Newman Keuls post-hoc test; *$p < 0.05$ when compared to the control group. (**c**) Illustration of calcium profiles obtained from cultured neurons pre-treated with either a vehicle solution, the TAT-GluN1C1 or the TAT-GluN1C1Δ peptides and loaded with the calcium probe Fluo-4. Calcium influx is induced by a stimulation with glutamate 10 μM added at the indicate time. (**d**) Diagram summarizing the results shown in (**a–c**). Adapted from [18]

2.2 Guideline on Storage and Use of Interfering Peptides Through the Example of D1R/GluN1 Disruption Using TAT-GluN1C1

1. *Preparation and storage of peptides*:

   TAT-GluN1C1 and TAT-GluN1C1Δ, synthesized with a purity >98%, were diluted in sterile deionized water as stock solutions of 1 mM. Since peptides are unstable molecules by nature (see **Note 4**), it is critical to avoid freeze-thaw cycle(s). The stock solution should be aliquoted in a volume corresponding to the need for one experiment and stored at −80 °C. One aliquot should be defrosted slowly on ice and used for a single experiment.

2. *Incubation of cultured neurons with TAT-interfering peptides.*

The incubation of primary cultured neurons with the TAT-GluN1C1 to D1R/GluN1 heteromers requires the following steps (see **Note 5**):

(A) Primary cultures of striatal neurons were performed from striata dissected out from 14-day-old embryos from pregnant Swiss mice as described in [30]. Neurons were then plated at a density of 1000 cells/mm^2 and maintained for 6 days in vitro (DIV) in a Neurobasal A culture medium supplemented with B27, 500 nM L-glutamine, 60 µg/ml penicillin G, and 25 µM β-mercaptoethanol in a humidified atmosphere of 95% air and 5% CO_2.

(B) At DIV 6, the culture medium was replaced by fresh medium and put back to the incubator in order to avoid the presence of endogenous glutamate that may alter the response of the cell to low doses of exogenous glutamate (see steps 2–4).

(C) At DIV 7, an aliquot of TAT-GluN1C1 and TAT-GluN1C1Δ stock solutions were defrosted on ice and added directly to the culture medium containing the neurons to reach a final concentration of 5 µM (see **Note 6**). Neurons were then put back at 37 °C in the incubator for 1 h.

(D) Neurons incubated in the presence of interfering peptides were then stimulated with 3 µM of the D1R agonist SKF 38393 together with 0.3 µM of glutamate to stimulate concomitantly D1R and NMDAR for 10 min at 37 °C in the incubator.

(E) At the end of the treatments, when immunostaining or PLA experiments were performed (Fig. 3a), the culture medium was removed and neurons were fixed for 20 min at room temperature with a solution of Phosphate Saline Buffer (PBS) containing 4% paraformaldehyde (PFA) and 4% sucrose and washed three times with PBS (see **Note 7**). To perform immunoblotting (see Fig. 2), the culture medium was removed and the culture dishes containing the neurons were placed on dry ice. Neurons were then homogenized by sonication in a boiling solution of 1% sodium dodecylsulfate (SDS) and 1 mM of the phosphatase inhibitor sodium orthovanadate diluted in water, and then placed for 5 min at 100 °C before being stored at −80 °C.

3. *Infusion of TAT-interfering peptides into the brain.*

To evaluate the functional roles of D1R/GluN1 heteromers in the brain, we performed intracerebral injection of the TAT-GluN1C1 peptide although other routes of administrations can be considered (see **Note 8**).

To deliver the TAT-GluN1C1 in the ventral part of the striatum (i.e., the nucleus accumbens), bilateral injections were performed on mice anesthetized with isoflurane. Stainless-steel guide cannulae of 22 gauge were implanted in the nucleus accumbens by using the following stereotaxic coordinates: antero-posterior: +1.5; medio-lateral: ±1.6; dorso-ventral: 4.1 with an angle of 13° and then fixed to the skull with dental cement. After 1 week of recovery, 5 pmol of peptide diluted in water were infused 1 h prior to any further pharmacological treatment. At the end of the experiments, the mice were perfused by intracardiac perfusion of 4% PFA in 0.1 M Na_2HPO_4-NaH_2PO_4 buffer, pH 7.5 delivered with a peristaltic pump. Brains were then removed and post-fixed overnight in the aforementioned fixative solution. The brains were then sliced in 30 μm-thick sections kept in a solution of 30% ethylene glycol, 30% glycerol, 0.1 M phosphate buffer. These sections can be used to detect receptor heteromers (and their disruption by an interfering peptide) by using techniques such as PLA [18, 22, 23]. When biochemical analyses were performed, mouse heads were immediately frozen in liquid nitrogen and punches were prepared from both hemispheres of the brain and homogenized as described in Sect. 2.4 for cultured neurons.

3 Testing of the Efficacy and Specificity of an Interfering Peptide-Based Strategy

To validate the specificity of an interfering peptide, the first step is to show the ability of the peptide to alter the interaction between the receptors of interest and confirm the absence of effect of the control peptide. In the case of D1R/GluN1 heteromers, cultured striatal neurons, which co-express both D1R and NMDAR [18], were pre-incubated with TAT-GluN1C1 or TAT-GluN1C1Δ and co-stimulated or not with a D1R agonist together with glutamate as described in Sects. 2-3 and 2-4. After 10 min of co-stimulation, neurons were fixed as described in Sect. 2-5. Endogenous D1R/GluN1 heteromers were detected by performing a PLA according to the manufacturer's instructions. For details on the PLA procedures, see [18, 22, 31]. As illustrated in Fig. 3a, D1R/GluN1 proximity visualized by PLA appears as a punctate pattern and a concomitant stimulation of both D1R and NMDAR for 10 min triggers an increase of the PLA signal in the presence of the control TAT-GluN1C1Δ. By contrast, in the presence of the TAT-GluN1C1, basal levels of D1R/GluN1 are decreased and their increased formation induced by the co-stimulation of both receptors is fully blocked, thus illustrating the efficacy and selectivity of the TAT-GluN1C1 as an inhibitor of D1R/GluN1 heteromers.

On a functional standpoint, the second validation step is to verify that the TAT-GluN1C1 peptide does not interfere nonspecifically with the functions of D1R and NMDAR independently on heteromerization.

The D1R is a G protein-coupled receptor that is positively coupled to adenylyl cyclase and cAMP production through the G_{olf} protein [32]. As shown in Fig. 3b, an Elisa-based assay was used to measure cAMP production in cultured neurons that have been co-stimulated with a D1R agonist and glutamate to confirm that the presence of TAT-GluN1C1, which binds to the t2 domain of D1R, did not alter cognate pathway downstream from D1R (see **Note 9**). On the other side, the impact of TAT-GluNC1 has also been tested on NMDAR-mediated calcium influx triggered by a purely glutamatergic stimulation with 10 μM of glutamate. Live calcium imaging on cultured striatal neurons loaded with the calcium probe Fluo-4 demonstrated that the calcium dynamics and amplitude were similar when glutamate was applied on neuron pre-treated with TAT-GluN1C1, TAT-GluN1C1Δ, or a vehicle solution (Fig. 3c).

Altogether, this series of controls showed that the TAT-GluN1C1 peptide blocks the binding of D1R to GluN1 subunits of NMDAR, while preserving the normal functioning of individual D1R and NMDAR (Fig. 3d). This interfering peptide-based strategy allowed the demonstration that D1R/GluN1 heteromers can be considered as molecular bridges by which DA facilitates both NMDAR-dependent signaling in D1R-expressing neurons and neuronal plasticity in the striatum [18].

4 Notes

1. The tethering of receptors at synaptic sites involves protein–protein interactions with numerous partners through interaction sites that may be contiguous or overlapping with the docking domains responsible for heteromer formation. Prior to the design of an interfering peptide sequence, an exhaustive literature screening is necessary to identify the location of known interaction domains of all binding partners of a given receptor of interest. This preliminary step is important in order to increase the chances to disrupt the receptor heteromer without altering the interaction between the receptors of interest and other protein partners, as it may change the function and trafficking of this receptor independently of heteromerization.

2. Alternatives to the cell-penetrating TAT sequence to deliver interfering peptides into the cell:

   (a) The penetrating sequence can be composed of a stretch of positively charged amino acids, such as arginines, or other known plasma membrane-permeant peptide sequence such as penetratin [33].

(b) For electrophysiological studies, the interfering peptide does not necessarily need to be coupled to a penetrating sequence as it can be filled in the patch pipet. The peptide will diffuse in the cytoplasm of the recorded cell and reach its target receptor [18].

(c) Virus-mediated expression of interfering peptides can be used and offers several advantages when compared to the use of synthetic peptides. The use of strong promotors such as CMV, CAG, and PKG to drive the expression the interfering sequence allows a continuous production of the peptide that can overcome the degradation of the peptide by endonucleases and exonucleases. Virus-mediated expression should be privileged in studies where a long-lasting blockade of receptor heteromer needs to be achieved. Furthermore, the injection of viruses expressing a floxed construct encoding the interfering sequence into the brain of a mouse line that expresses the enzyme cre recombinase driven by a cell-type specific promotor enables the blockade of receptor heteromers in specific cell subpopulations of interest.

3. The most common way to test for the specificity of an interfering peptide is to use a control peptide composed of the same amino acids placed in a random order (i.e., "scramble peptide"). However, as shown for several heteromers including D1R/GluN1 [29] or D2R/A2R [34], clusters of charged amino acids are critical for the formation of heteromers through electrostatic interactions. Hence, it is important to be aware that the randomized control peptide should display a distribution of charged amino acids that has to be as much as possible different from the one of the active interfering peptide.

4. To overcome the low stability of peptides against proteolysis, several chemical modifications of amino acids can be used. For instance, retro-inverso amino acids in which the N- and C-terminus are reversed and the natural L-conformation of the peptide is changed to the D-enantiomer form. Other modification such as introduction of N-alkylated amino acids or cyclization can also be considered [35].

5. The experimental procedures described in this section relate to the disruption of D1R/GluN1 heteromers with TAT-GluN1C1 striatal cultured neurons. However, the global procedure may be generalizable to the use of any TAT-coupled peptide and any cell type [36].

6. A dose–response curve has to be performed since the concentration of interfering peptide yielding a maximum inhibitory effect on the disruption of a receptor heteromer of interest may depend on the accessibility of the targeted receptor, its localization within the cell, and its expression level.

7. The penetration of the peptide can be confirmed by coupling the peptide to a fluorophore [36] or a molecule of biotin [18] or any kind of tag that can be detected by immunofluorescence. However, such labeled peptides may not be suitable for functional studies since the presence of the fluorophore may alter the ability of the peptide to reach the target receptor.

8. TAT peptides have the characteristic to cross the plasma membrane but they also cross the blood–brain barrier. As such, they can be injected intraperitoneally in order to reach neurons of the central nervous system [37].

9. The absence of effect of TAT-GluN1C1 on D1R-mediated cAMP production is in agreement with the fact that this peptide binds to the t2 domain of D1R, which does not overlap with the third intracellular loop of D1R that is involved in the coupling of D1R to the G protein [38].

References

1. Bastide MF, Meissner WG, Picconi B et al (2015) Pathophysiology of L-dopa-induced motor and non-motor complications in Parkinson's disease. Prog Neurobiol 132:96–168

2. Roze E, Cahill E, Martin E et al (2011) Huntington's disease and striatal signaling. Front Neuroanat 5:55

3. Laurelle M (2014) Schizophrenia: from dopaminergic to glutamatergic intervention. Curr Opin Pharmacol 14:97–102

4. Pauls DL, Abramovitch A, Rauch SL et al (2014) Obsessive-compulsive disorder: an integrative genetic and neurological perspective. Nat Rev Neurosci 15:410–424

5. Cahill E, Salery M, Vanhoutte P et al (2014) Convergence of dopamine and glutamate signalling onto striatal ERK activation in response to drugs of abuse. Front Pharmacol 4:172

6. Pascoli V, Cahill E, Bellivier F et al (2014) Extracellular signal-regulated protein kinases 1 and 2 activation by addictive drugs: a signal toward pathological adaptation. Biol Psychiatry 76:917–926

7. Wang M, Wong AH, Liu F (2012) Interaction between NMDAR and dopamine receptors: a potential therapeutic target. Brain Res 1476:154–163

8. Fuxe K, Borroto-Escuela DO, Romero-Fernandez W et al (2014) Moonlighting proteins and protein-protein interactions as neurotherapeutic targets in the G-protein-coupled receptor field. Neuropsychopharmacology 39:131–155

9. Borroto-Escuela DO, Carlsson J, Ambrogini P et al (2017) Understanding the role of GPCR heteroreceptor complexes in modulating the rain networks in health and disease. Front Cell Neurosci 1:37

10. Lee FJS, Xue S, Pei L et al (2002) Dual regulation of NMDA receptor functions by direct protein-protein interactions with dopamine 1 receptor. Cell 111:219–230

11. Pei L, Lee FJS, Moszczynska A et al (2004) Regulation of dopamine D1 receptor function by physical interaction with NMDAR receptors. J Neurosci 24:1149–1158

12. Cepeda C, Levine MS (2006) Where do you think you are going? The NMDA-D1 receptor trap. Sci STKE 2006:pe20

13. Scott L, Zeleniin S, Malmersjo S et al (2006) Allosteric changes of the NMDA receptor trap diffusible dopamine 1 receptors in spines. Proc Natl Acad Sci U S A 103:762–767

14. Liu XY, Chu XP, Mao LM et al (2006) Modulation of D2R-NR2B interactions in responses to cocaine. Neuron 52:897–909

15. Nai Q, Li S, Wang SH et al (2010) Uncoupling D1-N-Methyl-D-Aspartate (NMDA) receptor complex promotes NMDA-dependent long-term potentiation and working memory. Biol Psychiatry 67:246–254

16. Pascoli V, Besnard A, Hervé D et al (2011) Cyclic adenosine monophosphate-independent tyrosine kinase phosphorylation of NR2B mediates cocaine-induced extracellular signal-regulated kinase activation. Biol Psychiatry 69:218–227

17. Ladepeche L, Dupuis JP, Bouchet D et al (2013) Single-molecule imaging of the functional crosstalk between surface NMDA and dopamine D1 receptors. Proc Natl Acad Sci U S A 110:18005–18010

18. Cahill E, Pascoli V, Trifieff P et al (2014) D1R/GluN1 complexes in the striatum integrate dopamine and glutamate signalling to control synaptic plasticity and cocaine-induced responses. Mol Psychiatry 19:1295–1304

19. Fiorentini C, Gardoni F, Spano PF et al (2003) Regulation of dopamine 1 receptor trafficking and desensitization by oligomerization with glutamate N-Methyl-D-aspartate receptors. J BiolChem 278:20196–20202

20. Fiorentini C, Busi C, Spano PF et al (2008) Role of receptor heterodimers in the development of L-dopa-induced dyskinesia in the 6-hydroxydopamine rat model of Parkinson's disease. Parkinsonism Relat Disord 14:S159–S164

21. Zhang J, Xu TX, Hallett PJ et al (2009) PSD-95 uncouples dopamine-glutamate interaction in the DA/PSD-95/NMDA receptor complex. J Neurosci 29:2948–2960

22. Trifilieff P, Rives ML, Urizar E et al (2011) Detection of antigen interactions ex vivo by proximity ligation assay: endogenous dopamine D2-adenosine A2 receptor complexes in the striatum. BioTechniques 50:111–118

23. Borroto-Escuela DO et al (2018) Receptor-receptor interactions in the central nervous system. Chapter 19 Neuromethods. Springer, New York

24. Biezonski DK, Trifilieff P, Meszaros J et al (2015) Evidence for limited D1 and D2 receptor coexpression and colocalization within the dorsal striatum of neonatal mouse. J Comp Neurol 523:1175–1189

25. Bontempi L, Savoia P, Bono F et al (2017) Dopamine D3 and acetylcholine nicotinic receptor heteromerization in midbrain dopamine neurons: relevance for neuroplasticity. Eur Neuropsychopharmacol 27:313–324

26. He Y, Li Y, Chen M et al (2016) Habit formation after random interval training is associated with increased adenosine A2A receptor and dopamine D2 receptor heteromers in the striatum. Front Mol Neurosci 9:151

27. Soriano A, Ventura R, Molero A et al (2009) Adenosine receptor-antagonist/dopamine D2-agonist bivalent ligands as pharmacological tools to detect A2A-D2 receptor heteromers. J Med Chem 52:5590–5602

28. Ehlers MD, Zhang S, Bernhardt JP et al (1998) Inactivation of NMDA receptors by direct interaction of calmodulin with NR1 subunit. Cell 84:745–755

29. Woods AS, Ciruela F, Fuxe K et al (2005) Role of electrostatic interaction in receptor–receptor heteromerization. J Mol Neurol 26:125–132

30. Garcia M, Charvin D, Caboche J (2004) Expanded Huntingtin activates the c-jun N terminal kinase/c-Jun pathway prior to aggregate formation in striatal neurons in culture. Neuroscience 127:859–870

31. Bellucci A, Fiorentinin C, Zalteri M et al (2014) The "in situ" proximity ligation assay to probe protein-protein interactions in intact tissues. In: Ivanov A (ed) Exocytosis and endocytosis. Methods in molecular biology (Methods and protocols), vol 1174. Humana, New York

32. Corvol JC, Studler JM, Schonn JS et al (2001) Galpha(olf) is necessary for coupling D1 and A2a receptors to adenylyl cyclase in the striatum. J Neurochem 76:1585–1588

33. Dupont E, Prochiantz A, Joliot A (2015) Penetratin story: an overview. Methods Mol Biol 1324:29–37

34. Ciruela F, Burgueño J, Casadó V et al (2004) Combining mass spectrometry and pull-down techniques for the study of receptor heteromerization. Direct epitope-epitope electrostatic interactions between adenosine A2A and dopamine D2 receptors. Anal Chem 76:5354–5363

35. Gentilucci L, De Marco R, Cerisole L (2010) Chemical modifications designed to improve peptide stability: incorporation of non-natural amino acids, pseudo-peptide bonds, and cyclization. Curr Pharm Des 16:3185–3203

36. Lavaur J, Bernard F, Trifilieff P et al (2007) A TAT-DEF-Elk-1 peptide regulates the cytonuclear trafficking of Elk-1 and controls cytoskeleton dynamics. J Neurosci 27:14448–14458

37. Besnard A, Bouveyron N, Kappès V et al (2011) Alteration of molecular and behavioural responses to cocaine by selective inhibition of Elk-1 phosphorylation. J Neurosci 31:14296–14307

38. König B, Grätzel M (1994) Site of dopamine D1 receptor binding to Gs protein mapped with synthetic peptides. Biochem Biophys Acta 1223:261–266

Chapter 21

Super-Resolution Imaging as a Method to Study GPCR Dimers and Higher-Order Oligomers

Kim C. Jonas and Aylin C. Hanyaloglu

Abstract

The study of G protein-coupled receptor (GPCR) dimers and higher-order oligomers has unveiled mechanisms for receptors to diversify signaling and potentially uncover novel therapeutic targets. The functional and clinical significance of these receptor–receptor associations has been facilitated by the development of techniques and protocols, enabling researchers to unpick their function from the molecular interfaces, to demonstrating functional significance in vivo, in both health and disease. Here we describe our methodology to study GPCR oligomerization at the single-molecule level via super-resolution imaging. Specifically, we have employed photoactivated localization microscopy, with photoactivatable dyes (PD-PALM) to visualize the spatial organization of these complexes to <10 nm resolution, and the quantitation of GPCR monomer, dimer, and oligomer in both homomeric and heteromeric forms. We provide guidelines on optimal sample preparation, imaging parameters, and necessary controls for resolving and quantifying single-molecule data. Finally, we discuss advantages and limitations of this imaging technique and its potential future applications to the study of GPCR function.

Key words GPCR, Dimer, Oligomer, Super-resolution, PALM, Single-molecule imaging

---

1 Introduction

The ability of G protein-coupled receptors (GPCRs) to associate as dimers and higher-order oligomers with themselves (homomers), or distinct GPCRs (heteromers), has provided a mechanism to understand how these receptors can diversify their activity in different tissues, display altered pharmacological properties of distinct ligands and the side-effects of drugs targeting these receptors. Mechanistically, GPCR homomerization has been implicated in receptor trafficking and signal diversification and/or amplification [1, 2]. GPCR heteromerization, however, can essentially result in a unique receptor functional unit compared to their respective homomeric counterparts, with reported distinct cell surface targeting, pharmacology, G protein complement, and ligand-induced trafficking [2–5]. The road to studying these complexes, however,

Kjell Fuxe and Dasiel O. Borroto-Escuela (eds.), *Receptor-Receptor Interactions in the Central Nervous System*, Neuromethods, vol. 140, https://doi.org/10.1007/978-1-4939-8576-0_21, © Springer Science+Business Media, LLC, part of Springer Nature 2018

has been very rocky, and remains an area of debate, historically due to the lack of approaches applied to directly demonstrate a required function of these homomers and heteromers in vivo [6–9]. Thus, more recent studies have developed methodological strategies to demonstrate that these complexes do indeed exist in vivo and are functionally relevant. Discussing the impact of GPCR homo- and heteromerization to cell signaling and physiological/pathological function is outside the scope of this chapter, and thus we refer the reader to the following recent reviews that specifically describe the role of GPCR homomer and heteromers in receptor function in vivo and in disease [2, 4, 5].

Despite this progress there remain numerous outstanding questions on the molecular organization, and debate on the functional relevance of dimer versus oligomeric GPCR complex [10–13]. Indeed, with such a large family of receptors it is not inconceivable that distinct GPCRs may differ in their ability (or necessity) to form such complexes. Even for a single GPCR it is likely there are roles for monomer, dimer, and oligomer, rather than a model where a receptor is capable of only forming a specific numerical complex, e.g., only dimers. The ability to provide molecular evidence to address these questions requires cross-disciplinary technology that can unpick the molecular interfaces, the full landscape of receptor complexes that are formed, and the specific cellular and physiological roles of GPCR monomers, dimers, and oligomers. The development of a suite of single-molecule imaging approaches has contributed to unveiling the molecular organization of different GPCRs in cells and tissues, and its significance to cellular signaling. The approaches used have ranged from employing diffraction-limited imaging and ultra-low receptor densities, but with superior temporal resolution, to unveil the dynamics of receptor–receptor associations at the plasma membrane; to those that break the light diffraction barrier of 200 nm and employ super-resolution imaging at physiological densities of receptor, however, with as yet poor temporal resolution. The former approach has used total-internal reflection fluorescence microscopy (TIRF-M) combined with post-acquisition extrapolation of intensity data to visualize the dynamics of individual GPCRs associating as dimers and oligomers in live cells, in real time [14–17]. With such approaches, receptors must be expressed at low densities (<10 molecules/μm^2) to enable tracking and localization of individual molecules [18, 19]. This is due to the point spread function (PSF) of each molecule in typical diffraction-limited imaging. The fluorescence of a single molecule has a Gaussian-like intensity distribution; the PSF. PSFs have a radius of ~200–250 nm, and correspond to the uncertainty of localization, or degree of spreading/blurring of a single point object, and is also impacted by the wavelength of emission and the light collection capacity of the objective [20]. If the PSF is sufficiently separated from other emitting species by a distance greater than the resolution limit (~200–250 nm), this center can be located. This is why molecule density for diffraction-limited single-molecule imaging is critical as all fluorophores are detected simultaneously, thus any overlap of the PSFs precludes super-resolution imaging and loss of structural detail [18]. Therefore, for visualization of GPCR dimers via this approach, the density must be less than a few molecules per μm^2 to enable visualization of individual receptors. Indeed, it has been optimized for enabling kinetic studies of individual receptor interactions rather than the detection of the overall landscape of GPCR monomers/dimers/oligomers [14–17]. The very high temporal resolution this technique affords, reveals the dynamic nature of these associations for certain GPCRs that would not be detectable by other real-time methods used to study protein–protein interactions, such as biophysical real-time techniques of BRET and FRET. We also refer the reader to recent reviews discussing all these distinct techniques for the study of GPCR di/oligomerization [18, 21]. Instead we will give a brief overview of super-resolution imaging with focus on photoactivated localization microscopy (PALM).

1.1 Photoactivated Localization Microscopy (PALM) to Study GPCR Oligomerization

Super-resolution approaches can map single molecules of labeled proteins down to <10 nm resolution, compared to conventional light microscopy that achieves ~200 nm maximal resolution and encompasses a number of techniques including 2D and 3D localization microscopy techniques (PALM and stochastic optical reconstruction microscopy (STORM)), stimulated emission depletion (STED), structured illumination and light sheet microscopy, each of which have specific cellular applications in terms of spatial and temporal resolution capabilities, and ability for live-cell imaging. With respect to studying GPCR oligomerization, only PALM out of the super-resolution imaging approaches has been employed to date [22–24]. Localization microscopy approaches such as PALM and STORM provide the ability to image labeled molecules of interest due to stochastic switching of fluorescent probes, providing precisely localized single-molecule imaging [25, 26]. PALM involves the use of photoswitchable or photoactivatable fluorophores, which remain in the dark state until unmasked or activated by UV light, emitting fluorescence in a fluorophore-defined wavelength range and subsequently photobleached into the dark state (Fig. 1). This activation occurs in a stochastic manner, thus allowing for spatially separate detection of the activated molecules. These cycles are repeated until all fluorophores are activated and bleached into the dark state in order to ensure accurate and defined coordinate-specific detection of proteins (Fig. 1).

The labeling of proteins for PALM requires photoactivatable fluorophores, such as tagging of proteins with large photoswitchable proteins, e.g., Dendra, Dendra2, and mEosFP [25, 27]. For our studies utilizing PALM to study GPCR oligomerization at the

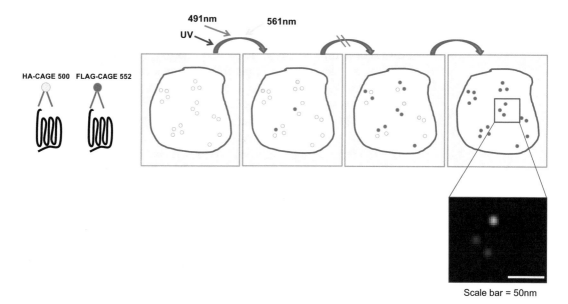

Fig. 1 Principles of PD-PALM. Schematic to demonstrate the principles of PD-PALM utilizing simultaneous dual-color imaging of CAGE 500- and 552-labeled receptors. CAGE 500 and 552 dyes are stochastically "uncaged" by UV, excited, and photobleached using 491- and 561-nm lasers, respectively. This is repeated through multiple cycles until all fluorophores are activated and bleached. A representative reconstructed PD-PALM image of an individual GPCR heterotrimer is highlighted. Scale bar = 50 nm. Adapted from [24]

cell surface, we employed photoactivatable dyes (PD) as they are brighter and more photostable than photoswitchable proteins, yet critically undergo irreversible activation and bleaching [28] to improve the accuracy of localizing and quantifying single receptors. An important distinction to highlight between prior single-molecule studies employing TIRF-M to localization microscopy approaches such as PD-PALM, is that the latter enables localization of receptors expressed at much higher densities (~200 molecules/μm^2 for PD-PALM vs 1–3 molecules for single-molecule imaging/tracking techniques). An argument for both techniques in studying GPCR oligomerization under physiologically relevant conditions has been made [14–18, 21]. For our prior studies with the luteinizing hormone receptor (LHR), we could localize and quantify between 2000 and 8000 receptors/cell, which are within the range previously reported for ovarian and testicular LHR (4000–20,000 receptors/cell) [24, 29, 30]. Certainly, there are GPCRs that are expressed at very low levels in vivo, although it is the temporal resolution of low-density/wide-field single-molecule tracking techniques not afforded by localization microscopy that enables the study of receptor dimeric association and dissociation kinetics. Receptor density will be discussed further below in relation to quantitation of PD-PALM. Overall, by employing a super-resolution localization microscopy technique to study and identify LHR complexes not only provided a methodology to resolve single molecules beyond the diffraction limit of conventional microscopy, but enabled identification of individual LHR molecules at equivalent levels to those previously reported in endogenous tissues, rather than necessitating under-expression, or low labeling of receptors to enable resolution of single molecules.

2 Materials

2.1 Antibody Labeled PD Dyes

Selecting the best methodology for labeling your receptor of interest to ensure single-molecule detection is an important and nontrivial task, as a number of factors dictate the choice of probe. A probe needs to be bright, emitting sufficient photons to be easily localized, and distinguished from background, which all impact localization precision values of single molecules and thus the integrity of data obtained [31]. If GPCR heteromers are to be imaged via simultaneous dual-channel imaging, the emission spectra of your chosen probes also need to be spectrally separated to ensure minimal overlap and the correct identity of the localized molecules. For these reasons, we selected PDs for PALM imaging and we will subsequently refer to it as PD-PALM. When labeling your antibody with PDs ensuring a 1:1 stoichiometric labeling of probe receptor for single-molecule imaging is an obvious essential. For our studies, PDs were directly conjugated to primary HA and FLAG antibodies (purchased from Covance and Sigma-Aldrich, respectively) were employed to visualize di/oligomers of two distinct mutant GPCRs at the cell surface. The CAGE 500 and CAGE 552 dyes, purchased from Abberior, are in the dark state and require uncaging via a UV light source to be activated. This uncaging is importantly irreversible. Antibodies were directly labeled with NHS-esters of these PDs, as directed by the manufacturer's instructions and protocols for labeling of antibodies/proteins with PDs (Abberior). It also is important to note that as the dye is in the dark state it is colorless and non-fluorescent. During handling and dye labeling it is very important to shield the dye from external light sources. Collect fractions before, during, and after elution of label for analysis of the degree of labeling. This can be determined from the absorption spectrum of the labeled antibody and calculated based on a derivation of the Beer–Lambert law [24] using a nanodrop spectrophotometer. For our antibodies, the degree of labeling was determined as 1.0 ± 0.2 for FLAG-CAGE 500 and 1.3 ± 0.1 for HA-CAGE 552. Therefore, achieving an approximate 1:1 labeling of receptor: PD-conjugated antibody.

2.2 Sample Preparation Materials

For PD-PALM analysis of cell surface organization of GPCRs, imaging is also carried out under TIRF with a 1.45 numerical aperture. Therefore, selecting the right glass coverslip for imaging is needed to ensure that the best quality images are obtained. A high quality, no. 1.5 thickness cover-glass with minimal variation in thickness is recommended for high numerical aperture objectives and super-resolution imaging. For our studies, we initially used 8-chamber well 1.5 borosilicate cover-glass slides (Lab-tek) but have also employed 35 mm dishes with 14 mm diameter glass inserts (MatTek).

We have also optimized distinct blocking agents to minimize background and nonspecific binding of the FLAG and HA CAGE-labeled antibodies including: combinations of fish skin gelatin (1%) that is used in immunogold labeling for electron microscopy to reduce background (although the amount used may impact labeling efficiency) and fetal bovine serum (FBS), FBS alone, goat serum, a combination of goat serum and FBS. In our hands, 10% FBS was used as both a blocking agent and to dilute the antibodies for incubation as this gave the least nonspecific background in HEK 293 cells.

PALM/PD-PALM employs fixed cells, although approaches to be able to use this technique in live cells are ongoing [32, 33]. This is because the photoactivatable dyes and photoswitchable proteins undergo stochastic excitation to ensure only a small subset of molecules are activated at any given time and repeated hundreds to tens of thousands of cycles in order to release and capture all unactivated molecules within the field of view. The fixative used for super-resolution imaging has been previously demonstrated to be very important as lateral diffusion of membrane proteins can still occur with conventional fixatives [34]. This study by Annibale and colleagues identified that simple addition of 0.2% glutaraldehyde to 4% paraformaldehyde exhibited the least lateral membrane diffusion of transmembrane proteins when compared to commonly used fixatives, such as paraformaldehyde alone and methanol. Once fixed, cells are stored at 4 °C and imaged in PBS containing 1% sodium azide until imaged.

2.3 Microscope Setup

For our published studies [19, 24], we used a custom-adapted Inverted Axiovert 200 manual inverted wide-field fluorescent microscope (Zeiss, Germany) fitted with a commercial TIRF condenser kit (TILL Photonics GmbH, Germany), with a 1.45 numerical aperture 100× oil immersion objective. To activate uncaging of PDs, a polychrome light source was used at 390 nm (Polychrome IV, TILL Photonics GmbH, Uckfield, UK). Simultaneous dual-channel imaging and photobleaching of activated FLAG-CAGE 500 and HA-CAGE 552 was conducted using 491- and 561-nm laser lines, respectively. Simultaneous imaging of CAGE 500 and 552 dyes was achieved using a beam splitter (Optosplit II, Andor) fitted with a T585lp dichroic and ET520–40 and ET632–60 emission filters (all Chroma). To minimize any environmental factors that may have impacted imaging, the microscope was enclosed in a plastic draft-proof system, always maintained at a constant temperature of 25 °C. Additionally, laser lines were always switched on at least 1 h before imaging to allow the system to stabilize and acclimatize prior to imaging. All of these measures ensured that minimal sample drift was observed when imaging. For capturing the images, a cooled electron multiplying charged-coupled device camera (EM-CCD; C9100–13, Hamamatsu) and Simple PCI software were used and time-lapse image series were taken, using an exposure time of 30 ms. For image registration and alignment of CAGE 500 and CAGE 552 generated images (required due to Optosplit-mediated simultaneous dual-channel imaging), a combination of fiducial markers and grid images was used for post-acquisition alignment in Fiji ImageJ software. The fiducial markers and grids acted as reference points to align the two channels post-acquisition.

There has been an increase in use of commercial systems for super-resolution imaging. We currently employ a Zeiss Elyra PS1 with an AxioObserver Z1 motorized inverted microscope. This microscope was custom designed to have two ECD cameras, to allow simultaneous dual-channel imaging without the necessity of an optisplit, and thus forgoing the need to align the two-channel images. Other advantages of this system include the automated nature of the TIRF and PALM imaging, plus certain features of the ZEN software that facilitate early image analysis, such as drift correction and signal to noise ratio values. In our hands, we would like to highlight that users must still be diligent to use sufficient laser power to ensure that PDs are fully bleached by the lasers within a single cycle, as our experience has been that more photoblinking/continuous PD activation over successive frames has been observed via commercial systems.

3 Methods

3.1 Cell Labeling and Image Acquisition

Cells are seeded ideally between 30% and 60% confluency to enable to locate a sufficient number of adherent cells suitable for imaging in the TIRF plane. We suggest the following conditions to be optimized for labeling density, minimize background fluorescence, and prevent antibody- and fixation-induced artifacts. For each labeled antibody perform dilution curves, vary antibody incubation time and temperature of labeling conditions, in order to ensure labeling reaches saturation and to prevent under-labeling of the target protein. For our studies, we observed minimal differences in the total percentage of detected associated molecules obtained when cells were fixed and subsequently labeled overnight at 4 °C, versus when PD-conjugated antibodies were incubated in live cells for 30 min at 37 °C and then subsequently fixed. These comparisons also demonstrate that the receptor di/oligomers could not be attributed to antibody-induced fixation. For assessing ligand-dependent changes, we therefore chose to label our cells live as steady-state labeling was reached using these conditions, and potential ligand-dependent changes could be carried out on pre-labeled cells [24]. Post-fixation, cells were imaged in an aqueous buffer or mounting fluid, such as PBS, which can aid achieving the optimal TIRF angle.

For ensuring the optimal TIRF angle is selected for imaging, we have found it easiest to locate the coverslip, at the most acute TIRF angle, and locate the cell membrane (and labeled receptors), using a combination of adjusting the focal plane, and adjusting the TIRF angle. For image acquisition, there are a few key parameters to consider. The laser power used will determine the number of fluorophores activated, and thus the spatial separation achieved. However, this will also impact on the efficiency of photobleaching, therefore the laser power selected is a fine balance between achieving spatially separated fluorophores, while ensuring rapid photobleaching occurs. Image speed is also another consideration to be made. Imaging too quickly will often result in the same fluorophores being detected in multiple frames; however, imaging too slow will miss activated fluorophores, and so will impact on the information collected. Imaging speed is also dictated by the resolution of the camera. For our studies, we have found that imaging at 30 frames/s provides a good balance for these parameters. In terms of selecting the number of photoactivation/bleaching cycles, this will be dictated by the number of fluorophores detected and localized, and is thus microscope, camera, and acquisition setup dependent. For our study, the number of photoactivation/photobleaching cycles selected was typically between 25,000 and 35,000 frames, to capture all labeled molecules and enable the assembly of the complete cell surface landscape of GPCR monomers/dimers/oligomers.

As discussed above, super-resolution imaging approaches can resolve higher densities of molecules, which for certain GPCRs is more physiologically relevant such as in our studies with LHR. However, whether imaging via diffraction-limited single particle tracking, or via super-resolution imaging, the density of receptor expression has been reported to impact the level and nature of certain forms of receptor complexes detected. We have reported that for LHR, increasing receptor density had little effect on the number of lower-order receptor oligomers observed (dimers, trimers, and tetramers), but rather translated to differences in the number of higher-order oligomers formed [19]. Likewise, prior studies using PALM and mEOS-tagged β2-adrenergic receptor found an increase in the number of receptor oligomers with increasing receptor expression [23]. A similar observation was observed with diffraction-limited single particle tracking and methodologies such as SpIDA (spatial intensity distribution analysis) with distinct GPCRs [16, 35]. Although they may be density dependent, significance of these higher-order forms

(>5 receptors) should not be overlooked. Some of these structures may represent signal microdomains, and certainly GPCR expression levels will dynamically change under both physiological and pathophysiological conditions. Further, given the reports of GPCR heteromeric complexes that contain three distinct GPCRs [36, 37], and if each of these receptors also self-associate, then this may manifest itself in formation of very high order complexes.

Additional controls that can be employed for both experimental conditions employed and the chosen data analysis parameters (providing confidence in the ability of a given GPCR to form dimers and oligomers), is the use of a non-clustering transmembrane protein, or receptor. For our published studies, we utilized a member of the receptor tyrosine kinase family that is activated by macrophage colony stimulating factor (M-CSF), that basally is primarily monomeric. PD-PALM imaging of FLAG-tagged M-CSF receptor revealed that 85% of this receptor were monomers, with the majority of associating receptors residing as dimers. Further, coexpression of M-CSF with WT LHR detected minimal (<2%) heterocomplexes, demonstrating the specificity of the GPCR complexes observed by PD-PALM [24].

3.2 Post-acquisition Analysis Part 1: Resolving Single-Molecule Data

Post-image acquisition, single molecules are assigned and identified via localization analysis to resolve the coordinates of individually activated CAGE-labeled molecules across all frames that are compressed into a single file. For our studies, individual receptors were identified using a freely available, open source ImageJ plug-in, QuickPALM [38]. The software identifies and localizes each fluorophore with subpixel accuracy across frames of the time-lapse series taken. Within the software, the parameters of analysis can be adjusted to refine the stringency of single molecules identified and localized, e.g., image plane pixel size (155 nm for our system), the minimum signal to noise ratio of each experiment (typically 8, but reanalyzed for each experiment) and the full width, half maximum of single molecules (3). Each identified single molecule is presented as a reconstruction map in image form, simultaneously generated as the analysis occurs, and a data set containing the localizations, or XY map coordinates, of each counted single molecule (Fig. 3). For our ongoing studies using the Zeiss Elyra setup, we have continued to use QuickPALM for generation of XY coordinate maps. However, there are other ImageJ plug-ins, such as ThunderSTORM that offer a good alternative for handling of super-resolution PALM and STORM imaging data, offering processing abilities such as data rendering and drift correction.

Localization precision is the standard measure used for assessing the accuracy of localizing each detected fluorophore, and calculated via the theory of Thompson et al. [39]. This takes into account several factors including the photon count for each molecule, noise, the full width half maximum of the observed PSF, and the camera pixel size. For our studies, localization precision was calculated to be approximately 20 nm. The resolution that we achieved was determined to be ~8 nm, based on the number of photons emitted by the CAGE dyes during activation, and the PSF of the activated dye. Commercial systems also contain software to calculate the signal to noise ratio and localization precision obtained.

3.3 Quantification of GPCR Monomers, Dimers, and Oligomers from PD-PALM Datasets

PD-PALM generates multiple large data sets of localized receptor molecules (up to <8000 data points for every cell), therefore we developed custom software to quantify the number of receptors existing as monomers, dimers, and oligomers, but also to distinguish different receptor populations participating in homo- versus heterocomplexes [19, 24]. We developed a JAVA-based software that employed an adaptation of Getis and Franklin's nearest neighborhood approach (termed PD-Interpreter). This software is freely available to download at www.superimaging.org. The software was designed to identify a single molecule and recursively searched within a chosen radius for further single molecules until no further molecules were found within the search radius (Fig. 2). The software can then inform not only monomeric receptor or those participating in a dimer/oligomer, but also if an oligomer, the number of molecules in that oligomer. For dual-color imaging and the study of heteromers, both information on homomeric and heteromeric complexes can be obtained (Fig. 2). The radius that the nearest neighborhood analysis is conducted over is user selected, ranging from 10 to 100 nm. For our studies with the LHR we selected 50 nm. Although the LHR is a Class A GPCR, where sizes of the members of this family is ~6 nm [40, 41], this receptor belongs to the glycoprotein hormone receptor subfamily containing a very large extracellular domain. Further, considering the localization precision of our imaging (20 nm), the size of the PD-labeled antibody, ~20 nm in size, and the nature of the labeling means that the position of the dye on the antibody is unknown and will be heterogenous. The search radius of 50 nm gave the highest number of lower-order dimers, trimers, and tetramers when compared to radii of 20–100 nm (Fig. 3). As an added level of stringency to our quantitation, events within a radius of 10 nm of a "parent"-activated fluorophore were excluded from the analysis; however, such molecules were infrequent and typically resulted in only discounting approximately <1% of activated molecules. CAGE dyes undergo irreversible activation; however, in commercial systems we have found that photobleaching may be suboptimal. As such, we have developed an add-on algorithm that we run after generation of the XY coordinate tables (and before interrogation on di/oligomeric status) to eliminate molecules that occur in more than one frame. Grouping of molecules activated across more

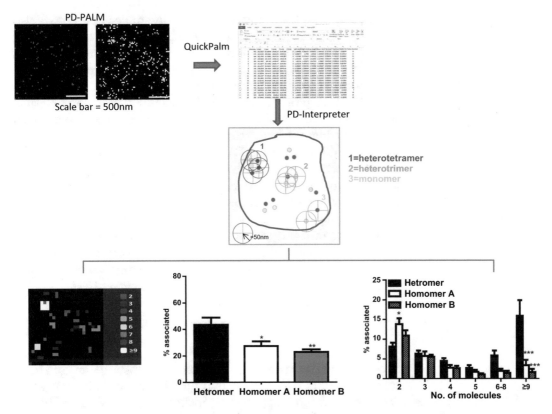

Fig. 2 Workflow of PD-PALM imaging and quantitation of GPCR oligomers. Following image acquisition, reconstruction and extraction of single molecule coordinates is carried out via the Fuji/ImageJ plug-in QuickPALM. Data tables containing *x-y* particle localization coordinates were generated, and two-dimensional coordinates were determined. To analyze the number of associated receptor molecules from the *x-y* particle localization coordinates, a custom Java application was designed (PD-Interpreter). A second order Getis Franklin neighborhood analysis was conducted, using defined search radii, to determine the degree of both homomeric associations within an individual channel and heteromeric associations across channels. To identify monomers, dimers, and oligomers, the software recursively searches at a specific radius (here 50 nm) from each associating molecule until no further associating molecules were identified within the allotted search radius. Once an associating group of molecules is assigned, the composition of the di/oligomer is identified and omitted from further searches, so that molecules were not double counted. Data can be represented in the form of heat maps, with individual colors depicting different numbers of associating molecules and outputted in Excel format for graph presentation. Adapted from [24]

than one frame is also possible to do using the data rendering ability of the Zen software that operates the Zeiss Elyra.

Data from PD-Interpreter is outputted in pictorial and Excel spreadsheet form depicting self-associating and co-associating molecules. The spreadsheets contain a summary page with the total of self-associating and co-associating molecules. A breakdown of the number of each complex type, specifically the number of molecules within each complex, is also detailed within this summary page. Additional tabs contain the individual localization information on each of the assigned self-associating and co-associating complexes, and non-associating molecules. Data is also outputted as several different individual image files, initially depicting all molecules in the localization data sets from the two channels. From the analysis, co-localization plots are generated using blue and yellow colors to differentiate the two channel populations. A multi-color plot depicts complex size and represents clusters with two or more molecules each represented by different colors (both omitting non-associating molecules) (Fig. 2).

When first quantifying PALM data sets, we strongly suggest the use of theoretical or simulated datasets as controls. Excel can generate randomly dispersed data sets where one can vary the total numbers to simulate comparable receptor densities observed in experimentally obtained data sets. Theoretical data sets can then be subjected to the same analysis via PD-Interpreter software. By using this control in our studies with LHR, we demonstrated that the total percentage of associated molecules in our experimental data sets was greater than that for a randomly dispersed sample set [24].

4 Conclusions

Super-resolution imaging has enabled scientists to image structures and proteins in cells beyond the diffraction limit of standard fluorescent imaging approaches, including TIRF-M, unveiling an unprecedented depth of information to unveil the "inner secrets" of how cells and proteins function. In the context of GPCR

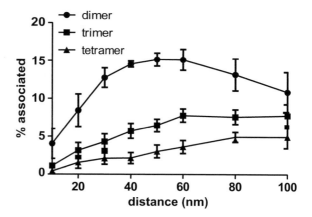

Fig. 3 Percentage of GPCR dimers, trimers, and tetramers at varying search radii. Following reconstruction of PD-PALM images in QuickPALM and extraction of *xy* coordinates for all single receptors imaged, the number of dimers and oligomers were quantitated with varying radii (20–100 nm). From this analysis, a search radius of 50 nm was selected as it identified the highest number of lower-order associating complexes, i.e., dimers/trimers/tetramers. Data obtained from HEK 293 cells stably expressing HA-tagged mouse LHR. All data points represent the mean ± S.E. of 10–12 individual cells, *n* = 3. From [24]

di/oligomerization, super-resolution imaging enables the study of these receptor complexes at densities to that found under physiological conditions. At present, there have been limited studies on the application of super-resolution imaging to study GPCR oligomerization, although these advanced imaging techniques are being increasingly applied to study other aspects of receptor function such as membrane trafficking [42–44]. Future applications of approaches such as PD-PALM that enable multi-channel imaging (>2) to visualize not only the receptor complexes, but their organization with the signaling machinery will enable assessment of current models that propose the asymmetric organization of receptor homo- and heteromers with, for example, different G proteins [1, 24, 45, 46]. The limitations at present of the current presented technique is the lack of temporal resolution, though these "snapshots" of GPCR organization provide the full detailed complement of receptor organization; improvements to apply live-cell PALM may require a reduction in the level of spatial resolution to achieve this. Further, extensions of PD-PALM to 3D PALM has the potential to analyze receptor complexes not just at the cell surface but also in distinct subcellular compartments, such as endosomes, Golgi, and mitochondria that all have been reported to exhibit active GPCR/G protein signaling [47–50]. 3D super-resolution imaging will also facilitate single-molecule resolution in tissue and multi-cellular in vitro systems such as organoids. Overall, we predict that this is only the start of the application of such techniques to the study of GPCRs, and in combination with other approaches such as molecular modeling, an unprecedented insight into how these receptors function at the nano-scale.

Acknowledgements

This research was supported by Biotechnology and Biological Sciences Research Council (BBSRC) project grant (BB/1008004/1), a BBSRC Sparking Impact Award, and a Society for Endocrinology Early Career Grant.

References

1. Ferre S et al (2014) G protein-coupled receptor oligomerization revisited: functional and pharmacological perspectives. Pharmacol Rev 66(2):413–434

2. Farran B (2017) An update on the physiological and therapeutic relevance of GPCR oligomers. Pharmacol Res 117:303–327

3. Franco R et al (2016) Basic pharmacological and structural evidence for class A G-protein-coupled receptor heteromerization. Front Pharmacol 7:76

4. Gomes I et al (2016) G protein-coupled receptor heteromers. Annu Rev Pharmacol Toxicol 56:403–425

5. Jonas KC, Hanyaloglu AC (2017) Impact of G protein-coupled receptor heteromers in endocrine systems. Mol Cell Endocrinol 449:21–27

6. Chabre M, le Maire M (2005) Monomeric G-protein-coupled receptor as a functional unit. Biochemistry 44(27):9395–9403

7. James JR et al (2006) A rigorous experimental framework for detecting protein oligomerization using bioluminescence resonance energy transfer. Nat Methods 3(12):1001–1006

8. Bouvier M et al (2007) BRET analysis of GPCR oligomerization: newer does not mean better. Nat Methods 4(1):3–4. author reply 4

9. Salahpour A, Masri B (2007) Experimental challenge to a 'rigorous' BRET analysis of GPCR oligomerization. Nat Methods 4(8):599–600. author reply 601

10. Lambert NA, Javitch JA (2014) Rebuttal from Nevin A. Lambert and Jonathan A. Javitch. J Physiol 592(12):2449

11. Lambert NA, Javitch JA (2014) CrossTalk opposing view: weighing the evidence for class A GPCR dimers, the jury is still out. J Physiol 592(12):2443–2445

12. Bouvier M, Hebert TE (2014) CrossTalk proposal: Weighing the evidence for Class A GPCR dimers, the evidence favours dimers. J Physiol 592(12):2439–2441

13. Bouvier M, Hebert TE (2014) Rebuttal from Michel Bouvier and Terence E. Hebert. J Physiol 592(12):2447

14. Hern JA et al (2010) Formation and dissociation of M1 muscarinic receptor dimers seen by total internal reflection fluorescence imaging of single molecules. Proc Natl Acad Sci U S A 107(6):2693–2698

15. Kasai RS et al (2011) Full characterization of GPCR monomer-dimer dynamic equilibrium by single molecule imaging. J Cell Biol 192(3):463–480

16. Calebiro D et al (2013) Single-molecule analysis of fluorescently labeled G-protein-coupled receptors reveals complexes with distinct dynamics and organization. Proc Natl Acad Sci U S A 110(2):743–748

17. Kasai RS, Kusumi A (2014) Single-molecule imaging revealed dynamic GPCR dimerization. Curr Opin Cell Biol 27:78–86

18. Scarselli M et al (2016) Revealing G-protein-coupled receptor oligomerization at the single-molecule level through a nanoscopic lens: methods, dynamics and biological function. FEBS J 283(7):1197–1217

19. Jonas KC, Huhtaniemi I, Hanyaloglu AC (2016) Single-molecule resolution of G protein-coupled receptor (GPCR) complexes. Methods Cell Biol 132:55–72

20. van der Merwe PA et al (2010) Taking T cells beyond the diffraction limit. Nat Immunol 11(1):51–52

21. Vischer HF, Castro M, Pin JP (2015) G protein-coupled receptor multimers: a question still open despite the use of novel approaches. Mol Pharmacol 88(3):561–571

22. Annibale P et al (2011) Quantitative photo activated localization microscopy: unraveling the effects of photoblinking. PLoS One 6(7):e22678

23. Scarselli M, Annibale P, Radenovic A (2012) Cell type-specific beta2-adrenergic receptor clusters identified using photoactivated localization microscopy are not lipid raft related, but depend on actin cytoskeleton integrity. J Biol Chem 287(20):16768–16780

24. Jonas KC et al (2015) Single molecule analysis of functionally asymmetric G protein-coupled receptor (GPCR) oligomers reveals diverse spatial and structural assemblies. J Biol Chem 290(7):3875–3892

25. Betzig E et al (2006) Imaging intracellular fluorescent proteins at nanometer resolution. Science 313(5793):1642–1645

26. Owen DM et al (2013) Super-resolution imaging by localization microscopy. Methods Mol Biol 950:81–93

27. Williamson DJ et al (2011) Pre-existing clusters of the adaptor Lat do not participate in early T cell signaling events. Nat Immunol 12(7):655–662

28. Belov VN et al (2010) Rhodamines NN: a novel class of caged fluorescent dyes. Angew Chem Int Ed Engl 49(20):3520–3523

29. Luborsky JL, Slater WT, Behrman HR (1984) Luteinizing hormone (LH) receptor aggregation: modification of ferritin-LH binding and aggregation by prostaglandin F2 alpha and ferritin-LH. Endocrinology 115(6):2217–2226

30. Dehejia A et al (1982) Luteinizing hormone receptors and gonadotropic activation of purified rat Leydig cells. J Biol Chem 257(22):13781–13786

31. Patterson G et al (2010) Superresolution imaging using single-molecule localization. Annu Rev Phys Chem 61:345–367

32. Manley S, Gillette JM, Lippincott-Schwartz J (2010) Single-particle tracking photoactivated localization microscopy for mapping single-molecule dynamics. Methods Enzymol 475:109–120

33. Deschout H et al (2016) Complementarity of PALM and SOFI for super-resolution live-cell imaging of focal adhesions. Nat Commun 7:13693

34. Annibale P et al (2011) Identification of clustering artifacts in photoactivated localization microscopy. Nat Methods 8(7):527–528

35. Ward RJ et al (2015) Regulation of oligomeric organization of the serotonin 5-hydroxytryptamine 2C (5-HT2C) receptor observed by spatial intensity distribution analysis. J Biol Chem 290(20):12844–12857

36. Sohy D et al (2009) Hetero-oligomerization of CCR2, CCR5, and CXCR4 and the protean effects of "selective" antagonists. J Biol Chem 284(45):31270–31279

37. Navarro G et al (2010) Interactions between intracellular domains as key determinants of the quaternary structure and function of receptor heteromers. J Biol Chem 285(35):27346–27359

38. Henriques R et al (2010) QuickPALM: 3D real-time photoactivation nanoscopy image processing in ImageJ. Nat Methods 7(5):339–340

39. Thompson RE, Larson DR, Webb WW (2002) Precise nanometer localization analysis for individual fluorescent probes. Biophys J 82(5):2775–2783

40. Mercier JF et al (2002) Quantitative assessment of beta 1- and beta 2-adrenergic receptor homo- and heterodimerization by bioluminescence resonance energy transfer. J Biol Chem 277(47):44925–44931

41. Gurevich VV, Gurevich EV (2008) GPCR monomers and oligomers: it takes all kinds. Trends Neurosci 31(2):74–81

42. Eichel K, Jullie D, von Zastrow M (2016) beta-Arrestin drives MAP kinase signalling from clathrin-coated structures after GPCR dissociation. Nat Cell Biol 18(3):303–310

43. Pons M et al (2017) Phosphorylation of filamin A regulates chemokine receptor CCR2 recycling. J Cell Sci 130(2):490–501

44. Cooney KA et al (2017) Lipid stress inhibits endocytosis of Melanocortin-4 Receptor from modified clathrin-enriched sites and impairs receptor desensitization. J Biol Chem 292:17731–17745

45. Navarro G et al (2016) Quaternary structure of a G-protein-coupled receptor heterotetramer in complex with Gi and Gs. BMC Biol 14:26

46. Ferre S (2015) The GPCR heterotetramer: challenging classical pharmacology. Trends Pharmacol Sci 36(3):145–152

47. Irannejad R et al (2017) Functional selectivity of GPCR-directed drug action through location bias. Nat Chem Biol 13(7):799–806

48. Irannejad R et al (2013) Conformational biosensors reveal GPCR signalling from endosomes. Nature 495(7442):534–538

49. Suofu Y et al (2017) Dual role of mitochondria in producing melatonin and driving GPCR signaling to block cytochrome c release. Proc Natl Acad Sci U S A 114(38):E7997–E8006

50. Godbole A et al (2017) Internalized TSH receptors en route to the TGN induce local Gs-protein signaling and gene transcription. Nat Commun 8(1):443

INDEX

Printed in the United States
By Bookmasters